China's Water Resources Management

Seungho Lee

China's Water Resources Management

A Long March to Sustainability

Seungho Lee
Graduate School of International Studies
Korea University
Seoul, Korea (Republic of)

ISBN 978-3-030-78778-3   ISBN 978-3-030-78779-0 (eBook)
https://doi.org/10.1007/978-3-030-78779-0

© The Editor(s) (if applicable) and The Author(s), under exclusive license to Springer Nature Switzerland AG 2021
This work is subject to copyright. All rights are solely and exclusively licensed by the Publisher, whether the whole or part of the material is concerned, specifically the rights of translation, reprinting, reuse of illustrations, recitation, broadcasting, reproduction on microfilms or in any other physical way, and transmission or information storage and retrieval, electronic adaptation, computer software, or by similar or dissimilar methodology now known or hereafter developed.
The use of general descriptive names, registered names, trademarks, service marks, etc. in this publication does not imply, even in the absence of a specific statement, that such names are exempt from the relevant protective laws and regulations and therefore free for general use.
The publisher, the authors and the editors are safe to assume that the advice and information in this book are believed to be true and accurate at the date of publication. Neither the publisher nor the authors or the editors give a warranty, expressed or implied, with respect to the material contained herein or for any errors or omissions that may have been made. The publisher remains neutral with regard to jurisdictional claims in published maps and institutional affiliations.

Cover credit: View Stock

This Palgrave Macmillan imprint is published by the registered company Springer Nature Switzerland AG
The registered company address is: Gewerbestrasse 11, 6330 Cham, Switzerland

Acknowledgments

There are many esteemed scholars, colleagues, and friends who have encouraged me to work on this book project. While pondering how I express my gratitude to them, I would like to first mention Professor Tony Allan at King's College London and SOAS, my mentor and great friend, who has been a guiding light for my research on water resources management and policy. Sadly, Tony passed away in April 2021. Although there is no way to have a chat with him about my book anymore, I would like to say big thanks for Tony's wisdom, insights, kindness, and interest in my research since I studied Shanghai's water pollution control policy back in the late 1990s.

I have been planning to write this book since I was fascinated by some of China's mega-scale water infrastructure projects, i.e., the Three Gorges Dam and the South North Water Transfer Project. I bumped into Mr. Jeremy Berkoff at one of Tony Allan's SOAS Water Seminars in London in 2000, who kindly invited me to his house at Cambridge and allowed me to skim through his unparalleled collection of China's water project documents. Dr. Richard Edmonds has helped me explore a wide range of environmental challenges in China based on his in-depth knowledge and insight.

China's water policy changes over the past two decades have always fascinated me, including institutional reform, water market, and megascale water infrastructure projects. As China's influence on the world grows, I start to become interested in China's involvement with its transboundary rivers, and I decided to investigate the case of the Lancang-Mekong River. My research on the topic has paved the way to not only publish several research papers and book chapters in English and Korean but also come to know excellent scholars on transboundary river issues. Dr. Eunsuk Hong at the School of Finance and Management of SOAS kindly arranged my affiliation with SOAS as a visiting scholar in 2016 and I thank Professor Steve Tsang, back then at the University of Nottingham and now the Director of the China Institute of SOAS, who took good care of me as a visiting fellow in 2016.

Dr. Eva Sternfeld, a good friend and China water expert, kindly invited me to the Berlin Workshop in June 2017 where I enjoyed discussions with Professor James Nickum, Editor in Chief of *Water International*, and Professor Shaofeng Jia at the Chinese Academy of Science. I thank both James and Shaofeng for their continuous interest in my research on China's transboundary rivers.

I particularly appreciate Professor Jia for his kind invitation to the conference in Beijing, September 2019 and gave me the opportunity to visit the Beijing end point of the Middle Route of the South North Water Transfer Project. I would like to express my gratitude toward Dr. Ma Miaomiao at the China Institute of Water Resources and Hydropower Research in Beijing who allowed me to have access to documents on the Three Gorges Dam and the South North Water Transfer Project.

I would like to thank the Korea Water Resources Association and the Chinese Hydraulic and Engineering Society for their support to my trip to Yichang City, Hubei Province, in October 2019, where I joined a fieldtrip to the Three Gorges Dam. Writing several chapters was not possible without the data and information on China's water market as well as a brilliant workspace and accommodation provided by the K-water Research Institute for a couple of weeks in September 2020, and special thanks go to Dr. Moonhyun Ryu at K-water. A good friend of mine as well as superb scholar, Professor Jongpil Chung at Kyunghee

University, has given me his insights and thoughtful comments on politics and international relations of China through numerous lunches and bottles of beer.

I owe a lot to my school, the Graduate School of International Studies, Korea University, for second-to-none facilities, library collections, efficient administrative support, and fantastic colleagues who are intellectually and emotionally the backbone of my work. I would like to thank my former postgraduate students, Dr. Yookyung Lee who kindly edited my final draft, and Mr. Jonghak Kim who helped draw the maps in the book.

This book has never come out without my family's patience and support. I would like to thank my little girl, Sooyoun Lee, who has been keen on what I am doing. This book is dedicated to my beloved wife, Hyegyoon Kim, who has always been supportive of my book project.

Contents

1	**Introduction**	1
	Background	1
	Purpose and Scope of the Book	3
	Major Issues	4
	Structure of the Book	9
	References	12
2	**Water and Development**	15
	Introduction	15
	Water and Sustainable Development	17
	Water for Economic Growth	26
	Water for Social Development and Ecosystems	37
	Conclusion	44
	References	45
3	**Overview of Water Resources**	51
	Introduction	51
	Water Resources	54
	Water Quality	74

	Flood and Drought	84
	Climate Change Impacts	93
	Conclusion	98
	References	100
4	**Water Plan and Governance System**	105
	Introduction	105
	Plans	106
	Governance Structure	116
	Laws and Regulations	136
	Conclusion	148
	References	149
5	**Sustainable Water Use**	153
	Introduction	153
	Water Shortage	155
	Water Pricing	158
	Water Rights Trading	167
	Virtual Water and Water Footprint	175
	Conclusion	183
	References	185
6	**Water Quality Management**	191
	Introduction	191
	General Framework	193
	River (Lake) Chief System	201
	Water Pollution Trading	210
	Case Study: Tai Lake Algae Incident in 2007	217
	Conclusion	223
	References	225
7	**Water Resources Development**	229
	Introduction	229
	Dam Development	231
	Water Transfer Projects	252
	Conclusion	283
	References	285

8	**Water and Wastewater Service Market**	293
	Introduction	293
	Development of Water and Wastewater Service Market	295
	Governance Structure and Regulatory Frameworks	302
	Market Analyses	310
	Implications	325
	Conclusion	331
	References	333
9	**Transboundary Rivers**	337
	Introduction	337
	Transboundary Rivers in China	339
	Water Diplomacy Strategies of China	345
	China and the 1997 UN Water Convention and Its Transboundary River Cases	348
	Case Study: The Lancang-Mekong River Basin	352
	Conclusion	369
	References	371
10	**Conclusion**	377
	References	381
Index		383

List of Figures

Fig. 1.1	Structure of the book (*Source* Author)	11
Fig. 1.2	Key themes of the book (*Source* Author)	12
Fig. 2.1	Substantial impacts of SDG 6 water and sanitation on the other SDGs (*Source* Author)	19
Fig. 3.1	Trend of per capita water resources in China from 1998 to 2019 (*Source* MWR [2018] and NSB [2020])	56
Fig. 3.2	Spatial distributions of water resources, population, river basins and farmland in China (Unit: %) (*Source* Modified based on Wang et al. [2017])	60
Fig. 3.3	Location of seven major rivers in China (*Source* Author)	66
Fig. 3.4	Water supply trends in China from 2000 to 2019 (*Source* MWR [2018, 2019, 2020] and NSB [2019])	68
Fig. 3.5	Water supply in water resources districts of China, 2019 (*Source* MWR [2020])	69
Fig. 3.6	Water use trends by agriculture, industry, household and ecology from 2010 to 2019 (*Source* MWR [2018, 2019, 2020], NSB [2019])	73
Fig. 3.7	National water quality assessment results from 2015 to 2019 in China (*Source* MEE [2015, 2016, 2017, 2018b, 2019])	75

List of Figures

Fig. 3.8	Water quality of major river basins in China, 2019 (Unit: %) (*Source* MEE [2019])	76
Fig. 3.9	Poor water quality of major river basins (Grade IV, V, and V+) in China, 2019 (Unit: %) (*Source* MEE [2019]; *Remarks* Zhe-Min indicates rivers in Zhejiang and Fujian Provinces)	77
Fig. 3.10	Annual increase of wastewater discharge in China from 2000 to 2015 (*Source* NSB and MEE [2019])	81
Fig. 3.11	Trends of total direct economic losses by flood in China from 1990 to 2017 (*Source* MWR [2018])	87
Fig. 3.12	Death toll caused by flood in China from 1990 to 2017 (*Source* MWR [2018])	89
Fig. 3.13	Trends of drought damage in China between 1980 and 2017 (*Source* MWR [2018])	91
Fig. 4.1	Institutional arrangements of water resources management in China (*Source* Modified based on Xie and Jia [2018])	121
Fig. 4.2	Development of the Ministry of Ecology and Environment in China (*Source* Modified based on Chan and Xu [2018] and Chang and Li [2019])	127
Fig. 4.3	Development of the Ministry of Natural Resources in China (*Source* Modified based on Chan and Xu [2018] and Voita [2018])	128
Fig. 4.4	Responsibilities of the Ministry of Emergency Management in China (*Source* Liao [2019])	129
Fig. 4.5	The administrative structure of flood and disaster management in China. *Remarks* CCP means the Chinese Communist Party, and NDRC stands for the National Development and Reform Commission (*Source* Modified based on Ding [2016])	129
Fig. 5.1	Water tariff for household in 36 major cities in China, 2020 (*Source* H2O Website. http://h2o-China.com [accessed 25 September 2020])	165
Fig. 5.2	Major water policy developments in China from 2000 to 2012 (*Source* Modified based on Jiang et al. [2020])	170
Fig. 5.3	Major water policy developments in China between 2013 and 2017 (*Source* Modified based on Jiang et al. [2020])	173

Fig. 6.1	Organization of the river and lake system (*Source* Modified based on Liu [2019])	204
Fig. 6.2	Workflow of the water pollution trading scheme in China. *Remarks* EEBs mean Ecology and Environment Bureaus (*Source* Modified based on Shen et al. [2017])	211
Fig. 6.3	Structure of the water pollution control system in the Tai Lake Basin (*Source* Modified based on Zhang et al. [2012])	213
Fig. 6.4	Tai Lake Basin (*Source* Author)	218
Fig. 7.1	Hydropower installed capacity (GW) of the top 10 countries in 2020 (*Source* IHA [2020])	233
Fig. 7.2	Current governance structure of hydropower generation in China (*Source* Modified based on Habich [2016], Mertha [2008], and Tilt [2015])	239
Fig. 7.3	**a.** Façade of the Three Gorges Dam. **b.** The Ship Lock of the Three Gorges Dam. **c.** Transmission Lines from the Three Gorges Dam (*Source* Author [photo taken in October 2019])	240
Fig. 7.4	Location of the Three Routes in the South North Water Transfer Project (*Source* Author)	262
Fig. 7.5	Roadmap of the eastern route of the South North Water Transfer Project (*Source* Author)	264
Fig. 7.6	Roadmap of the middle route of the South North Water Transfer Project (*Source* Author)	268
Fig. 7.7	Roadmap of the western route of the South North Water Transfer Project (*Source* Author)	274
Fig. 8.1	Ministries and bureaus at the national level involved in water resources management in China since 2018 (*Source* Updated based on Lee [2010] and World Bank [2006])	308
Fig. 8.2	Bureaus involved at the local level in Water Resources Management in China (*Source* Updated based on Lee [2006, 2010])	310

Fig. 8.3　Diverse models in China's Water and Wastewater Public Private Partnership (PPP) Projects. *Remarks* BOT means Build-Operate-Transfer, TOT, Transfer-Operate-Transfer, ROT, Renovate-Operate-Transfer, and BROT, Build-Rehabilitate-Operate-Transfer. WFOE stands for Wholly Foreign Owned Enterprise. CJV indicates Cooperative Joint Ventures, EJV, Equity Joint Ventures (*Source* Modified based on Choi and Lee [2010])　317

Fig. 9.1　Lancang-Mekong River Basin (*Source* Author)　354

List of Tables

Table 2.1	Targets for the Sustainable Development Goal 6 and related actions by China	23
Table 3.1	Renewable water resources of China in 2019 (Unit: billion m^3)	55
Table 3.2	Precipitation in 10 water resources districts of China in 2019 and comparison with precipitation in 2018 and previous years	57
Table 3.3	Surface water resources in water resources districts of China in 2019 and comparison with surface water resources in 2018 and previous years	58
Table 3.4	Total amount and spatial distribution of groundwater resources in China, 2019 (Unit: billion m^3)	62
Table 3.5	General situations of seven major rivers in China	65
Table 3.6	Changes in population, economy and water use in China from 1980 to 2010	72
Table 3.7	Surface water quality standard 6-grade rating system in China	75
Table 4.1	Future water planning targets—the Three Red Lines	111
Table 4.2	Water Ten Plan measures	115

xvii

List of Tables

Table 4.3	Roles of government agencies in China involved in water pollution control under the Water Ten Plan 2015	117
Table 4.4	Ministries involved in water resources management of China	124
Table 4.5	Major state-level laws and regulations, and rules on water resources management in China	145
Table 5.1	Increasing block tariffs in Beijing since January 2018	164
Table 6.1	Water quality enhancement targets for the 13th Five Year Plan (FYP) (2016–2020)	197
Table 7.1	Summary of 13 hydropower bases in China	234
Table 7.2	China's hydropower development plan from 2020 to 2050	236
Table 7.3	Technical specifications of the Three Gorges Dam	244
Table 7.4	Siltation in the Yangtze River in the different time periods	250
Table 7.5	Construction plan of each phase of the three routes of the South North Water Transfer Project	258
Table 7.6	Specifications of the three routes in the South North Water Transfer Project	261
Table 7.7	Water transfer volume in the three phases of the eastern route (Unit: billion m^3/year)	263
Table 7.8	Water transfer volume in the two phases of the middle route (Unit: billion m^3/year)	267
Table 7.9	Water transfer volume of the three phases of the western route (Unit: billion m^3/year)	274
Table 8.1	Four development stages of water and wastewater public-private partnership projects in China	300
Table 8.2	Laws and regulations on public private partnership water and wastewater projects in China	306
Table 8.3	Investment models in the Chinese Water Market	316
Table 8.4	Types of water and wastewater public-private partnership projects in China from 1994 to 2019	318
Table 8.5	Four development stages and classification of private players depending on the nationalities of water companies in China	319
Table 8.6	The World's Top 50 Private Water Operators (Chinese water companies highlighted in gray)	321

Table 8.7	2019 Top 10 Most Influential Water Companies in China	325
Table 9.1	15 Major transboundary rivers in China	342

1

Introduction

Background

China's socioeconomic development since the late 1970s is unprecedented in its history. This development has entailed the fast pace of urbanization, industrialization, and population growth, which have accelerated the transformation of Chinese society into a wealthy, industrialized, and urbanized society. Such an eye-catching metamorphosis of Chinese society has been, to a large extent, attributed to a maximum exploitation of the environment, including water resources. It is undeniable to note a salient role of water in China in terms of a variety of sociopolitical, economic, and environmental aspects, i.e., water and sanitation for public health and education, industrialization, food security, energy generation, and the growth of large-scale urban centers, to name a few.

Many of the most influential and grand empires of ancient times succeeded in controlling river systems and developing vast irrigated areas for agricultural production, including Chinese, Mesopotamian, Egyptian, or Maya empires (Molle et al. 2009). Water has served as a catalyst to create ancient civilization in North China, particularly contributing to the establishment of hydraulic bureaucracy for developing irrigation projects through the mobilization of innumerable people. Such phenomena simply tell the weight of water issues for top leaders throughout history and encapsulates the political imperativeness of water resources management. The Chinese ancient civilization began adjacent to the Yellow River, more than a few thousand years ago, and since then, ups and downs of significant political, economic, social, and environmental events have often occurred centered on major river basins, including the Yellow or Huang River, the Hai River, the Huai River (3H Rivers), and the Yangtze River. Peoples in China recognized the significance of rivers, lakes, reservoirs, and groundwater for their livelihoods, and competed for securing fertile land as well as the places in which farmers have better access to the most imperative component for food production, water.

Although the phenomena of China are not necessarily unique, this backdrop leads to a plethora of historical records and anecdotes on monumental and outstanding achievements of emperors or empresses, kings and queens, governors, and leaders at localities in harnessing flood events and providing substantial amounts of water for irrigated fields. Embankments along the Yellow River for harnessing flood events have been regarded as some of essential tasks for rulers in the river basin, and the Grand Canal between Hangzhou and Beijing and the irrigation system in Dujiangyan are the examples that manifest themselves for the magnitude of water resources management in Chinese history (Wang et al. 2018).

The open-door policy of Deng Xiaoping prompted China's return to the world stage in 1978 and ignited China's unstoppable willingness to become rich and powerful. The country has shown outstanding records of economic growth to become the largest manufacturing country, to compete with the US as an influencing export and import country, and to burn substantial amounts of fossil fuels, thereby being the largest Green House Gas (GHG) emitter in the world (Wu and Edmonds 2017). All these facts imply the two different sides of a coin in China's socioeconomic development, not only a positive side but also a worrying part, ecological degradation. A huge toll at the environment has been taken in society, and among them, water is one of the most damaged and degraded elements in the country due to its extensive and far-reaching impacts on ecosystems and biodiversity. In the course of the remarkable socioeconomic development since the late 1970s, the concept of sustainable development has been introduced to the country, and the incorporation of sustainability into water resources management has come into being in the context of multiple uses for economic purposes (Wang et al. 2018; Xie and Jia 2018).

It is worth having a close investigation on the roles and contributions of water for China's modernization efforts in the reform era because the investigation helps understand the contribution of water resources management to China's sustainable development. Socioeconomic and political responses of the Chinese government to environmental challenges, such as water shortage, water pollution, and water-related disasters, have disclosed the institutional capacity of the government to adequately tackle water and environmental issues based on both top-down and bottom-up approaches.

Purpose and Scope of the Book

This book aims to evaluate one of the most challenging issues for China's development, water resources management, with a special focus on the period after the new millennium. The rapid modernization in the reform era has brought about the revolutionary transformation of political, socioeconomic, and environmental governance systems, and the water sector is not an exception. China suffers from water shortage, water pollution, and water-related disasters including flood and drought, and degradation of water and ecosystems. These issues are intertwined with structural problems that have stood out during the reform era, such as urbanization, population growth, industrialization, and climate change.

Confronted with these challenges, the Chinese government has endeavored to establish effective water resources management systems over the last few decades for achieving sustainable development. The path to sustainability in China's water resources management since the 1980s has been a long and winding road like the Long March of the Chinese Communist Party to the hinterland in 1934–1935 escaping from the attack of the Nationalist Government (Salisbury 1987). An array of policy shifts favoring sustainable development have been introduced in water resources management, e.g., the Three Red Lines in 2011, the Water Ten Plan in 2015, and the Ecological Red Lines in 2016. These macro- and national-level policies encapsulate a new architecture of policies, plans, and strategies in water resources management that helps envisage the future look of China's water and environment. Ecological civilization for achieving 'Beautiful China' has strongly been promoted since 2016 when the 13th Five Year Plan (FYP) was officially launched, and more emphasis is placed on ecological and environmental matters than pro-growth policies and agendas in the country. China's revolutionary march to modernization has shifted its direction from a high-speed to high-quality and green development, which is reiterated in the newly announced the 14th FYP (2021–2025) in March 2021 (CCICED 2019; Tan 2016; Xinhua 2021).

The book investigates the extent to which the Chinese government has been effective in tackling a complexity of problems in water resources management, highlighting institutional frameworks, including plans, laws and regulations, and economic instruments, and physical infrastructures, such as multi-purpose and agricultural dams, aqueducts, and water and wastewater treatment facilities. Attention is paid to the overall framework of water resources management and sectoral approaches of diverse ministries, agencies, and different levels of local governments together with the contribution of private sector players to water and wastewater services since the late 1990s. An interesting aspect of the book is to broaden the geographical scope of water resources management beyond China's territorial areas, exploring transboundary river issues.

Major Issues

The study investigates the role of water for China's modernization from socioeconomic, political, and environmental perspectives in the reform era with special reference to the two decades since 2000. The study sheds light on socioeconomic and political development in the country that have been facilitated and affected by water resources management and policy. Discussions are on the linkage between water and sustainable development, demonstrating water's multifaceted role for achieving Sustainable Development Goals (SDGs). With the recognition of China's commitment to achieving SDGs until 2030, the study focuses on the contribution of water to accelerating economic development and improving social well-being and ecosystems.

China's per capita water resources in 2019 were 2,078 m^3, which is equivalent to less than one quarter of the world's average and one-sixth of the figure in the US, and therefore, one of the critical challenges for top leaders in the country is to secure sufficient amounts of water. The uneven distribution of water resources between regions, seasonal and temporal variations of precipitation, and a disparity in spatial distribution of water demand and availability all affect water shortage in China. A striking mismatch in terms of water resources between regions in the country is that more than 80% of water resources in China are available in South with more than 50% of the people but 60% of the arable land belong to North China with less than 20% of water and 46% of the total population.

Surface water resources account for 80% of the total amount of water supply, groundwater 19%, and others 1%, which indicates that China should strive to develop unconventional water resources via water reuse and recycle, rainwater harvesting, and desalination. Whereas the agriculture sector has shown a slow growth rate of water use over the past four decades, the household and industrial sectors have increased water use in the same period. Nevertheless, the largest water user in the country is still the agricultural sector, accounting for 61%, the industrial sector, 20%, the household sector, 14%, and the environment, 4% in 2019.

Water quality of major rivers, lakes, and groundwater resources has improved to a great deal over the past four decades, and in 2019, the sections of Grade I-III in freshwater bodies amounted to 74.9%, those of Grade IV, 17.5%, those of Grade V, 4.2%, and those worse than Grade V, 3.4% (MEE 2019). More works are necessary for controlling Non-Point Source (NPS) pollution in suburban and rural areas together with more attention to the quality of groundwater resources.

Floods and droughts have wreaked havoc to millions of people in China every year. Some of the deadliest flood events in the world have badly hit the country, triggering tremendous scales of human and economic losses, often concentrated in the Yangtze and the Yellow River Basins. Droughts accompany long-spelled pains and devastation in food production, water shortage, and damage to public health. Structural and non-structural measures have been introduced for coping with water-related natural disasters, which have proven to be effective for reducing human and economic losses. However, climate change and its adverse impacts have made these disaster prevention efforts sometimes in vain. More detailed and specific strategies and plans should be prepared for coping with new norms driven by climate change.

The two main ways to tackle water shortage are either to supply more water, supply management or to restrict water consumption, leading to a consumption pattern change, demand management. China has experimented a series of demand management measures, i.e., adequate water pricing, water rights trading, and the virtual water and water footprint approach. Sustainable water use has been encouraged through the promotion of an adequate level of water tariffs in Chinese cities over the past four decades. The Beijing Municipality has been a leading city to introduce the cost benefit principle into the water pricing mechanism and put an appropriate level of water pricing concerning the consumption of different water users. Many cities in China have endeavored to follow the case of Beijing although it is still a long way to go for the establishment of the sustainable level of water pricing mechanisms in many urban areas of China. Water rights trading has been practiced more than a decade in the country, and the trading appears to contribute to water saving and sustainable water use between rural communities. The virtual water and water footprint approach provides food for thought related to true water consumption between regions within the country as well as its contribution to global water supply and consumption, which demonstrates China's hidden but significant linkage with global water issues.

The Water Law of China (1988 and 2002) has helped providing basic and fundamental guidelines on how to manage water resources, and subsequent sets of water-related laws have been enacted following the changing needs of society. Numerous laws and regulations have also been established at different levels of local governments, which have served as useful foundations to monitor and regulate water intake, water pollution, illegal actions against water and ecosystems and develop preventive measures against water-related disasters. There is a consensus that China has made good progress in terms of the number of laws and regulations at the central and local levels. However, different judgments are often made pertinent to law enforcement on non-complying public and private units.

Water resources development has been a major task of water engineers in China. Mega-scale water engineering projects have been the manifestation of hydraulic and engineering legacy. The state has been able to secure water, generate hydropower, and protect people from water-related disasters through the Three Gorges Dam and attempts to slake the thirst of North China through various inter-basin water transfer projects, including the South North Water Transfer Project. The dominance of the central government in water policy is often found in many authoritarian regimes in the world, including China, and technocrats of the Ministry of Water Resources, working in the center of power, Beijing, are still having a far-reaching influence on the overall water governance in China. The two cases of water resources development epitomize the top-down, centrally controlled, and supply management-driven solutions which Chinese hydraulic engineers have resorted to among many other options related to demand management and non-conventional water sources, e.g., water reuse and recycling, rainwater harvesting, and desalinated water.

Special attention will be placed on state's interests and hydraulic bureaucracies in water resources development in China, which are shared by four power factors: the Chinese Community Party (CCP), construction companies (state-owned enterprises, mainly); techno-elites (water bureaucrats); and development banks. Such 'synergistic relationships' unveil the way the flows of water are engineered by water infrastructures and the way these infrastructures are intertwined with power and influence (Molle et al. 2009). The 2018 administrative reform has seemed to shake the conventional authority of the MWR over water resources management, giving more mandate to the hands of the Ministry of Ecology and Environment, however, it needs more time to investigate if there would be a fundamental shift of power between water-related ministries.

Increasing levels of water pollution in rivers, lakes, and coastal waters have threatened livelihoods of Chinese people, and the unraveling achievements of socioeconomic development in China can be jeopardized owing to the intensity of water pollution. Numerous water pollution accidents are often reported in rural villages, nature reserves, urban areas, and major rivers, i.e., the outbreak of blue-green algae in Tai Lake, Jiangsu Province in 2007. Primary causes are bundled with the weak level of law enforcement by local environmental authorities, non-compliance of business units, and the dominance of pro-growth policies in localities, which occasionally de-emphasize the strict law enforcement of water and environmental regulations toward local business units. The government has introduced the water pollution trading scheme and the River (Lake) Chief system. The water pollution trading scheme has been implemented for more than three decades, however, the scheme has not worked well due to a myriad of institutional challenges unlike the similar scheme, i.e., water rights trading. The River (Lake) Chief system was introduced after the Tai Lake blue-green algae incident in 2007 and has played a significant role in improving water quality in rivers and lakes with its strong leadership, law enforcement, and the integrated approach to addressing water pollution.

While the central government in Beijing has kept its predominant position in lieu of water policymaking and implementation, the key to success of water resources management and relevant projects depends upon local governments at the provincial, prefectural, municipal, and village levels, which are increasingly collaborating with private sector players. Public private partnership in water and wastewater service projects have prevailed in many parts of the country since the late 1990s. The fast development of private sector participation in water and wastewater services in China has been phenomenal in urban areas, which is attributed to the rapid urbanization, and has demonstrated the strenuous efforts of local governments to improve the quality of public utility services as well as resolve budget constraints. Multinational companies and foreign investors were dominating the newly created water market of China at the beginning, in the late 1990s, however, the Chinese water market has been led by competent Chinese companies in the fields of construction, technology, investment, and consulting since the start of the new millennium.

China's growing quest for water resources had led the country to pay more attention to transboundary rivers beyond its exploitation of domestic water resources. There are 15 major international river basins in China, and the country is the most powerful upstream hegemon in most of the transboundary rivers except for the Yalu-Amnok, the Tumen-Dooman, and the Heilong-Amur Rivers. This strategic position of China has provided little incentives for the government to cooperate with other riparian countries. Even though China has signed various types of agreements and treaties with other riparian countries, the general tendency of these transboundary frameworks seems to show the inactive and reluctant engagement of China with its neighbors.

It is imperative for the country to articulate its strategies dealing with riparian countries as the Belt and the Road Initiative (BRI) is incrementally having profound impacts on different parts of the world. For instance, China has transformed its non-cooperative to more cooperative attitude toward the Lower Mekong countries, i.e., Thailand, Laos, Vietnam, and Cambodia, in recent few years because of political, economic, and security reasons. The Lancang-Mekong Cooperation (LMC) mechanism, which was established by China in 2015, has played a pivotal role in creating a new style of transboundary river governance in the Lancang-Mekong River against the disappointing cooperative regime, the Mekong River Commission.

Structure of the Book

The book begins with the exploration of the interconnectedness between water resources management and its contribution to socioeconomic development and environmental sustainability in China. Chapter 2 focuses on the contribution of water resources management to sustainable development in the country by highlighting the country's pursuit of Sustainable Development Goals following global norms as well as implementing its own tailored targets.

Chapter 3 describes the overall situations of water resources in China, focusing on the up-to-date information on water resources and the features of relevant trends of water supply and use, water quality, and water-related disasters. Attention will be placed on water resources management plans and governance system in Chapter 4, which evaluates major plans for water resources, such as the Three Red Lines in 2011 and the Water Ten Plan in 2015, the administrative reform in 2018, and water-related national laws and regulations.

Chapter 5 sheds light on the demand management methods and policies for sustainable water use in China. In particular, the chapter discusses adequate water pricing in urban areas, the introduction and development of water rights trading, and the applicability of the virtual water and water footprint theories to the Chinese context. Chapter 6 investigates the development of water quality management, exploring the continuous efforts of the government to abate water pollution in freshwater bodies with special reference to the water pollution trading scheme and the River (Lake) Chief System. The chapter also provides the in-depth case study of the blue-green algae incident in Tai Lake, 2007.

Chapter 7 evaluates the trajectory of water resources development in the country, highlighting the two primary topics, large dam development and inter-basin water transfer projects. Large dams have benefited Chinese society in various ways, including agricultural water supply, flood prevention and control, hydropower generation, and inland navigation. Special attention will be paid to the Three Gorges Dam for assessing achievements and challenges. Inter-basin water transfer projects have made substantial contributions to delivering water to water-scarce areas in China, especially in North China, and critical assessment will be conducted for the case of the South North Water Transfer Project.

Chapter 8 is dedicated to exploring the development of water and wastewater service market in China since the late 1990s. Private sector participation in water and wastewater service sectors has developed for more than two decades, and major foci will be on the evolution of related policies and interactions between the public sector (central and local governments) and the private sector (foreign and Chinese companies).

Transboundary river management is the key issue in Chapter 9, which draws readers' attention out of China's territory and to its neighboring countries and appraises the extent to which China has been involved in transboundary river issues. The case study on the Lancang-Mekong River Basin unfolds China's gradual change of attitude toward its riparian countries with new rules of the game through the establishment of the LMC in 2015. Figure 1.1 describes the overall structure of the book coupled with major points on each chapter and Fig. 1.2 summarizes core

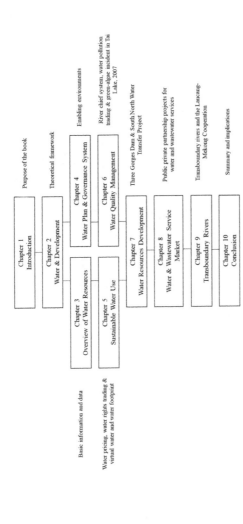

Fig. 1.1 Structure of the book (*Source* Author)

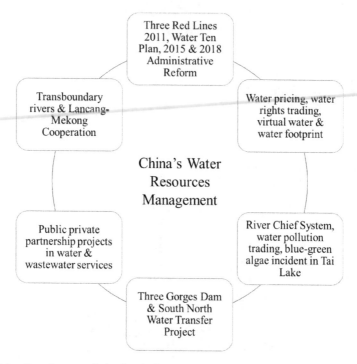

Fig. 1.2 Key themes of the book (*Source* Author)

themes discussed in the book.

References

China Council for International Cooperation on Environment and Development (CCICED). 2019. The Shift to High-Quality, Green Development. CCICED Issus Paper 2019.

Ministry of Ecology and Environment (MEE). 2019. China Ecology and Environment Situation Bulletin 2019 (中国生态环境状况公报). Beijing, Ministry of Ecology and Environment of the People's Republic of China.

Molle, Francois, Peter Mollinga, and Philippus Wester. 2009. Hydraulic Bureaucracies and the Hydraulic Mission: Flows of Water, Flows of Power. *Water Alternatives* 2 (3): 328–349.

Salisbury, Harrison. 1987. *The Long March: The Untold Story*. McGraw-Hill.
Tan, Debra. 2016. Beautiful China 2020: Water and the 13th Five Year Plan. *China Water Risk*. 16 March.
Wang, Xiao-jun, Jian-yun Zhang, Juan Gao, Shamsuddin Shahid, Xing-hui Xia, Zhi Geng, and Li. Tang. 2018. The New Concept of Water Resources Management in China: Ensuring Water Security in Changing Environment. *Environment, Development and Sustainability* 20 (2): 897–909.
Wu, Fengshi, and Richard Edmonds. 2017. Chapter 7: Environmental Degradation. In *Critical Issues in Contemporary China*, ed. Czeslaw Tubilewicz, , 2nd ed., 105–119. London: Routledge.
Xie, Lei, and Shaofeng Jia. 2018. *China's International Transboundary Rivers*. Abingdon: Routledge.
Xinhua. 2021. 14th Five Year Plan of the National Economic and Social Development and 2035 Long Term Goal (中华人民共和国国民经济和社会发展第十四五年规划和2035年远景目标纲要). Xinhua, 12 March.

2

Water and Development

Introduction

This chapter explores the close linkage between water resources management, socioeconomic development, and environmental sustainability in China since the onset of the open-door policy. Water has played a fundamental role in facilitating China's socioeconomic development. Whereas a large body of literature points out negative impacts of the rapidly developing economy on water resources, it is imperative to highlight positive roles of water resources management as a backbone to support the remarkable transformation of contemporary Chinese society into the society with socioeconomic well-being.

Sustainable development epitomizes a shift of the global community from the modernity-centered to the post-modernistic world that is more concerned about the environment in parallel with economic growth and social development. This concept was suggested from the publication of 'Our Common Future' in 1987, and relevant discourses on sustainable development began to draw attention of global communities at the United Nations World Summit on Sustainable Development in 1992. Since then, the brand-new approach to global development has had a far-reaching impact on leaders in many countries, including China. China has taken an active part in an array of international conferences on sustainable development from Rio 1992, Rio+10 (2002), to Rio+20 (2012). The country has incrementally transformed itself from a country shying away from collective responsibilities for the environment to the country which puts sustainable development as a critical policy agenda and national strategy to pursue.

Amongst the many agendas in sustainable development, water is one of the basic and fundamental environmental elements to be safeguarded not only for the environment but also for economic growth and social well-being. In the long history of China, political leaders have striven to harness mighty rivers for daily living, livelihoods, and public health of ordinary people, consolidating their political power, and beefing up agricultural production and industrialization.

In recent few decades, the fast pace of urbanization, industrialization, and population growth in China has entailed a complexity of challenges in water resources management, such as water shortage, water pollution, and water-related disasters, i.e., flood and drought, which are often compounded by climate change. Together with these direct impacts by societal and economic transformations of contemporary China, ensuing natural and anthropogenic problems regarding water issues have followed, including soil erosion, land contamination, a reduction of arable land, and a threat to public health due to drinking unclean water, to name a few.

It is necessary to note that no development in contemporary China has been feasible without critical contributions of water. Water is vital for human survival as generally recognized and serves as an essential component for agricultural and industrial production and energy generation. For example, the agricultural sector in China requires no less than 60% of the total renewable water resources, and thermal or nuclear power plants can produce energy based on vast amounts of cooling water consumption. Keeping up more than 6% of the annual GDP growth rate is a key to legitimizing the ruling of the Chinese Community Party (CCP) over China at the moment, and that is the reason why Premier Li Keqiang pinpointed water as one of the three environmental elements to focus on for creating the 'Beautiful China', i.e., blue sky (air), green land (ecosystems), and clear water, in the 12th National People's Congress when the 13th Five Year Plan was officially launched in 2016 (Tan 2016).

Investigations on these aspects in this chapter will shed light on the three sectors, economy, society, and the environment, and the three sectors should be balanced for achieving sustainable development associated with water resources management. The first part of the chapter focuses on the relationship between water and sustainable development in China by appraising the linkage between water and socioeconomic development. The 2030 development agenda, namely the Sustainable Development Goals (SDGs), will be explored with special reference to SDG 6 Water and Sanitation from global as well as Chinese perspectives.

The second part spotlights the close relationship between water and economic growth, and particular attention will be placed on the extent to which economic growth has been achieved through the optimal use of water resources. Social dimensions of water are explored at the third part, highlighting progress in water and sanitation services of China. Critical roles of water for ecosystems are also investigated, exploring the situations of management of forestry and groundwater resources which remain fragile in the country.

Water and Sustainable Development

Water's contribution to achieving sustainable development is found in many fields, i.e., clean water and sanitation services, water resources conservation and development, sustainability for ecosystems, shared water resources and cooperation between riparian countries, and poverty eradication. Such contributions of water demonstrate the role of water as a medium and linkage for other development agendas (Ait-Kadi 2016). SDG 6 Water and Sanitation reiterates the importance of water as the foundation for the achievement of sustainable development and takes into consideration the impacts of hydrological cycle affected by climate change for development projects (SIWI 2017).

Efficient and sustainable water resources management for society is the starting point and the foundation for poverty eradication, food and energy security, healthy life, sustainable human settlements, the reduction of water-related disasters, and the response to climate change. All these aspects are embedded in 17 SDGs, and therefore, water is a key to success of the 2030 sustainable development agenda.

The connectivity of water for other SDGs the other SDGs is directly described in the following SDGs, such as SDG 3 Good Health and Well-being, SDG 11 Sustainable Cities and Communities, SDG 12 Sustainable Consumption, and SDG 15 Life on Land. With regard to SDG 3 Good Health and Well-being, the two sub-goals of SDG 3 directly mention the roles of water, i.e., 3.3 reduction of water-borne diseases, and 3.9 reduction of water pollution and decrease of relevant diseases and death. SDG 11.3 describes the task of reduction of water-related disasters, e.g., typhoons or cyclones, flashflood, and tsunami.

SDG 12 Sustainable Consumption includes the roles of water referring to 12.4 reduction of chemical pollution from various water bodies, including rivers and lakes, and SDG 15 Life on Land explicitly touches upon the roles of water indicating the task of 15.1 protection from the intrusion of alien species in freshwater ecosystems and 15.8 conservation and rehabilitation of freshwater ecosystem services (UNDESA 2015).

Discussions on the direct linkage between water and the other SDGs display the imperative roles of water for sustainable development. ICSU and ISSC (2015) maintain that SDG6 directly and indirectly relates the other SDGs, and safeguarding multi-dimensions of water can have a far-reaching impact on the achievement of the other SDGs which do not mention the roles of water directly, including SDG 2 Zero Hunger, SDG 5 Gender Equality, SDG 7 Energy, SDG 13 Climate Change and SDG 14 Life below Water. Figure 2.1 summarizes the far-reaching impacts of SDG 6 on the other SDGs, which are also closely tied to the success of water resources management.

Let us turn attention to China's commitment to achieving sustainable development through water resources management. As a new beacon of China's modernization, Premier Li Keqiang presented the idea of 'Beautiful China' at the fourth session of the 12th National People's Congress in 2016. The propaganda advocates the constant commitment of the government to revitalizing its environment by making the sky blue, the land green, and the water clear. This approach was neatly entrenched in

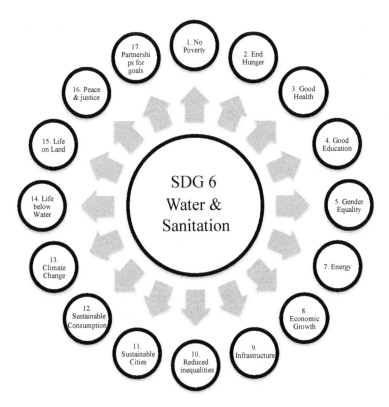

Fig. 2.1 Substantial impacts of SDG 6 water and sanitation on the other SDGs (*Source* Author)

the 13th Five Year Plan (FYP) (2016–2020) with a particular emphasis on environmental protection and ecosystem rehabilitation (Tan 2016).

The reality is far from the situation of picturesque landscapes with high degrees of biodiversity and plenty of clean water resources. Regarding water resources, the State of the Environment Report by the Ministry of Environmental Protection, 2016 succinctly illustrated grim situations of the quality of major water bodies, with 76% of groundwater resources being assessed as 'poor and very poor', and around 29% of major rivers, Grade IV, V, and worse than V, being unsuitable for human contacts owing to heavy pollution (MEP 2017).

Details of targets for water resources management in the 13th FYP mirror which fields or sectors the Chinese government should focus on. As for water use efficiency, the plan embraces the target of the reduction of water consumption per RMB 10,000 of GDP at 5.1% even though it is expected to witness a surge of water consumption in absolute terms thanks to the GDP growth in the region of 6.5–7% per annum. It seems that the country has shown its commitment to the structural optimization of economy by shifting its development focus from the over-exploitation of natural resources to more efficient utilization of resources, which can bring about a significant change in the water consumption pattern (Hao et al. 2019; Tan 2016).

Sustainable water resources management can be achieved through the deployment of an integrated approach. There are several aspects the central government should consider, and there should be a balanced distribution of sufficient amounts of water resources between rural and urban areas in China. Urban centers have achieved a great deal of contribution to China's rapid socioeconomic development. The trajectory of China's urban development has demonstrated an upsurge of megacities in many parts of the country, such as Beijing, Shanghai, Guangdong, and Chongqing and an emergence of innumerable cities with more than several million residents. More water demands have been observed in such cities and urban centers by both households and industrial sectors.

It is also important to notice that approximately half of the total cultivated area has been allocated for irrigated farming, which accounts for 64% of all water withdrawals. The agricultural sector takes responsibility for food security and livelihoods in rural communities that are closely associated with complex socioeconomic issues, i.e., migration between rural and urban areas, the divide between rural and urban areas in terms of income levels as well as living standards, and the ideological (socialist) backbone for the Chinese Communist Party (CCP). Whilst the contribution of the agricultural sector to GDP has decreased since the late 1970s, the magnitude of food security appears to be more significant than before considering adverse impacts of climate change that can engender extremely long spelled and intensified droughts. Soaring water demands from the urban and rural sectors may give tremendous pressure to water resources. It is imperative to consider integrated approaches in allocating water (GWP 2015).

The philosophy of sustainable development has increasingly been entrenched in China's national policies, including water resources management policies. Currently, Chinese society fully embraces the concept of ecological civilization, putting the environment ahead of other policy priorities, even socioeconomic development, and the SDGs have been regarded as a salient global norm which China should follow not only as a member of the global community but also the country that successfully achieves the goals better than any other countries.

Right after the establishment of the SDGs in 2015, the Chinese government took an urgent action to set up the National Plan on Implementation of the 2030 Agenda for Sustainable Development in 2016. The plan encompasses guiding thoughts, a general roadmap, and specific plans for implementing the SDGs. Furthermore, the central government will cling to the basic national policy of resource conservation and environmental protection, make a civilized development path, and achieve green development (Dai 2019; Government of PRC 2016).

China's basic plan for SDG 6 Water and Sanitation encompasses specific actions conforming to the sub-goals of SDG 6. Whilst following the SDG 6 targets, the government has specified its own action plans. For instance, for achieving 6.1 universal and equitable access to clean drinking water, the government pledged to embark on the project of consolidating and improving the safety of rural drinking water. This reflects the commitment of the central government to enhancing general conditions of water supply services for rural communities, which have lagged compared with urban communities. Specific targets are set. By 2020, more than 85% of rural areas will have access to centralized water supply, and universal and equitable access to safe and affordable drinking water will be guaranteed by 2030.

Sanitation and hygiene issues are emphasized for rural areas, too. For achieving 6.2 adequate and equitable to sanitation and hygiene for all, the government is committed to achieving a full coverage of water hygiene infrastructure. By 2030, investments and policy measures will be devoted to renovating rural household toilets. Pertaining to 6.3

improving water quality, the government explicitly points out the magnitude of the Water Pollution Prevention and Control Action Plan (Water Ten Plan) 2015 for safeguarding good water quality in major river basins and coastal water areas.

Water use efficiency is the key focus in SDG 6.4, and water demand and consumption management will be improved in China coupled with close monitoring and control of total volume and intensity of water resource consumption. Specific figures are presented for enhancing water use efficiency through the introduction of water efficiency assessment systems, for instance, the water use amount per RMB 10,000 of GDP. Effective use of irrigation water and the reduction of water consumption are promoted in accordance with the Three Red Lines in 2011 which includes future water planning targets.

SDG 6.5 Integrated Water Resources Management and Transboundary Water Management are linked to the government's push for strengthening river basin management based on the system of the Seven River Commissions and the administrative reform in 2018. Unfortunately, there is no specific plan to tackle transboundary water issues from the Chinese perspective, which may imply the conventional low priority of transboundary rivers for China's water resources management. The protection and restoration of water-related ecosystems, SDG 6.6, are closely associated with China's recent pledge of ecological civilization, and the government shows its commitment to restoring water-related ecosystems and ameliorating the abuse of groundwater resources.

SDG 6.a, international cooperation, is linked to the Belt and Road Initiative for infrastructure development as well as non-physical infrastructure development, e.g., capacity building for resource conservation, climate change mitigation, and green and low-carbon development. China has already undertaken a series of policies and programs related to SDG 6.b participation of local communities, such as the establishment and running of water users' association in rural areas and pledged to support the improvement of local communities' participation in water and sanitation management (Government of PRC 2016) (see Table 2.1).

Progress on the overall implementation of the SDGs has been reported by the Chinese government since 2016, and the most recent national report was published in September 2019. Good progress has been made

Table 2.1 Targets for the Sustainable Development Goal 6 and related actions by China

SDG6. Ensure availability and sustainable management of water and sanitation for all

SDG Target	China's Action Plan
6.1 By 2030, achieve universal and equitable access to safe and affordable drinking water for all	Launch the project of consolidating and improving the safety of rural drinking water By 2020, centralized water supply rate will exceed 85% and tap water coverage rate will exceed 80% in rural areas By 2030, achieve universal and equitable access to safe and affordable drinking water for all
6.2 By 2030 achieve access to adequate and equitable sanitation and hygiene for all; and end open defecation, paying special attention to the needs of women and girls and those in vulnerable situations	Work toward full coverage of water hygiene infrastructure. By 2030, complete revamping of rural household toilets and achieve access to adequate equitable sanitation and hygiene for all
6.3 By 2030, improve water quality by reducing pollution, eliminating dumping, and minimizing release of hazardous chemicals and materials, having the proportion of untreated wastewater, and substantially increasing recycling and safe reuse globally	Implement the Water Pollution Prevention Control Action Plan to increase the proportions of good quality water in key river basins and coastal water areas and qualified treatment of sewage water. Intensify monitoring of key functional water zones and sewage discharge outlets, and strengthen categorized and tiered management of functional water zones
6.4 By 2030, substantially increase water-use efficiency across all sectors and ensure sustainable withdrawals and supply of freshwater to address water scarcity and substantially reduce the number of people suffering from water scarcity	Build a water-saving society in a holistic manner by enforcing the strictest water resources management system, strengthening water demand and water consumption management, and exercising dual control of total volume and intensity of water resources consumption

(continued)

Table 2.1 (continued)

SDG6. Ensure availability and sustainable management of water and sanitation for all

SDG Target	China's Action Plan
	Establish water efficiency assessment systems such as the water consumption quantity per RMB 10,000 of GDP, and continuously improve water efficiency across all sectors
	By 2020, increase the effective use of irrigation water to above 0.55 nationwide, and reduce water consumption per RMB 10,000 of GDP and per RMB 10,000 of industry added value by 23% and 20%, respectively
6.5 By 2030, implement integrated water resources management at all levels, including through transboundary cooperation as appropriate	Improve the water resources management system that combines river basin management and administrative area management, and enhance the role of comprehensive river basin management in water governance
6.6 By 2020, protect and restore water-related ecosystems, including mountains, forests, wetlands, rivers, aquifers, and lakes	Build a national ecological security framework to protect and restore water-related ecosystems by managing the overuse of groundwater in some areas. By 2030, endeavor to improve the national overall water quality and generally restore functions of water ecosystems
6.a By 2030, expand international cooperation and capacity-building support to developing countries in water- and sanitation-related activities and programs, including water harvesting, desalination, water efficiency, wastewater treatment, recycling, and reuse technologies	Actively advance the South-South Cooperation on water- and environment-related areas; help other developing countries strengthen the capacity building for resource conservation, climate change mitigation, and green, low-carbon development; and provide them with assistance and support within China's capacity

(continued)

Table 2.1 (continued)

SDG6. Ensure availability and sustainable management of water and sanitation for all	
SDG Target	China's Action Plan
6.b Support and strengthen the participation of local communities improving water and sanitation management	Continue to exercise working mechanism that involves water users' participation and support; strengthen and urge the participation of water users and local communities in improving water and sanitation management

Source Government of PRC (2016)

pertinent to access to drinking water and water use efficiency for both urban and rural residents. Substantial contributions of the South North Water Transfer Project had been made for easing water stress in Beijing and northern parts of the country with more than 23 billion m^3 of water diverted from the Yangtze River. The proportion of centralized water supply and tap water in the rural areas increased from 82 to 88% and 76 to 81% between the end of 2015 and the end of 2018, respectively.

As for water quality enhancement, by the end of 2018, 4,332 wastewater treatment plants had been constructed in cities and towns with a daily capacity of 195 million m^3. The 13th FYP for the Comprehensive Improvement of the Rural Environment pushed forward diverse programs to protect drinking water sources for rural areas and treat household waste and wastewater, and in 2018, environmental improvement projects were undertaken in 25,000 administrative villages at the national level, including the enhancement of toilet facilities.

Strict observation of the Three Red Lines in 2011 is praised for effective control of water consumption. Particularly, a variety of water saving technologies, incentives, and management had culminated in the increase of water use efficiency. For instance, the effective utilization coefficient of farmland irrigation water surged from 0.536 in 2015 to more than 0.55 in 2018. Water consumption per RMB 10,000 GDP and per RMB 10,000 industrial added value in 2018 was 18.9% and 20.9% lower, respectively.

Technological advancement for river basin management had optimized water resources management, and national and local efforts had been made for restoring water-related ecosystems. The River (Lake) Chief system has contributed to ensuring water quality enhancement, coordinating diverse works between environmental agencies, and supervising environmental regulatory works effectively. The National Water and Soil Conservation Plan (2015–2030) was introduced, and in 2018, additional 64,000 km^2 of soil erosion damaged land were treated, and 345 ecologically clean small watersheds were identified.

The government demonstrated an increase of cross-border cooperation. With an emphasis of the South-South cooperation, the government had conducted numerous capacity building programs in areas such as water resources planning, flood control and disaster reduction, water-saving irrigation, and small hydropower development, to name a few. The Lancang-Mekong Cooperation Mechanism (LMC) is China's symbolic transboundary water cooperation vehicle for Mainland Southeast Asia, and the Lancang-Mekong River Environment Cooperation Center was created as part of the mechanism.

Despite such achievements, the top priority areas for improvement in China's SDG 6 are threefold. First, more progress is necessary for achieving the basic water and sanitation target for rural residents. Second, the government should enhance the legal system for water safety and law enforcement, undertaking scientific and technological innovation, and to improve urban sewage treatment services together with the promotion of reclaimed water. Third, priorities should be given to improve water distribution in major river basins, to control the overuse of groundwater resources in North China, to facilitate water conservation, and to be continuously committed to undertaking water-related South-South cooperation and transboundary river cooperation (MFA 2019).

Water for Economic Growth

A common notion on the relationship between water and economic growth is water as the vector for socioeconomic development. Cox

(1987) maintains the significant contribution of water resources management to economic development referring to incremental improvement of socioeconomic conditions in various developing countries. A body of water and development literature highlights the key roles of water infrastructure for society in terms of facilitating water supply, sanitation services, agricultural water supply, flood protection, and inland navigation (HSBC 2012; Min et al. 2013; Sunsik 2015; Tropp 2013). A group of scholars in South Korea pays close attention to the relationship between water and economic growth, arguing that Korea's rapid economic growth has been possible thanks to efficient and sustainable water resources management (Lee et al. 2018).

The role of water for China's economic growth is neatly reflected in numerous historical records, particularly supporting agricultural production and promoting complicated systems of irrigation in every possible plot of land in the country. The increase of agricultural production via irrigation projects in the northern part of China has made politically powerful governing groups and sophisticated bureaucratic systems possible together with the rise of rich and wealthy merchants. Such an anecdotal evidence on the relationship between water and economic growth becomes more conspicuous in the reform era.

In the 1980s, China's water resources management primarily focused on agricultural water supply and made tremendous efforts for the rapid achievement of socioeconomic well-being. At the advent of the 1990s, more weight was laid on developing water resources for industrialization and the initiation of industrialization of water and sanitation services opting for cost recovery and reasonable profits for future investment. Economic diversification and industrial restructuring in the new millennium accompanied the growing awareness of environmental issues, including water and ecosystem rehabilitation (Shen and Wu 2017). A series of water policy initiatives, such as the Three Red Lines in 2011, the Water Ten Plan in 2015, and the Ecological Red Lines in 2016, epitomize recent commitments of the central government to addressing water challenges more seriously under the framework of sustainable development.

China's economic development since the late 1970s has been phenomenal with the remarkably high annual GDP growth rate. Between 1978 and 2016, the annual GDP growth rate averaged at 9.5%, which has made China an upper-middle-income country with a per capita GDP of US$ 8,640 in 2017, measured by market exchange rates, from a per capita GDP of US$ 155 in 1978 (Lin and Shen 2018). Despite the remarkable socioeconomic development in the past four decades, the socialist market economy of China is at a crossroads. The era of the almost two-digit GDP growth has been over due to endogenous and exogenous factors, such as an upsurge of debt of local governments, lower employment rates, and the US-China Trade War. China's average GDP growth rate per annum was recorded at 6.6% between 2016 and 2019, for instance (Lin and Shen 2018; NBS 2021).

Against this backdrop, the CCP opts for doubling the 2010 GDP and per capita personal income by 2020 but this has not been achieved, recognizing the GDP growth of 2.3% in 2020 because of the unexpected global pandemic of COVID-19 (Cheng 2020; NBS 2021). Can this ambitious goal be achievable with the paucity of sufficient amounts of water resources? The lack of water resources can trigger detrimental influences on agricultural and industrial productions in major agricultural and industrial bases such as the North China Plain.

A conventional approach to the exploration of the roles of water for economic growth is to appraise the linkage between water consumption and economic growth. For instance, GWP (2015) stresses that the value of industrial production from each cubic meter of water consumption in China is equivalent to one-third of the world average, and the GDP production from each cubic meter of water consumption is at best one-fifth of the world average. A low level of water use efficiency serves as a hindrance for China's continuous economic growth, and for instance, 108 m^3 of water is required per US$ 1,600 (RMB 10,000) industrial added value. It is estimated that irrigation efficiency reaches about 48% on average. A leakage rate of piped water networks in many cities of China is as high as 20%, and the rate of industrial water reuse reaches only about 60%.

This approach, however, does not seem to be straightforward as it looks, because there would be more than one factor, the measurement of water consumption, to influence the contribution of water to economic growth. Hao et al. (2019) stress that an adequate empirical analysis should require several variables for addressing the roles of water in achieving economic growth, not only water consumption, but also per capita water resource, real per capita GDP, population density, trade openness, the ratio secondary industry value added to GDP, and per capita real capital stock. There are three major variables that mainly affect the demand for water resources: (1) population scale; (2) trade openness; and (3) industrial structure. It is common to note that population growth in China would require more water resources thanks to the growing number of newborn people as well as migration from rural to urban areas. The variable of trade openness is instrumental, which displays the ratio of the export and import trade to the current GDP and the extent of trade liberalization. This would be one of the determinants to engender the difference in demands for water resources.

The reason why the variable of the ratio secondary industry value added to GDP is necessary reveals the fact that the early phase of economic development would necessitate more demands of water resources due to a heavy reliance on the agricultural and industrial sectors whilst service and information-focused industries would require less water resources. The study results of Hao et al. (2019) demonstrate the N-shaped relationship between water use and economic growth, which implies that more economic growth and accumulated wealth would not guarantee a decrease of water consumption. In order to make such a pattern into an inverted-N shaped one, the Chinese government should be committed to enhancing water use efficiency by altering population, trade and industrial policies and introducing cutting-edge technologies for water saving.

In China, market-based and regulatory programs have culminated in a reduction of water consumption since the 1980s whereas GDP has continued to increase in the same period. Wang et al. (2018) provide a decoupling analysis of water use from economic growth in some of the most industrialized and developed urban centers in China, namely, Beijing, Shanghai, and Guangzhou. Shanghai is the most developed city with the economic value of RMB 1,895.3 billion (US$ 261.5 billion) in 2015 from RMB 789.2 billion (US$ 121.4 billion) in 2005, which shows an average annual rate of increase at 9.16%. Guangzhou demonstrates the higher growth rate of its economy than that of Shanghai's, estimated at 11.78%, increasing from RMB 515.42 billion (US$ 79.3 billion) in 2005 to RMB 1,569.44 billion (US$ 241.4 billion) in 2015. The third place is Beijing, whose GDP growth rate was estimated at 9.31% from RMB 463.82 billion (US$ 71.3 billion) in 2005 to RMB 1,129.56 billion (US$ 173.7 billion) in 2005.

The trend of water use and wastewater discharge of these cities largely epitomizes the broken linkage between economic development and water use or wastewater discharge, thereby showing a good practice of decoupling. The pattern of water consumption in Shanghai from 2005 to 2015 indicates a decreasing trend from 12.128 billion to 10.38 billion m^3, Guangzhou from 8.361 billion to 6.614 billion m^3. However, Beijing's water consumption demonstrates an upward trend, from 3.45 billion to 3.82 billion m^3. These results imply the successful decoupling of water consumption from economic growth in Shanghai and Guangzhou whereas Beijing requires multifaceted policy efforts to achieve decoupling.

There are developed economies with little endowment of water resources that have made impressive socioeconomic development, such as Japan, France, Germany, and the UK after transforming their industrial structure from the agriculture and manufacturing to tertiary or knowledge-based structure. Such a success is possible thanks to the import of water-intensive products, in other words, the virtual water trade with larger external water footprints. Although the per capita GDP of China is much lower that of the developed economies, China manages to feed its own people and to provide water resources for its own diverse needs from household, agriculture, industries, and the environment. In 2009, China exported virtual water to overseas countries, amounting to 153.3 billion m^3, which were in the forms of textile, clothing, electrical and telecommunication equipment, and passenger transport items (Hou et al. 2018).

An array of five-year socioeconomic development plans of the Chinese government has laid out the priorities and targets for public and private sectors, which encapsulate a centralized notion of what China should achieve, its vision for the foreseeable future, and the way the country should make policies and investments for accomplishing its goals. The five-year plans have evolved from economic growth-centered approaches to symbiotic policies between economic growth and ecological conservation, particularly water resources. However, the supposedly parallel and balanced relationship between the two sectors has not necessarily worked well.

The 13th FYP (2016–2020) provided an urgent call for greater efficiency in developing and using water resources and in requiring a dramatic decrease of the discharge of major pollutants in water bodies. In tandem with the environmental objectives, the plan advocates the promotion of the socioeconomic development of the Beijing-Tianjin-Hebei Region and the Yangtze River Economic Belt and accommodates the country's ambition to implement the Silk Road Economic Belt and the 21st-Century Maritime Silk Road (or the Belt and Road Initiative). Special attention is placed on urban–rural integration, broadened space for rural development, and carefully designed measures for poverty alleviation and elimination. Ecological rehabilitation plans are also included for forests, rivers, lakes, wetlands, oceans and their ecosystem functionality.

There is a growing recognition that water can be the most salient resource bottleneck in the continued growth pathway of China in the next 10–15 years. Water shortage and water pollution for a few decades have cost China about 2.3% of GDP per annum, and the calculation of all types of pollution damages range from 6 to 9% of GDP. Even though the central government envisages that 60% of all Chinese people would live in a 'moderately prosperous society' in cities by 2020, the mounting tension between China's continued medium–high economic growth and the sluggish recovery of its damaged water systems serves as a major hindrance. These phenomena can be compounded by mediocre levels of service delivery and poor coordination among diverse institutions, agencies, and bureaus in charge of water resources management (ADB 2018). It is maintained that China's growth and prosperity of economies, cities, and population will depend upon sustainable water resources management.

The Three Red Lines was proposed in 2011 for capping water consumption, increasing water use efficiency, and improving water quality until 2030 and demands the whole country to be determined to save invaluable water resources (World Bank 2018). Thanks to this plan, public authorities in China have produced quite a few technical and engineering as well as economic instruments for saving water and improving water use efficiency. This unprecedented policy commitment is undeniably useful, especially allocating responsibilities to diverse sectors and introducing water saving technologies.

More innovative and ground-breaking policies and programs should be considered, such as: (1) the promotion of various economic instruments for providing incentives and levying penalties related to water consumption; (2) cropping pattern changes for less water-intensive products and the upgrade of industrial structure toward tertiary industries; (3) more import of water-intensive products and less export of water-intensive products; and (4) the encourage of water reuse, recycling, and water saving (China Water Risk and MEP 2016).

Leaders of the CCP seek for means to bolster the national economy with three national strategies: (1) the Belt and Road Initiative; (2) development of the Beijing-Tianjin-Hebei Region; and (3) development of the Yangtze River Economic Belt. The strategies summarize the government's thrust for vertical and horizontal economic axes which comprise economic belts along sea, river, and border through the overall strategy of regional development. The changing rhetoric from economic belts to economic axes implies the transfer of China's regional development into integration and coordination. The three national strategies will serve as a guide and catalyst to integrate the eastern, central, western, and northeastern regions. Whilst China's outward engagement continues through the BRI, the national integration through regional development strategies will usher in a wide opening along sea, river, and border transport routes to new domestic, regional and international markets. Effective water resources management through the construction of ecological civilization will be a prerequisite for such an ambitious proposal (ADB 2018).

The BRI is officially dubbed as the Silk Road Economic Belt and the 21st-Century Maritime Silk Road, which was announced by President Xi Jinping during his state trip throughout Central Asia and Southeast Asia in 2013. Specifically, the Silk Road Economic Belts embrace inland from China to Europe through Central Asia and Russia, to the Persian Gulf and the Mediterranean Sea through Central and West Asia, and to Southeast Asia, South Asia, and the Indian Ocean. Coastal ports of China are connected to the Indian Ocean and to Europe, and to South Pacific through the 21st-Century Maritime Silk Road.

As of 2018, 72 countries were involved in various infrastructure projects, including water, energy, food, and transport sectors. Motivations for the BRI are described as the enhancement of connectivity, openness, and innovation. In addition, sustainable development is one of the significant motivations for the initiative, which serves as a useful vehicle to achieve SDGs. Water conservation is one of the critical fields coupled with energy and food security which are inextricably associated with water security. China's water sector will contribute to the BRI through water diplomacy which plans to promote bilateral and multilateral cooperation with countries along the new Silk Road and belt. In particular, Chinese experiences and technologies will be instrumental for planning, survey, design, construction, and technology of hydraulic infrastructures such as hydropower dams (ADB 2018; OECD 2018).

Regional development strategies have been proposed at the national level since the late 1990s, such as the Western Development Strategy in 1999, the Beijing-Tianjin-Hebei Region in 2014, the Yangtze River Economic Belt in 2016, the Pearl River Delta Area 2018, and the upcoming strategy for the north-eastern region. Particular attention has been placed on the promotion of coordinated development of the Beijing-Tianjin-Hebei Region and the Yangtze River Economic Belt in the 13th FYP (ADB 2018; Xinhua 2019).

The Beijing-Tianjin-Hebei Region has been the center of politics, economy, culture, science and technology and regarded as an important growth pole in the eastern China and a major socioeconomic development engine. Whereas flood control and water supply systems have neatly been established in the region, the concentration of population and industries over a few decades has culminated in giving tremendous pressure on water resources and the environment. As a result, various water challenges occur, e.g., water shortage, aquatic ecosystem degradation, heavy water pollution, and inter- and intra-provincial conflicts for water use.

An unsustainable pattern of water supply and use in the region is observed in the over-exploitation of groundwater resources. It is interesting to note that the portion of agricultural water use is rather high with less water use from the tertiary sector and for the environment in Tianjin and Beijing. Tianjin's agricultural water use in 2019 amounted to 32.5% of the city's total water use whereas the household sector accounted for 26.4%, the industrial sector 19.2%, and environmental flow 21.9%. Groundwater resources contributed to almost 20% of the total water supply in the city. Non-conventional water resources, e.g., reclaimed, brackish, or desalinated water, appear to be under-utilized. For instance, desalinated water supply in the city was responsible for 11.2% of the total surface water supply in 2019 (Tianjin Water Authority 2019).

As the government unveils its primary blueprint of coordinate development in the region, Beijing may streamline its administrative functions, which leads to shifts in demand, changed means, and readjusted structure of water use. In addition, the city has relieved its water stress based on the transferred water through the middle route of the South North Water Transfer Project since 2014, which has, to some extent, enhanced the carry capacity of the city, and nearby areas within the region enjoy similar benefits through the transfer project. Nevertheless, water will continue to be a precious resource thanks to continuous pressure from socioeconomic development in the region. The circumstances will prompt the introduction of new types of agriculture, industry and commerce, which demands low water-intensive industries, primarily high-tech and high-value industries, and low water-use agriculture. Water use efficiency of industries in the region should be enhanced for adequately controlling the over-exploitation of groundwater resources, rehabilitating river and lake ecological systems, and seeking for water supply alternatives through the promotion of water reuse and recycle (ADB 2018).

Core policy directions in the Yangtze River Economic Belt (YREB) were explicitly addressed by President Xi Jinping in January 2016 with a particular emphasis on holistic ecological protection and green development along the Yangtze River. The launch of the YREB Development Plan encapsulates China's continuous commitment to enhancing ecological conditions and environmental sustainability in parallel with economic growth (China Water Risk and MEP 2016).

The Yangtze River in the YREB runs over 6,300 km, whose size of the river basin is as large as 1.8 million km^2. The river provides around 36% of freshwater resources for the country and received approximately 45% of the total amount of wastewater discharge in 2016, especially contributed by heavy industries, such as steel mills and petrochemical factories (Davies and Westgate 2019).

There are 10 provinces in the river basin, including Qinghai, Tibet, Yunnan, Sichuan, Hubei, Hunan, Jiangxi, Anhui, Jiangsu, and Guizhou together with the two centrally administered municipalities, Chongqing and Shanghai. The areas belonged to the YREB are slightly different from the provinces included in the river basin, which are Jiangsu, Zhejiang, Anhui, Hubei, Hunan, Sichuan, Jiangxi, and Guizhou alongside the Chongqing and Shanghai Municipalities. In the belt, more than 580 million people are living, and the river has been the most significant inland navigation route in the country, dubbed as 'the golden waterway' for transporting goods between cities and villages. Different kinds of agricultural and industrial products are shipped from inland areas to major port cities such as Chongqing and Wuhan, and then are transported to the other parts of China through railways and motorways. Shanghai and Ningbo, which are located downstream and some of the largest ports in the world, export and import products (China Water Risk and MEP 2016; Davies and Westgate 2019).

Pertaining to water resources management, the belt functions as a testing ground for how to sustainably manage water use and allocation, and water quality control in parallel with economic development. One of the primary causes for the central government to designate such a vast region as the special economic belt is to beef up local economies against the dwindling GDP growth rate in recent few years. The other imperative motive is to safeguard the environment, which explicitly indicates the government's policy shift for striking the balance between economic growth and environmental protection.

It is crucial to investigate the situations of local economies upstream, largely spearheaded by heavy industries, for lessening pollution burdens downstream, which can enhance polluted water bodies downstream in the Yangtze River Estuary. The river plays a key role in augmenting water supply for the thirsty North China via the South North Water Transfer Project and will be expected to contribute more to energy generation via hydropower and nuclear power plants.

In particular, the role of the river for energy generation underscores the magnitude of policy implications for the Water-Energy-Climate nexus that has a far-reaching impact on China and beyond, especially for the Hindu Kush Himalayan region. The region is also called, 'the Water Towers of Asia', and some of the largest rivers in the world start to flow from the high mountains and valleys in the region, i.e., the Heng-Ganges, the Indus, the Yarlung Tsangpo-Brahmaputra, the Nu-Salween, the Lancang-Mekong, the Yangtze, the Yellow and the Tarim Rivers (Scott et al. 2019). The latter four rivers originate from China so that impacts of China's socioeconomic development and ecological protection activities along the rivers can be influential domestically as well as globally. In this sense, China's sustainable water resources management in line with economic development influences the success of sustainable development in Asia and beyond.

As the socioeconomic powerhouse, the YREB occupied 43% of the total population of the country and was responsible for RMB 28 trillion (US$ 4.3 trillion) in Gross Regional Product in 2014, which accounted for approximately 42% of the total national GDP. A close look at the industrial structure of the YREB unveils the surprising level of its contribution to the overall socioeconomic development. As of 2014, the region produced 65% of rice, 58% of chemical pesticides, and 51% of fertilizers in the agricultural sector and generated 40% of electricity and 73% of hydropower in the energy sector. In addition, in the building material sector, the region took responsibility for producing 48% of cement, 40% of primary plastic, and 35% of crude steel. Around 81% of chemical fibers and 59% of cloth in China were produced in the region (China Water Risk and MEP 2016).

Despite such an impressive contribution to diverse industries, the YREB has its limitations, i.e., the lower level of water use efficiency for industries. Industries in the YREB have turned out to be uncompetitive compared with counterparts in other areas of China. For instance, water use efficiency per unit of industrial value added of the YREB in

2014 was estimated 24.9% higher than the national average.[1] This figure, however, does not include the contribution of agriculture and service sectors so that the actual water use per unit of GDP in the YREB was even 4.6% lower than the national average. It is necessary for the leaders of the YREB to take serious consideration the improvement of water use efficiency in industries (China Water Risk and MEP 2016).

Water for Social Development and Ecosystems

Social aspects of water resources management are related to the overall enhancement of the quality of life and living standards, particularly, regarding the level of universal access to water and sanitation at the national level. The newly coined motto, 'Beautiful China', embraces the social dimension regarding health as the root of happiness and the urgency to promote social well-being in 2016. The Chinese government announced that no less than 300 million people in rural areas had gained access to safe drinking water in the period of the 12th FYP (2011–2015). More comprehensive efforts will be made in order to prompt the enhancement of social well-being in rural areas by ensuring more than 80% of rural population in the country having good access to tap water supply during the 13th FYP (2016–2020).

Such policies should accompany sheer scales of investment in rural water supply and sanitation services alongside special attention to the improvement of water quality in local streams and rivers. Environmental authorities are committed to decreasing the rate of Chemical Oxygen Demand (COD) over ammonia nitrogen by 10% in the 13th FYP, which is one of major Non-Point Source (NPS) pollutants. In addition to these, a list of toxic pollutants needs more attention, i.e., endocrine disruptors (environmental hormones), antibiotics, and other persistent organic pollutants in Chinese freshwater bodies (Tan 2016).

---

[1] Water use efficiency is defined as the value added of a given major sector divided by the volume of water used. Available Online: http://unstats.un.org (accessed 31 March 2021).

According to the Joint Monitoring Programme for Water Supply, Sanitation, and Hygiene in 2019, 93% of the total population of the country had access to clean drinking water basic services as of 2017. The figure demonstrates sound circumstances at the national level, including 92% of urban areas with safely managed services[2] and 5% with basic services, and 86% of rural areas with basic services. As for sanitation services, situations are somewhat different. The same data source indicates that 85% of the total population of the country enjoyed basic sanitation services, including 84% of urban areas with safely managed services[3] and 7% with basic services and 72% of rural areas with safely managed services and 13% with basic services (WHO and UNICEF 2019). These statistics summarize that China has made impressive progress in several decades in terms of providing clean water supply and adequate sanitation services and is recognized as one of the successful cases in which a good level of water and sanitation services has been achieved in a short period of time.

There is still a gap between urban and rural areas, developed and under-developing areas, i.e., the divide between East and West, the rich and the poor in many provinces, prefectures, and municipalities. Even though scholars and experts can praise what has been done in the sectors of water and sanitation services in China, the fundamental challenge is how to narrow the gap between such divisions in various aspects.

Another issue that should not be ruled out is the social safety net system which guarantees universal access to water and sanitation for all Chinese people. This notion sounds familiar to those who have been following global discourses on water and development problems, and

---

[2] According to WHO and UNICEF (2019), safely managed drinking water services, an indicator, 6.1 of Sustainable Development Goal (SDG) 6, are defined as use of an improved drinking water source which is accessible on premises, available when needed and free from contamination. Information on the use of improved drinking water sources is combined with information on the accessibility, availability, and quality of drinking water. Basic services of drinking water do not include the elements mentioned above.

[3] According to WHO and UNICEF (2019), safely managed sanitation services, an indicator, 6.2 of SDG6, are defined as use of an improved sanitation facility which is not shared with other households and where excreta are disposed in situ or transported and treated offsite. Information on use of different improved sanitation facilities types (sewer connections, septic tanks, and latrines and other) is combined with information on containment, emptying, transport, and treatment. Basic services of sanitation services do not include the elements mentioned above.

the phrase is the exact wording for water challenges in the 2030 Sustainable Development Agenda, SDG 6. In addition, to achieve this policy goal means part of the ultimate goal of Chinese revolution, which has primarily been aimed at making the masses meet basic needs with the full responsibility of the CCP. It is true that the party should provide basic needs for the masses from the beginning so that water and sanitation services for all Chinese people have been one of the top priorities that should be fulfilled against all odds. Such an egalitarian approach is still valid, and there are various ways to support the ideological foundation for water resources management, imposed by the central government, particularly the level of water pricing.

Water pricing for the poor or the marginalized is widely discussed not only in developing countries but also in developed countries. The water affordability scheme is carefully designed for guaranteeing the basic right to have universal access to clean water supply and adequate sanitation services in many countries. Some countries offer water vouchers for the poor whilst other countries, such as England and Wales, have introduced legal settings for ensuring poor families' access to clean water and proper sanitation services with social tariff, called 'WaterSure Plus'. This tariff ensures that low-income households (about US$ 21,000 per annum) enjoy a discounted rate on their water bills, for instance, at 50% in the year between 2019 and 2020 (Thames Water 2020). The case of England and Wales is a useful benchmarking practice since many Chinese cities are increasingly depending upon private sector players' water and sanitation services.

Shen and Wu (2017) present a detailed overview of water pricing reform from the late 1970s to 2014 and conclude that continuous reform efforts of water pricing in China have been the outcome of sociopolitical concerns rather than economic considerations. Comparing a set of newly introduced regulations in water supply and wastewater treatment sectors with economic development trends, the authors point out that the increase of water and sanitation service fees has occurred not because of deteriorating water shortage or an upsurge of water pollution levels (internal factors) but because of economic policy and development and social stability (external factors). The evidence of the study reiterates the salience of social stability and concerns that are often the most significant factor for the national and local leaders without the consideration of economic feasibility of water and sanitation facilities and services.

Preventing water-related disasters including flood and drought not only save human lives but also protect tangible and invaluable economic assets. Water-related disasters account for more than 90% of various types of natural disasters at the global level every year, and these disasters often cause massive scales of human and economic losses and decimate multi-year achievements of a society as seen from numerous typhoons, cyclones, tsunamis, and flashflood cases in different countries. In 2018, there were 315 climate-related and geophysical disaster events, affecting 68 million people with the death toll of 11,804. Out of the 315 disaster events, 273 events, approximately 87%, were associated with water-related disasters, which include floods (127 events), storms (95 events), droughts and extreme temperatures (41 events), and wildfires (10 events) (CRED 2019a, b).

China is not an exception, since the country experiences several typhoons every summer, which typically traverse areas along the middle reaches of the Yangtze River after landing in eastern coastal provinces, such as Fujian Province, and result in flood and water logging and killing hundreds of people and destroying valuable economic assets, such as roads, railways, housings and buildings. In 2015, water logging and flood events cost RBM 98 billion (US$ 15 billion) of direct economic losses, and typhoons battered China six times, which was relatively more frequent than previous years, usually two or three times (Tan 2016).

Droughts do not necessarily draw attention of top leaders and water managers compared with flood events, because this type of disaster occurs comparatively in a longer period and has a slow impact on the livelihoods of people. In addition, the impact of droughts is not as conspicuous as that of flood events, and instead, droughts affect diverse parts of society in a sluggish pattern, which makes socioeconomic damage of droughts difficult to measure. Droughts trigger serious damage to primarily agricultural sectors, which often becomes a risk factor for economic losses in agricultural industries, such as crop and livestock production.

The period between 2009 and 2010 was recorded as the driest period for the southwestern part of China (Yunnan, Guizhou, Guangxi, and

Sichuan Provinces) and the second driest in North China (Hebei, Shanxi, and Liaoning Provinces), which shows approximately 25% below the average precipitation in the hydrological year. The 2009–2010 drought wreaked havoc on more than 25 million people, 18 million livestock by the lack of water, and damaged 8 million hectares of arable land, triggering total economic losses at approximately US$ 3.5 billion. Hydropower generation in the affected southwestern part of China, which accounts for about 70% of the concentration of hydropower plants in the country, had to be reduced in order to allow more water to flow downstream (Barriopedro et al. 2012; Qiu 2017).

A typical approach to the increase of resilience against water-related disasters and climate change by Chinese authorities has been building various infrastructures to prevent flood events such as dams, barrages, reservoirs, embankments, and retention areas. To cope with droughts, Chinese authorities have built reservoirs, aqueducts for water transfer, and extension of water pipelines. These efforts are conventional measures to resolve water-related disasters, namely supply-side management or hardware-centered approach. In addition to these, the central government of China officially launched the Catastrophe Insurance in 2016 after testing the scheme for about two years since June 2014 in Shenzhen, Ningbo, and Chongqing. The early version of the program encompassed only earthquakes and typhoons. However, insured fields would include droughts, other flood events, and snowstorms in due course.

Although this policy initiative is useful, a low quality of national and local level data about catastrophes, little regulatory frameworks, and little access to risk mapping and hazard information hinders private insurance companies from being actively engaged in the insurance market. In addition, the low level of catastrophe insurance penetration in China appears to be a great challenge. Insurance payout against total direct economic losses related to the Wenchuan 512 earthquake in 2008 was estimated only 0.2% and below 1% related to the Ya'an earthquake in 2013, which contrasts with the rate of catastrophe insurance indemnity in developed countries, i.e., 21% in Europe, 38% in North America, 33% in Australia and New Zealand. Considering around 9% in Asia, China's insurance penetration rate for catastrophes is particularly low. A breakthrough in the catastrophe insurance sector would be possible if better cooperation took place between government agencies, legislative departments, commercial insurance companies, and financial institutions (Qiu 2017; Tan 2016).

Water is imperative for maintaining ecosystems as sound ecosystems are the foundation for healthy water environments. Natural forests, for instance, are habitats and serve as conduits of providing water for flora and fauna. Various types of precipitation, such as rainfall and snow, are absorbed by trees and plants as well as land, which become part of underground water tables and aquifers, which recharge local streams and major rivers. Therefore, it is vital to adequately protect water sources away from any unnecessary anthropogenic harms, such as over-exploitation and pollution. The significance of ecosystems has well been recognized by Chinese environmental authorities over a few decades. Nevertheless, a rampant over-exploitation and illegal logging of natural forests have not been suspended regardless of continuous regulatory efforts to ban unauthorized logging activities in major natural reserves, such as the upper reaches of the Yangtze River.

According to the 13th FYP, commercial logging in natural forests should be forbidden in order to achieve the percentage of forest coverage up to 23.04% by 2020 (Tan 2016). As of 2016, a total coverage of forest ecosystem in the country's landmass was estimated at approximately 22.35% according to the data of the World Bank. It would be possible for the Chinese government to reach the target by 2020 if regulatory efforts for the protection of forests and wetland areas were properly implemented at the national and local levels, especially centered upon the National Nature Reserves (NNRs) and local Nature Reserves (NRs) (Guo and Cui 2015).

Groundwater resource management is a prerequisite for the revitalization of damaged or degraded ecosystems throughout China. The situation of groundwater management, however, is rather grim and bleak. A wide-spread contamination and over-exploitation of groundwater resources are easily found in the country, in particular, in the North China Plain in which numerous farmers have desperately required water for their irrigation schemes and have fiercely been in competition with urban water users in Beijing, Tianjin, and other adjacent cities in Hebei Province.

Degradation of groundwater resources has entailed a series of negative impacts, such as land subsidence and sink holes, desertification of grasslands, and soil pollution and erosion. Furthermore, decreasing levels of groundwater resources have spawned less water flow in local streams, rivers, and freshwater lakes, which has had detrimental influences on local ecosystems.

To make things worse, the rapid pace of urbanization at the national level has posed a grave threat to the sustainability of water and ecosystems. China's unprecedented urbanization rate was recorded at 56% in 2015 and surged up to 60.6% at the end of 2019, and the total number of urban population reached 848.43 million (Tan 2016; Xinhua 2020). The natural hydrological cycle has seriously been hampered by human settlements of China, especially in urban areas, which has resulted in ecosystem disruption, thanks to the increasing built environment over the last four decades. Such a vicious cycle of urbanization, the encroachment of forests, and degradation of ecosystems should be put to an end.

Confronted with the complexity of development challenges, the central government embarked on the Sponge City Initiative in 2014 which was aimed at reversing the patterns of anthropogenic disasters against ecosystems into the reinvigoration of hydrological cycle in urban settings. Cities, which have been criticized as a major culprit to induce environmental degradation, are supposed to be transformed as the new media that can collect, drain, and purify rainwater run-off for water reuse, flood prevention, and water quality enhancement and play a significant role in facilitating the hydrological cycle through the construction of eco-friendly and 'sponge-like' buildings, housings, roads, and green spaces. The fast and large-scale expansion of urban areas across the country has culminated in larger volumes of wastewater discharged. A recovered urban hydrological cycle can recharge groundwater tables and aquifers, thereby increasing the capacity of climate resilience (Chan et al. 2018; Tan 2016).

Conclusion

This chapter has examined the roles of water for the overall development in China, highlighting water and its interconnectedness with economic growth, social development, and ecosystems which comprise three significant pillars leading to the achievement of sustainable development. The chapter serves as an important platform to lead to in-depth discussions of various subjects in water resources management for the following chapters by portraying the contributions of water to sustainable development in China. Particular attention has been placed on the most recent five-year plan of China, the 13th Five Year Plan (2016–2020) which accommodates paramount strategies and plans leading to ecological civilization under the flag of 'Beautiful China'.

Discussions have been made on the exploration of the interconnectivity between water and development with special reference to SDGs. Water influences a variety of sectors in development, and no development plan or project can work without sustainable water resources management. The Chinese government has been determined to contribute to achieving 17 SDGs by establishing its own strategies and plans and recognizes the centrality of water for the successful achievement of all SDGs by 2030.

Water has played a significant role in the rapid socioeconomic development of China over the four decades. China's global strategy, the Belt and Road Initiative, encompasses an array of investment projects, including water infrastructure projects. At the national level, more attention has been placed on the Beijing-Tianjin-Hebei Region and the Yangtze River Economic Belt as regional development strategies, and sustainable water resources management will be the key to success.

Whereas water and sanitation and other water services have been relatively well implemented in urban areas of China, rural communities have comparatively been marginalized and need more prioritization. There has been a gap between urban and rural areas in water and sanitation services, and Chinese water authorities should adequately tackle this problem in the coming decades. Social safety net systems have to be developed further, such as specific institutions for the poor family with regard to water bills as seen from England and Wales. Flood and disaster risk management should be strengthened for safeguarding social development in China, not only through physical infrastructure, such as dams, embankments, and barrages, but also institutional settings, including insurance systems.

Water contributes to sustaining healthy and sound ecosystems. The size of China's forests has increased thanks to good forestry management as well as restoration of water environments. Groundwater resources need more attention for sustainable use and protection. One of the recent policy efforts in water resources management, the Sponge City Initiative in 2014, can be a game changer for restoring the capacity of hydrological cycle, preventing floods, and providing additional amounts of water for urban life.

China's plans and strategies for water and development appear to have evolved for more green and environmentally friendly ones, demonstrating reflexive attitudes about pro-growth-centered drives and malignant consequences of industrialization over the past four decades. Detailed discussions will come in the following chapters on the extent to which the government has been committed to achieving sustainable water resources management whilst the goal of modernizing China is never ignored. The path to strike the good balance between economic growth and environmental protection in China, including water, is seemingly bumpy, however, collective efforts in Chinese society are necessary for ensuring sustainable water resources management.

References

Ait-Kadi, M. 2016. Water for Development and Development for Water: Realizing the Sustainable Development Goals (SDGs) Vision. *Aquatic Procedia* 6: 106–110.

Asian Development Bank. 2018. *Managing Water Resources for Sustainable Socioeconomic Development: A Country Water Assessment for the People's Republic of China*. Manila: Asian Development Bank.

Barriopedro, David, Celia Gouveia, Ricardo Trigo, and Lin Wang. 2012. The 2009/10 Drought in China: Possible Causes and Impacts on Vegetation. *Journal of HyDrometeorology* 13: 1251–1267.

Center for Research on the Epidemiology of Disasters (CRED). 2019a. Disasters 2018: Year in Review. CRED, USAID and UCLouvain.

Center for Research on the Epidemiology of Disasters (CRED). 2019b. Natural Disasters 2018. CRED, USAID and UCLouvain.

Chan, Faith Ka Shun, James A. Griffiths, David Higgitt, Shuyang Xu, Fangfang Zhu, Yu-Ting Tang, Yuyao Xu, and Colin R. Thorne. 2018. "Sponge City" in China—A Breakthrough of Planning and Flood Risk Management in the Urban Context. *Land Use Policy* 76: 772–778.

Cheng, Evelyn. 2020. China Says Its Economy Grew 2.3% in 2020, But Consumer Spending Fell. *CNBC*, January 17, 2021. Available Online: https://www.cnbc.com/2021/01/18/china-economy-release-of-fourth-quarter-full-year-2020-gdp.html (accessed 25 January 2021).

China Water Risk and Ministry of Environmental Protection (MEP). 2016. Water-Nomics of the Yangtze River Economic Belt: Strategies and Recommendations for Green Development Along the River. China Water Risk and Foreign Economic Cooperation Office, the Ministry of Environmental Protection, June 2016.

Cox, W., ed. 1987. The Role of Water in Socio-Economic Development. Part 1 of IHP-II Project C1. Prepared for the International Hydrological Programme by the Working Group of Project C1 (IHP-II). Paris: UNESCO.

Dai, Liping. 2019. *Politics and Governance in Water Pollution Prevention in China*. Cham, Switzerland: Palgrave Macmillan.

Davies, Paul, and Andrew Westgate. 2019. China Kick-Starts Investment in the Yangtze River Economic Belt. Latham and Watkins LLP, June 10, 2019.

Global Water Partnership. 2015. China's Water Resources Management Challenge: The 'Three Red Lines'. A Technical Focus Paper, Global Water Partnership.

Government of the People's Republic of China. 2016. China's National Plan on Implementation of the 2030 Agenda for Sustainable Development. Available Online: http://www.chinadaily.com.cn/specials/China'sNationalPlanon implementationofagenda(EN).pdf (accessed 24 January 2021).

Guo, Zhiliang, and Guofa Cui. 2015. Establishment of Nature Reserves in Administrative Regions of Mainland China. *PLOS ONE* 10 (3): 301196540. Available Online: https://doi.org/10.1371/journal.pone.0119650.

Hao, Yu, Xinlei Hu, and Heyin Chen. 2019. On the Relationship Between Water Use and Economic Growth in China: New Evidence from Simultaneous Equation Model Analysis. *Journal of Cleaner Production* 235: 953–965.

Hou, Siyu, Yu Liu, Xu Zhao, Martin Tilotson, Wei Guo, and Yiping Li. 2018. Blue and Green Water Footprint Assessment for China—A Multi-Region Input-Output Approach. *Sustainability* 10 (2822): 1–15.

HSBC. 2012. Exploring the Links Between Water and Economic Growth. A Report Prepared by HSBC and Frontier Economics. Executive Summary. June 2012.

ICSU and ISSC. 2015. Review of Targets for the Sustainable Development Goals: The Science Perspective. Available Online: http://www.icsu.org (accessed 20 March 2019).

Lee, Seungho, Donghoon Kim, Joo-Heon Lee, and Ilpyo Hong. 2018. Water and Economic Development—Analysis on the Linkage Between Water Resources Management and Socio-economic Development in the Republic of Korea (물과 경제개발 - 한국의 수자원관리와 경제개발의 연계성 분석). An unpublished report. Goyang City, Geonggi Province, Korea Institute of Civil Engineering and Building Technology.

Lin, Justin, and Zhingkai Shen. 2018. Chapter 7: Reform and Development Strategy. In *China's 40 Years of Reform and Development 1978–2018*, ed. Ross Garnaut, Ligang Song, and Cai Fang, 117–134. Acton: Australian National University Press.

Min, K., T. Shin, H. Cho, W. Song, J. Kim, and S. Hong. 2013. Water and Green Growth; Beyond the Theory for Sustainable Future. Volume 2, 2015. Daejeon, K-water, Ministry of Land, Infrastructure and Transport and the World Water Council.

Ministry of Environmental Protection (MEP). 2017. 2016 Report on the State of the Environment in China. Ministry of Environmental Protection, People's Republic of China.

Ministry of Foreign Affairs (MFA). 2019. China's Progress Report on Implementation of the 2030 Agenda for Sustainable Development (2019). September 2019. Beijing: Ministry of Foreign Affairs of the People's Republic of China.

National Bureau of Statistics (NBS). 2021. Statistical Communique of the People's Republic of China on the 2020 National Economic and Social Development. 28 February. Available Online: http://www.stats.gov.cn (accessed 2 March 2021).

OECD. 2018. China's Belt and Road Initiative in the Global Trade, Investment and Finance Landscape. In *OECD Business and Finance Outlook 2018*. Paris: OECD Publishing.

Qiu, Tom. 2017. A Decade of Direct Economic Losses from NatCat Perils in China: What Does the Future Hold? General Reinsurance AG September Newsletter. September 2017.

Scott, Christopher, Fan Zhang, and Aditi Mukherji. 2019. Chapter 8: Water in the Hindu Kush Himalaya. In *The Hindu Kush Himalaya Assessment*, ed. Philippus Wester, Arabinda Mishra, Aditi Mukherji, and Arun Shrestha. Cham, Switzerland: ICIMOD, HIMAP and Springer Open.

Shen, Dajun, and Juan Wu. 2017. State of the Art Review: Water Pricing Reform in China. *International Journal of Water Resources Development* 33 (2): 198–232.

SIWI. 2017. Water a Success Factor for Implementing the Paris Agreement. Policy Brief. Stockholm Water Institute, March 2017.

Sunsik, J. 2015. Economic Growth and Resource Use: Exploring the Links. UN University Institute for Integrated Management of Material Fluxes and of Resource (UNU-FLORES).

Tan, Debra. 2016. Beautiful China 2020: Water and the 13 FYP. *China Water Risk*, 16 March.

Thames Water. 2020. WaterSure, WaterSure Plus, and WaterHelp Schemes. Available Online: http://thameswater.co.uk (accessed 22 January 2020).

Tianjin Water Authority. 2019. Tianjin Water Resources Bulletin 2019 (天津市水资源年鉴 2019). Tianjin: Tianjin Water Authority.

Tropp, H. 2013. Chapter 3: Making Water a Part of Economic Development: The Economic Benefits of Improved Water Management and Services. In *Investing in Water for a Green Economy*, ed. M. Young and C. Esau. London: Earthscan.

UNDESA. 2015. The Critical Role of Water in Achieving the Sustainable Development Goals: Synthesis of Knowledge and Recommendations for Effective Framing, Monitoring and Capacity Development. 6 February.

Wang, Qing, Rui Jiang, and Rongrong Li. 2018. Decoupling Analysis of Economic Growth from Water Use in City: A Case Study of Beijing, Shanghai, and Guangzhou of China. *Sustainable Cities and Society* 41: 86–94.

WHO and UNICEF. 2019. Joint Monitoring Programme for Water Supply, Sanitation and Hygiene. Updated June 2019. Available Online: http://washdata.org (accessed 22 January 2020).

World Bank. 2018. *Watershed: A New Era of Water Governance in China—Thematic Report*. Washington, DC: World Bank.

World Bank Data on China's Forest Coverage. Available Online: https://data.worldbank.org/indicator/AG.LND.FRST.ZS?locations=CN (accessed 24 February 2020).

Xinhua. 2019. China's Regional Development Strategies Set Example for World. *Xinhua*, 6 October.

Xinhua. 2020. China's Urbanization Rate Hit 60.6 pct. *Xinhua*, 19 January.

3

Overview of Water Resources

Introduction

This chapter discusses the data and information on water resources, water supply and use, water quality, flood and drought coupled with climate change impacts in China. Prior to delving into a variety of water resources management issues, this chapter serves as a useful foundation to help have a better understanding of general facts and figures regarding China's water resources. The information on water resources will include the total amount of water resources available in the country, an annual mean volume of precipitation, major river and freshwater lake basins, and groundwater resources. The uniqueness of China's water resources is explored, including the uneven distribution of water resources between regions, seasonal and temporal variations of precipitation, a disparity in spatial distribution of water demand and availability and fragile water environment and ecosystems. North China's water resources are much less than South China's although North China contributes substantially to food production with more arable land, which intensifies water shortages in the region.

Groundwater resources are of great importance for North China even though more groundwater resources are available in South China. Intensive groundwater use in North China has culminated in chain effects, such as land subsidence and other geological hazards. Among more than 350 rivers with the river basin size of 10,000 km^2 or above in the country, seven rivers (the Yangtze, the Yellow, the Songhua, the Liao, the Pearl, the Hai, and the Huai Rivers) draw major attention of water managers in China thanks to their profound impacts on political, socioeconomic, and environmental aspects. China boasts almost 3000 lakes that function as primary sources for various water uses, including Poyang, Dongting, and Tai Lakes.

Water supply in China heavily depends on surface water bodies (rivers and lakes), 82%, groundwater resources, 16%, and other sources, 2%, i.e., wastewater reuse, rainwater, and desalinated water, in 2019. China's water use increased from 1980 to 2009 by 34% thanks to an increase of demands from household and industrial sectors whereas the agricultural sector use increased less by 5%, which implies a successful outcome of the enhancement of water use efficiency in the agricultural sector. In 2019, the agricultural sector is still the largest water user, responsible for using 61% of the total amount of water supply, the industrial sector, 20%, household 14%, and environmental flow, 4%.

The overall water quality of China's surface and groundwater resources in 2019 unveils that the ratio of Grade I–III freshwater bodies was 74.9%, Grade IV, 17.5%, Grade V, 4.2%, and worse than Grade V, 3.4%. These figures appear to be better than those in the 1980s and the 1990s, however, the water quality of small streams and tributaries has not been adequately monitored and has not drawn attention of authorities, and many lakes are highly subject to either eutrophication or oligotrophication, including Tai, Dianchi, and Chao Lakes.

Vast amounts of wastewater are discharged from households and industrial units in China, which are the two major pollution sources, and since 2007, household wastewater discharge has overpassed industrial wastewater discharge, partly affected by Non-Point Source (NPS) pollution as well as the growth of Township Village Enterprises (TVEs) in suburban and rural areas. The overuse of fertilizers and pesticides in rural areas has triggered pervasive water pollution despite some nationwide campaigns, such as the Action Plan for Zero Growth of Fertilizer and Pesticide.

Water-related disasters such as flood and drought have wreaked havoc on China for many years. To cope with floods, the government has introduced structural measures, i.e., flood control levees, reservoirs (dams), and diversion of waterways, and non-structural measures, i.e., flood protection policy, catastrophe insurance pilot projects, and the introduction of relevant laws. Drought risk management has actively been implemented in China for protecting people, agricultural commodities, and economic assets. Drought-prone areas are southern, central, and eastern provinces. The mixture of structural and non-structural measures is essential for tackling droughts, such as construction of reservoirs, laws and regulations for drought prevention and control and the introduction of new technical standards.

Climate change in China induces an increase of temperature, a geographical and seasonal change in water distribution, and the sea level rise. The government has established national-level plans and strategies to deal with climate change impacts, including the National Assessment Report on Climate Change in 2015. In addition, the country's specific responses to climate change have been neatly entrenched in its 12th Five Year Plan (2011–2015) and the 13th Five Year Plan (2016–2020), and water resources management projects, such as the South North Water Transfer Project, have demonstrated the government's proactive response to adverse impacts caused by climate change, water shortage.

The first part of the chapter sheds light on the general situations of total water resources, and attention is paid to water quality, investigating the water quality of rivers, lakes, and groundwater in recent years as well as trends. Flood and drought issues are highlighted in the third part, which leads to the exploration of climate change impacts on China.

Water Resources

Total Water Resources and Climate

Renewable water resources are generally measured based on the availability of surface and groundwater resources. The average volume of renewable water resources in China was estimated at 2775 billion m^3 per annum from 1998 to 2017 (MWR 2018). In 2019, the total volume of renewable water resources in the country was 2904.1 billion m^3, which was slightly more than the one in 2018 by 4.8%. Surface water resources were estimated at 2799.33 billion m^3, and groundwater resources, 819.15 billion m^3 with the non-overlapped amounts of surface and groundwater resources, 104.77 billion m^3. There are ten water resources districts in the country, and renewable water resources in the four South Water Districts are almost four times more than those in the six North Water Districts, which signifies the unevenness of water availability between North and South China. The largest amounts of renewable water resources in the country are available in the Yangtze River District, 1054.97 billion m^3, which are almost five times more than those in the Songhua River District, the largest water district in North China, 222.32 billion m^3 (MWR 2020). Table 3.1 describes each water district's renewable water resources.

An average water availability per capita is 2104 m^3 from 1998 to 2017 if the total volume of renewable water resources is evenly distributed to every single person in China. The level of per capita water resources has been up and down in the same period, the lowest, 1726 m^3 in 2011 and the highest, 2723 m^3 in 1998. The most recent estimations of it were 1972 m^3 in 2018 and 2078 m^3 in 2019 whose levels are even less than the average in the last two decades (MWR 2018; NSB and MEE 2019; NBS 2020) (see Fig. 3.1). This result is equivalent to less than one quarter of the world average and one-sixth of the figure in the US (GWP 2015; Yang et al. 2012).

In 2019, the annual average precipitation in China reached 651.3 mm based on the data collected from more than 18,000 precipitation monitoring stations, which decreased by 4.6% compared with precipitation in 2018 and by 1.4% compared with precipitation in previous years.

Table 3.1 Renewable water resources of China in 2019 (Unit: billion m³)

Water resources district	Surface water resources	Groundwater resources	Non-overlapped amounts of surface and groundwater resources	Total renewable water resources
Nation	2799.33	819.15	104.77	2904.10
6 North Districts*	471.3	256.37	89.78	561.08
4 South Districts**	2328.03	562.78	14.99	2343.02
Songhua River District	193.51	62.84	28.81	222.32
Liao River District	30.57	19.51	10.19	40.76
Hai River District	10.45	19.04	11.7	22.14
Yellow River District	69.02	41.59	10.72	79.75
Huai River District	32.81	27.48	17.9	50.72
Yangtze River District	1042.76	258.05	12.21	1052.97
*Tai Lake Basin***	20.42	4.41	2.16	22.58
Southeast Rivers District	247.5	54.2	1.36	248.85
Pearl River District	506.58	119.84	1.42	508
Southwest Rivers District	531.2	130.7	0	531.2
Northwest Rivers District	134.94	85.92	10.47	145.4

Remarks * 6 North Districts are Songhua River District, Liao River District, Hai River District, Yellow River District, Hai River District and Northwest Rivers District. ** 4 South Districts indicate Yangtze River District (including Tai Lake River Basin), Southeast Rivers District, Pearl River District, and Southwest Rivers District. *** Tai River Basin is specially mentioned due to its significance in water environments and socioeconomic development.
Source MWR (2020)

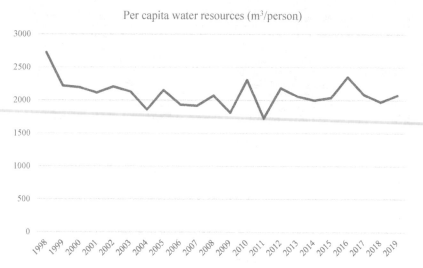

Fig. 3.1 Trend of per capita water resources in China from 1998 to 2019 (*Source* MWR [2018] and NSB [2020])

Eight out of the ten districts in 2019 showed higher levels of precipitation than those of the annual mean precipitation in previous years but compared with the statistics in 2018, only four districts show an increase of precipitation, the Songhua, the Liao, the Southwest, and the Pearl River Districts. Overall, China had a relatively wet year in 2019 although the level of precipitation in 2018 was dwarfed by that in 1998, almost 750 mm, when the highest level of precipitation was recorded since 1956. It will be expected to have the statistics on the unusually high levels of precipitation in 2020 due to heavy rainfalls in the central and southern parts of China in the summer of 2020. The most conspicuous fact observed from the data on precipitation in China is the wide gap between North and South China, which demonstrates that South China received precipitation three times more than North China (MWR 2020) (see Table 3.2).

The total volumes of surface water resources of China in 2019 were estimated at 2799.33 billion m^3, which decreased by 6.4% compared with those in 2018 and was more than the multi-year average level by 4.8%. Surface water resources in six out of ten water resources districts

Table 3.2 Precipitation in 10 water resources districts of China in 2019 and comparison with precipitation in 2018 and previous years

Water resources district	Precipitation (mm)	Compared with precipitation in 2018 (%)	Compared with average precipitation in multiple years (%)
Nation	651.3	−4.6	1.4
6 North Districts	346	−8.7	5.5
4 South Districts	1192.30	−2.3	−0.6
Songhua River District	603.4	5.9	19.7
Liao River District	557.9	9.1	2.4
Hai River District	449.2	−16.9	−16
Yellow River District	496.9	−9.9	11.4
Huai River District	610	−34.1	−27.3
Yangtze River District	1059.80	−2.4	−2.5
Tai Lake Basin	1261.80	−8.7	6.4
Southeast Rivers District	1844.90	14.8	11
Pearl River District	1627.50	1.7	5.1
Southwest Rivers District	1013.60	−11.6	−6.7
Northwest Rivers District	183.2	−10.2	13.8

Source MWR (2020)

were estimated more than the multi-year average, including three North Districts (the Yellow, the Northwest, the Liao River Districts) and three South Districts (the Yangtze, the Southwest, and the Pearl River Districts). In particular, the Songhua River District showed more surface water resources than the multi-year average by 49.9%. The Hai River and the Huai River Districts showed a steep decrease of surface water resources compared with the multi-year average by 51.6% and by 51.5%, respectively. These areas have been some of the worst regions hit by water shortage so that the pressure about more water resources will intensify. More details are described in Table 3.3.

Thanks to the sheer size of its territory, China has a variety of climate zones and is largely dominated by the monsoon system, which results in

Table 3.3 Surface water resources in water resources districts of China in 2019 and comparison with surface water resources in 2018 and previous years

Water resources district	Surface water resources (billion m^3)	Compared with surface water resources in 2018 (%)	Compared with average surface water resources in multiple years (%)
Nation	2799.33	6.4	4.8
6 North Districts	471.3	−2.4	7.7
4 South Districts	2328.03	8.4	4.3
Songhua River District	193.51	34.2	49.9
Liao River District	30.57	−0.7	−25.1
Hai River District	10.45	−39.9	−51.6
Yellow River District	69.02	−8.6	12.9
Huai River District	32.81	−57.4	−51.5
Yangtze River District	1042.76	12.9	5.8
Tai Lake River Basin	20.42	0.1	27.5
Southeast Rivers District	247.5	64.4	24.6
Pearl River District	506.58	6.4	7.6
Southwest Rivers District	531.2	−11	−8
Northwest Rivers District	134.94	−2.3	15.1

Source MWR (2020)

the clear demarcation between wet summers and dry winters. The range of climate variety shows cold temperature, i.e., Northeast, to tropical, i.e., Southwest, and precipitation patterns and amounts in different parts of the country display large gaps depending upon regions. Whereas arid or semi-arid conditions are common in the northwestern part of China, an average rainfall of the areas along the eastern coastal line is usually estimated between 500 and 1500 mm per annum.

The isohyet of 350 mm is found diagonally across the country from Northeast to Southwest, which divides rain-fed farming areas in the south and the east and irrigated farming in the north and the west. The feature puts water managers in a difficult position for water allocation. The Yangtze River contributes to around 80% of the country's water resources but provide water to only 36% of the country's land resources. On the contrary, in North China, approximately 60% of the farming areas are located, but less than 20% of the country's water resources are available (ADB 2018; GWP 2015).

In the period between 1951 and 2017, the mean annual surface temperature had increased up by 0.24 degrees Celsius every 10 years, which is higher than the global average temperature rise. The situation of temperature rises varies depending upon regions in the country, and it was the Qinghai-Tibetan Plateau region that showed the highest rise in the period according to the China Meteorological Administration. The trend of annual precipitation from 1961 to 2017 demonstrates little change, but the average precipitation level since 2012 had been estimated above the average. In 2017, the total amounts of annual precipitation in the country reached 641.3 mm, which is 1.8% higher than the annual average (Xinhua 2018).

Uniqueness of Water Resources in China

The uniqueness of water resources in China is summarized fourfold. First, water resources in the country are unevenly distributed spatially. Large amounts of surface water resources are concentrated in South China that includes the landmass under the southern parts of the Yangtze River whereas North China is endowed with much less water resources. For instance, 81.5% of water resources are concentrated in the heavily industrialized southern China compared with 18.5% of water resources in the agrarian north. North China occupies 64% of the national territory and has 46% of the people and 60% of the country's cultivated areas (ADB 2018; GWP 2015).

The Yellow (Huang), the Hai, and the Huai River Basins, which are called the 3H Region, encompass the average precipitation from 300 to 700 m^3 per annum. Arid and semi-arid climates in the inland regions make human settlements in severe water shortages, however, the low population density thanks to less people and the vast territory leads to the large volume of water resources per capita.

Most of the southern parts of China boast abundant water resources thanks to large rivers and freshwater lakes, such as the Yangtze and the Pearl Rivers, Dongting and Tai Lakes, and numerous local streams that are connected through complicated networks between cities and villages. In particular, the South-West regions are endowed with vast amounts of water resources in rivers and lakes and several transboundary rivers, including the Lancang-Mekong, the Nu-Salween, and the Yarlung Tsangpo-Brahmaputra Rivers (Yang et al. 2012).

Second, the seasonal and temporal variations of precipitation are generally large because of the climate regime of the East Asian monsoon, which pours tremendous amounts of rainfall in the summer months, typically from June to August and accompanies typhoons. The country receives little precipitation in the rest of the year that entails the seasonal imbalance of water resources and long-spelled drought conditions (Yang et al. 2012). The third distinctive feature is associated with a disparity in spatial distribution of water demand and availability. The amount of water availability per capita in the northern river basins are just one-third of that in the southern river basins (Fig. 3.2).

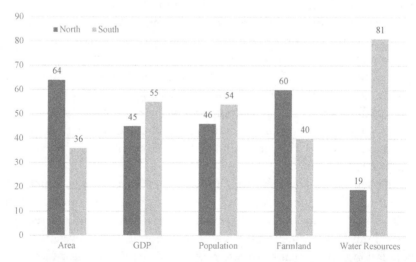

Fig. 3.2 Spatial distributions of water resources, population, river basins and farmland in China (Unit: %) (*Source* Modified based on Wang et al. [2017])

Fourth, water environment and ecosystems are fragile. It is estimated that soil erosion has caused economic losses by about 3.5% of the national GDP. The reduction of forest coverage, wetland and swamps, an acceleration of desertification, soil erosion, and the salinization of grasslands are some examples which are closely tied to the deterioration of water environment and ecosystems (Wang et al. 2017).

Soil erosion triggers land degradation and naturally takes place when the rate of soil erosion has developed much faster than that of soil build-up. According to the first National Water Resources Census in 2013, approximately one-third (31%) of the total land area in China were influenced by soil erosion. Among any other areas in the country, the Yellow River Valley, namely the Loess Plateau, has badly been affected by serious soil erosion, and an amount of 1200 million to 1600 million tons of sediment per annum has been delivered toward Yellow Sea.

A striking fact is that additional one-third of the total land mass of China is highly subject to desertification, because arid (less than 250 mm of precipitation per annum) and semi-arid (from 250 to 500 mm per annum) areas account for one-third of China's total land, and about 40% of these areas are undergoing the process of desertification (MWR and NBS 2014).

The soundness of water and ecosystems is measured in several ways, and one of the common ways is the ratio of water withdrawal to water availability, which demonstrates the intensity of water use and the pressure on the ecosystem. One of the worst performing areas in China is the 3H Region in which the Hai River Basin, for instance, indicates the ratio of water withdrawal to water availability at 124%. The basin has overused water resources, especially groundwater resources through deep boreholes, with little precipitation and little amounts of desalinated water. More than 90 billion m^3 of groundwater resources in the North China Plain had been overexploited between 1990 and 2010, which entailed soil erosion, seawater intrusion, and ecosystem disruption. Another common phenomenon is that the Yellow River could not reach its estuary in many days of the year due to the lack of water flow in the late 1990s (Yang et al. 2012).

Groundwater Resources

Groundwater resources have made substantial contribution to socioeconomic development and ecosystems in China for recent few decades. In particular, the magnitude of groundwater for the overall development of the North China Plain and the northwestern part of China is indisputable, because these areas are arid or semi-arid with little precipitation and surface water resources.

In 2019, the total amount of groundwater resources in China was 819.15 billion m^3, which is 28% of the total water resources of the country (2904.1 billion m^3) (MWR 2020). Groundwater resources are unevenly distributed, and almost 70% of groundwater resources are available in South China (38% of the country's total land area) whereas about 30% of them found in North China (62% of the total land area). Considering more demands of groundwater in the northern parts of China due to less surface water and precipitation, the uneven distribution of groundwater resources is posing a threat to water security in North China. In addition, 80% of the total amounts of groundwater is stored in the mountainous areas, and 20% in plain areas (Liu and Zheng 2016; MWR 2020). Table 3.4 shows the total amount and spatial distribution of groundwater resources in China.

Groundwater exploitation in China has escalated over the past few decades with an average rate of 2.5 billion m^3/year, and in the 1970s, 57 billion m^3/year was pumped up, 75 billion m^3/year in the 1980s, and 111 billion m^3 by 2011. Groundwater consumption accounted for 16.8% of total water supply in 2017, 101.6 billion m^3 (Liu and Zheng 2016; MWR 2018).

Table 3.4 Total amount and spatial distribution of groundwater resources in China, 2019 (Unit: billion m^3)

Area	Total	
Nation-wide	819.15	
Plain area	171.48	20%
Mountain area	677.96	80%
Overlapped amounts	30.29	

Source MWR (2020)

Groundwater is often used as a major water supply source for hundreds of cities in China, and more than 400 (61%) out of the 657 cities in China use groundwater as their primary water source. Rural communities rely on groundwater for their drinking as well as irrigation, and 65% of household water use, 50% of industrial water use, and 33% of irrigated water use stem from groundwater (Liu and Zheng 2016).

The intense exploitation of groundwater resources in North China has engendered 72 cones of depression[1] that would occupy 61,000 km^2 and has resulted in ground subsidence and other geological hazards and environmental challenges. Many aquifers in the region have been over-exploited, and in particular, the whole areas of Hebei Province and the cities like Beijing, Tianjin, Shenyang, Haerbin, Jinan, Taiyuan, and Zhengzhou have over-pumped groundwater resources (Liu and Zheng 2016; Wang et al. 2017).

The huge boom of groundwater development in the country for more than four decades has culminated in millions of wells being drilled, and about 4.5 million wells have been drilled only in North China. These wells are interconnected across multiple aquifers, and therefore, the wells can serve as a catalyst to spread pollution through shallow and deep aquifers.

The fact that deeper aquifers would contain high-quality water because of its less exposure to surface level activities can be a myth. Under the current groundwater pollution control system, there is almost no effective mechanism for well licensing and maintenance system. There is no guarantee of water quality in shallow or deep groundwater resources, and cross contamination can be unavoidable (Han et al. 2016; Shen 2015).

Declining groundwater levels due to overexploitation in China have caused the shrinking or disappearing of wetlands and degradation of vegetation coverage. Land subsidence has occurred in more than 40 cities due to groundwater overexploitation, and the maximum accumulative land subsidence by more than two meters have been reported in Shanghai, Tianjin, and Taiyuan. Coastal areas have also been affected

---

[1]The process called 'cone of depression' indicates a concentric pattern of water table drawdown because of over-pumping of groundwater resources.

by groundwater overdraft, including Dalian, Qinhuangdao, Changzhou, Qingdao, and Beihai, which leads to sea water intrusion and its impact on groundwater quality in an area of almost 1000 km^2. Aquifer salinization is indirectly caused by groundwater overdraft thanks to intense irrigation practices in the North China Plain (Liu and Zheng 2016).

Rivers

China has a land area of more than 9.6 million km^2 (NSB 2019) (World Bank Data in 2018, 9,388,210 km^2), and there are innumerable rivers that connect different parts of the country. The number of rivers varies depending upon how large a catchment area is. For instance, there are 46,796 rivers considering the number of river basins with the size of up to 50 km^2 and above, and the length of all the rivers is as long as 1,514,592 km. If the rivers with the size of river basins of up to 10,000 km^2 and above are only considered, there are 362 rivers in the country whose length reaches 136,721 km (MWR 2018).

Among the rivers, special attention is paid to seven major rivers, namely the Yangtze River, the Yellow River, the Songhua River, the Liao River, the Pearl River, the Hai River, and the Huai River. These rivers are of great significance to Chinese politics, economy, society, and environment. The Yangtze River Basin is the largest river basin in China with the size of 1,808,500 km^2. The 3H Region, including the Huang (Yellow), the Hai, and the Huai River Basins, have made substantial contributions to food production, such as maize, soybean, and wheat, with the population of approximately 340 million (MWR 2018). More detailed information on the seven major rivers is as follows in Table 3.5, and the location of the rivers is shown in Fig. 3.3.

Lakes

China's lakes have increasingly been affected by both climate change and anthropogenic activities. The number of lakes in China in 2018 is 2865, including 1594 freshwater lakes (56%), 945 saltwater lakes (33%), 166 salt lakes (6%), and 160 other types (5%). The total surface

3 Overview of Water Resources

Table 3.5 General situations of seven major rivers in China

River	River basin area (km²)	Annual runoff (billion m³)	Mean Average (1949–1988)						
			Mean annual runoff (billion m³)	Population (100 million persons)	Cultivated land (1000 ha)	Annual runoff per capita (m³/person)	Annual runoff per Mu of farmland (m³)	Total yield of grain production (10,000 ton)	Average yield per Mu of cultivated land (kg)
Yangtze	1,808,500	951.3	928	3.79	23,467	2449	2636	14,334.42	279
Yellow	752,443	66.1	62.8	0.92	12,133	683	345	2757.98	160
Songhua	557,180	76.2	73.3	0.51	10,467	1437	467	2920.97	224
Liao	228,960	14.8	12.6	0.34	4400	371	191	1770.64	308
Pearl	453,690	333.8	336	0.82	4667	1098	4800	2195.58	236
Hai	263,631	22.8	28.8	1.1	11,333	262	169	3731.24	221
Huai	269,283	62.2	61.1	1.42	12,333	430	330	6122.13	247

Source MWR (2018)

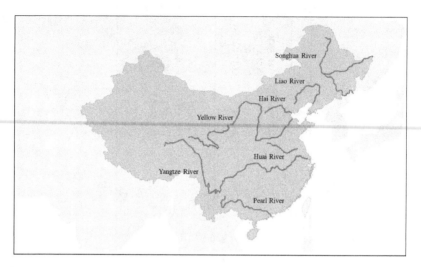

Fig. 3.3 Location of seven major rivers in China (*Source* Author)

area of all the lakes is 78,007.1 km^2, and freshwater lakes occupy 35,149.9 km^2, saltwater lakes 39,205 km^2, salt lake, 2003.7 km^2, and others, 1648.6 km^2. Among the 31 provincial level administration units, the Tibet Autonomous Region has the largest number of lakes, 808, followed by Inner Mongolia Province, 428, Heilongjiang Province, 253, and Hubei Province, 224. The lakes mentioned here are larger or equal to 1 km^2, and there are 40 trans-provincial lakes (MWR 2019).

There are five lake regions in the country: (1) the Eastern Plain; (2) the Mongolia-Xinjiang (Mong-Xin) Plateau; (3) the Northeast Plain; (4) the Yunnan-Guizhou (Yun-Gui) Plateau; and (5) the Tibetan Plateau lake regions. These five lake regions have had far-reaching impacts on ecosystems and livelihoods of local people in different ways. For instance, the Mong-Xin Plateau is significant for the grasslands, residents, and endangered species. In the Eastern Plain, there are all five of the largest freshwater lakes in China according to the size of surface area (Poyang, Dongting, Tai, Hongze and Nansi Lakes), which provide the most biologically diverse wetlands in China together with the Yangtze River Basin (Ma et al. 2010; Tao et al. 2020).

The statistics on China's lakes are not static but changing. The intensification of population growth, industrialization, and urbanization has threatened the sustainability of those lakes, which has been compounded by climate change since the 1980s. Ma et al. (2010) analyzed the change of lakes in the country based on the data from the 1960s to the 1980s and from 2005 to 2006 and argued that whereas 60 lakes increased in the Tibetan Plateau, 243 lakes vanished, and 254 lakes shrank out of 2928 lakes. Tao et al. (2020) have produced similar results based on the evaluation of the change of lakes from the 1980s to 2015. Lakes in the Tibetan Plateau increased from 1065 to 1184 (+119), and the lake area increased from 38,596 km^2 to 46,831 km^2. The Eastern Plain showed a dramatic decrease of lake surface area from 19,384 km^2 to 18,162 km^2 (-1222) even though the number of lakes was similar between the 1980s and 2015, from 621 to 618. The other suffering area was the Mong-Xin Plateau with a net loss of 111 lakes from 531 to 420 and the decrease of lake surface area from 8173 km^2 to 7243 km^2.

It is important to discuss the cause of such remarkable changes for China's lakes over the last three decades. Climate change can be blamed as a major cause for such changes, but this argument is applied to the Tibetan Plateau whereas changes for lakes in the Eastern Plain and the Mong-Xin Plateau were more relevant to anthropogenic factors. For instance, the Mong-Xin Plateau has been influenced by reclamation activities for agricultural purposes, which require large volumes of water for cropping. The plateau is semi-arid with limited rainfall so that water diversion and withdrawal have intensified, which has seemed to give detrimental impacts on lakes in the region. In addition, the plateau embraces one of the world's largest coal-mining regions, and mining damaged groundwater aquifers, depleted rivers and rapidly dried lakes around the mining sites.

The Eastern Plain includes the middle and lower reaches of the Yangtze River, some of the most highly industrialized, developed, and populous areas, and hydraulic engineering projects have been rampant, i.e., dam construction, river regulation, and lake impoldering that means the process of converting lakes into croplands and urban areas. Lake impoldering has occurred for many years in this region, which has intensified in recent few decades. Poyang, Dongting, and Tai Lakes have been victimized by the rapid impoldering, and the size of the lakes has gradually decreased (Tao et al. 2020).

Water Supply

Water supply in China is heavily dependent upon surface water resources with little contribution of groundwater resources and negligible amounts of non-conventional water resources, such as wastewater reuse, rainwater harvesting, and desalinated water. Figure 3.4 shows the trend of water supply from 2000 to 2019, and in 2000, surface water resources contributed to the total amounts of water supply by 80%, followed by groundwater resources, 19%, and others, 1%. In almost 10 years, such statistics have not much changed, surface water resources accounting for 82%, groundwater resources, 16%, and others, 2% in 2019. Although Chinese authorities have made many efforts to increase the contribution of non-conventional water resources, it is still difficult to observe noticeable outcomes yet (MWR 2018, 2019, 2020; NSB 2019).

China's renewable water resources are distributed unevenly, and similar patterns are observed with regard to water supply. The total amounts of

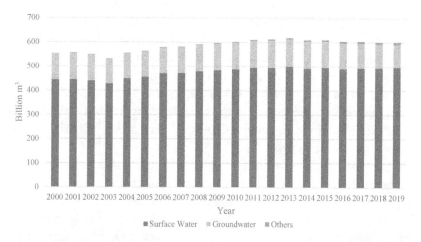

Fig. 3.4 Water supply trends in China from 2000 to 2019 (*Source* MWR [2018, 2019, 2020] and NSB [2019])

water supply in the Yangtze River subregion are equivalent to those of water supply in the five North China subregions (Songhua, Liao, Hai, Yellow, and Huai Rivers) (see Fig. 3.5). The notion of 'too much water in South and too less water in North' is applicable to the discussion of water supply at the national level.

The heavy reliance of North China on groundwater resources is conspicuous, particularly in the Songhua, the Liao, and the Hai Rivers compared with the other areas in South China. This implies the urgent needs of sustainable groundwater management including both quantity and quality in North China. As groundwater tables are dropping rapidly and groundwater resources are depleting in a speedy fashion in North China, it is imperative to consider the active promotion of alternative water resources, such as wastewater reuse, water recycling, and rainwater harvesting in both urban and rural areas (MWR 2019).

Special attention is placed on the general situations of water supply in 2019. Surface water supply was undertaken in four different ways: (1) reservoir projects, 31.2%; (2) water diversion projects, 32.4%; (3) water pumping projects, 30.7%; and (4) inter-basin water transfer projects, 5.7%. Water transfer projects were active between the Hai River, the Yellow River, the Huai River, the Yangtze River, the Pearl River, and the Northwest Rivers Districts, and the total volume of water transfer

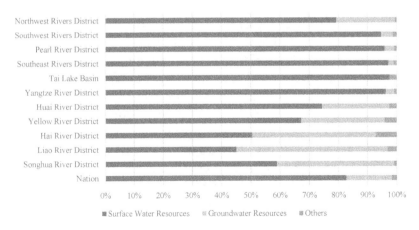

Fig. 3.5 Water supply in water resources districts of China, 2019 (*Source* MWR [2020])

projects in these districts was as large as 20.05 billion m³ in 2018. This sketches active inter-basin water transfers in the country including the vast scale of the South North Water Transfer Project.

As for water supply through groundwater resources, shallow aquifer groundwater accounts for 95.4%, deep aquifer groundwater, 4.2% and brackish water, 0.4%. Other types of water supply include wastewater reuse, 83.6% and rainwater harvesting, 9.2%. Seawater contributes to water supply, to some degree. At the national level, seawater has been used for various purposes, particularly cooling water for thermal and nuclear power generation. The total amounts of seawater use in 2018 reached 112.58 billion m³, and the provinces for substantial amounts of seawater use were Guangdong (39.19 billion m³), Zhejiang (23.63 billion m³), Fujian (21.01 billion m³), Liaoning (7.22 billion m³), Shandong (7.06 billion m³), Jiangsu (5.51 billion m³), and Hainan (3.85 billion m³) Provinces in coastal areas in 2018. Other coastal provinces also use certain amounts of seawater for various purposes (MWR 2019, 2020).

Water Use

Since the onset of the open-door policy in the late 1970s, insatiable demands for food have prompted the rapid expansion of irrigated fields, which have induced an upsurge of agricultural water use. Industrialization has engendered growing demands of industrial water use, and household water use has rapidly increased thanks to urbanization, rising income levels, and consumption pattern changes.

The total amounts of water use in China increased by 34% between 1980 and 2009, and the primary cause for the upward trend of water use was attributed to the contribution of household and industrial sectors although the agricultural water sector was the largest water user. The agricultural sector showed a slow growth of water use by 5% from 391.2 billion m³ in 1980 to 372.3 billion m³ in 2009. An interesting fact is that the total irrigated areas in China were enlarged at 34% from 1980 to 2010, from 44.89 million ha to 60.35 million ha despite the increase of agricultural water use. It is plausible to maintain that the enhancement of water use efficiency in the agricultural water sector had contributed to such a trend. An average water use per ha of irrigated plots of land diminished from 8240 m^3/ha (549 m^3/mu) to 6280 m^3/ha (417 m^3/mu)^2 (Yang et al. 2012).

In the period between 1980 and 2010, the total amount of water use had grown by 32.9%, from 444 to 590 billion m^3. The agricultural sector accounted for around 68% of the total water consumption in the country, and the overall demands of the agricultural water sector had soared up from 330.9 billion to 369 billion m^3 by 10.3%. The irrigation sector, which accounts for 90.3% of the total agricultural water demand, had witnessed the decrease of water demand by 38.1 billion m^3 in the three decades owing to the enhancement of water productivity (Wang et al. 2017). In 2014, farmland irrigation accounted for 56% of total water use and 88% of total agricultural water use in China. Forestry, fruit gardens, pasture irrigation, and fishponds accounted for 12% of agricultural water use. Water supply for rural households had been improving but the availability of supply and quality of water still needs more investments (ADB 2018).

The household water demands had increased from 6.8 billion to 77 billion m^3 by more than ten times from 1980 to 2010. Urban household water use was 58.3 billion m^3 in 2014, occupying about a third of the total urban water use and almost doubled between 2000 and 2014 from 28.4 billion to 58.3 billion m^3. The rate of increase in this sector started to decline in 2010 (ADB 2018; Wang et al. 2017; Yang et al. 2012).

The industrial water sector consumed water resources at 45.7 billion m^3 in 1980 and 139.1 billion m^3 in 2009, which increased more than three times. The industrial water demand had grown up from 45.7 billion to 145 billion m^3 (68.4%). Industrial water use declined from 145 billion m^3 in 2010 to 135.6 billion m^3 in 2014, which accounted for approximately 22% of total water use at the national level. Such a declining trend in industrial water use may be attributed to improved water use efficiency or economic restructuring. Spatial differences are noted in terms of industrial water use in 2014, and

---

^{2}The Chinese measure of land area, 'mu' is equivalent to 1/15 ha.

for instance, industrial water use in the Yangtze River Basin reached 70.8 billion m^3 compared with merely one billion m^3 in the southwest rivers.

Water use for thermal and nuclear power generation was soaring, amounting to 35% of total industrial water use in 2014 (47.8 billion m^3). Water abstraction for thermal power generation demonstrated an upward trend in 2000 and sluggish growth rates after 2007 when inefficient and small coal-fired power plants were shut down in the 11th FYP (2006–2010). Thermoelectric power generation was responsible for 29% of total industrial water use in 2010 with an increase to 39% in 2012 (ADB 2018; Wang et al. 2017) (see Table 3.6).

A significant feature observed from Table 3.6 is the trend of water use efficiency, which is closely related to water productivity. Water productivity from 1980 to 2010 had increased from US$ 0.55/m^3 to US$ 8.7/m^3, almost 15 times more. Such a level of water productivity of China is, however, not good enough compared with that of the G20 countries, which is equivalent to US$ 42.9/m^3. Water productivity of paddy irrigation in China reaches RMB 1.0 (US$ 0.15)/m^3 whereas that of vegetables can be as high as RMB 12.3 (US$ 1.89)/m^3, which is as good as half of the level of developed countries. The manufacturing sector in China shows a higher level of water productivity, which is RMB 21.3 (US$ 3.27)/m^3, but this level is still low. To cope with growing demands of water, the Chinese authorities have continued

Table 3.6 Changes in population, economy and water use in China from 1980 to 2010

Item	1980	2010
Population (million)	987.1	1340.90
Urban population (%)	19.4	52.1
GDP (US$)	302.3	9012.90
Total water demand (billion m^3)	444	590
Agricultural water demand (billion m^3)	330.9	369
Industrial water demand (billion m^3)	45.7	145
Household water demand (billion m^3)	6.8	77
Water productivity (US$/m^3)	0.55	8.7

Source Wang et al. (2017)

to build numerous dams, drill boreholes, and undertake water transfer projects (MWR 2018; Wang et al. 2017).

Looking into various water users of China in 2019, the largest water user was the agricultural sector, which accounted for 61%, the industrial sector followed, occupying 20%, and the household was consuming 14%. Water resources for ecological uses were allocated by 4% (MWR 2018, 2019, 2020; NSB 2019). Figure 3.6 demonstrates more updated trend of water use in China from 2010 to 2019. The trend of water use in this period is different from that from 1980 to 2009, which displays a status quo in terms of the total amounts of water use. Although there are various explanations about this phenomenon, rigorous efforts of the government have seemed to be paid off regarding water saving policies and campaigns, especially pertaining to the largest water use sector, agriculture. Continuous campaigns and regulatory works have been undertaken for industrial users that have been encouraged to adopt water saving industrial process, technologies, and practices.

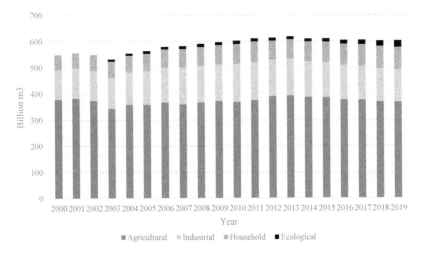

Fig. 3.6 Water use trends by agriculture, industry, household and ecology from 2010 to 2019 (Source MWR [2018, 2019, 2020], NSB [2019])

Overall, the amounts of water use in China have slightly decreased from 602.2 billion to 601.12 billion m^3 from 2010 to 2019. The agricultural water use in 2010 amounted to 61% of the total water use in the country but has shown a status quo, 61% in 2019. The portion of industrial water use in 2000 reached 24% whereas that in 2019 has reduced to 20%, and the household water use accounted for 13% in 2010 whereas the figure has not much changed, slightly increasing to 14% in 2019. A per capita water use decreased from 456 to 431 m^3 from 2013 to 2019 (MWR 2020). It seems that Chinese authorities allocated a minimum amount of water for ecological use from 2003, and in 2010, the ecological water use amounted to 2%. Such an ignorance of water allocation for ecological purposes appears to have changed since 2015 thanks to the recent policy shift favoring ecological rehabilitation and protection. Water for ecological water uses accounted for more than 4% in 2019 (MWR 2018, 2019, 2020; NSB 2019).

Water Quality

Overall Situation

The water quality of major surface and ground water bodies in China is assessed by the Ministry of Ecology and Environment, and the ministry publishes the annual ecology and environment bulletin. The water quality of freshwater bodies in recent five years is described in Fig. 3.7 below. The sections of Grade I–III have gradually increased from 64.5% in 2015 to 74.9% in 2019, and those of Grade IV–V have decreased from 26.7% to 21.7%. The good news is the conspicuous reduction of the sections of worse than Grade V, from 8.8% in 2015 to 3.4% in 2019. This signifies positive outcomes of the government's efforts to wage war against water pollution.

According to the 2019 Ecology and Environment Situation Bulletin, 1931 water quality assessment sections at the national level were evaluated. The sections of Grade I–III accounted for 74.9%, which increased by 3.9% compared with the ratio in 2018, and those of worse than Grade V, 3.4%, which reduced by 3.3% compared with that in 2018.

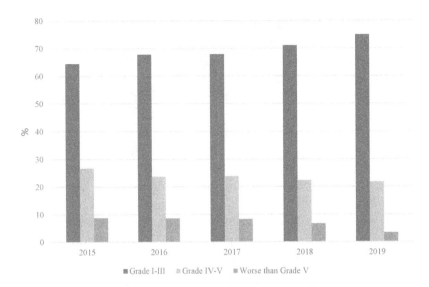

Fig. 3.7 National water quality assessment results from 2015 to 2019 in China (*Source* MEE [2015, 2016, 2017, 2018b, 2019])

The sections of Grade IV accounted for 17.5% and those of Grade V, 4.2%. China's surface water quality assessment is based on the six-grade rating system as described in Table 3.7.

Among the 1610 monitored major river reaches in China, 21% of the reaches met Grade IV or worse, which is unsuitable for drinking or

Table 3.7 Surface water quality standard 6-grade rating system in China

Grade	Classification/applicable uses
I	Pristine water source (e.g., river headwaters and protected natural catchment areas)
II	Class A water source protection areas for centralized drinking supply
III	Class B water source protection areas for drinking supply and recreation
IV	Industrial water supply and recreational water with no direct human contact
V	Limited agricultural water supply
VI	Essentially useless

Source Han et al. (2016)

human contact. Water quality in 6.3% of the river reaches was recorded worse than Grade V, which means no consumptive use or human contact of the water from them. The river reaches' water quality of the northwest, Zhejiang and Fujian Provinces, the southwest and the Yangtze River were regarded as 'very good,' and those of the Pearl River as 'good.' The five major river basins in North China, e.g., the Yellow, the Liao, the Hua, the Hai, and the Songhua Rivers include more river reaches classified as Grade IV and V compared with the other major river basins. Primary pollution indicators are Chemical Oxygen Demand (COD), Total Phosphorus (TP), Permanganate, and Ammonia Nitrogen (MEE 2019). Figure 3.8 shows the water quality of major river basins in China, and the seriousness of water pollution at the national level is delineated in Fig. 3.9, showing the percentage of the major river sections classified as Grade IV, V, and worse than Grade V.

Coastal areas should closely be monitored and regulated, because river systems discharge domestic, industrial, and agricultural wastewater into these coastal areas, which cause chain effects on coastal seawater pollution and disrupt marine ecosystems. In autumn 2014, coastal water's quality was characterized as 'poor' or 'very poor' (Grade IV or V), and

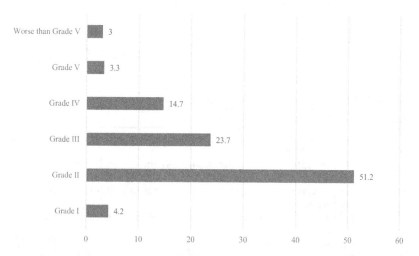

Fig. 3.8 Water quality of major river basins in China, 2019 (Unit: %) (*Source* MEE [2019])

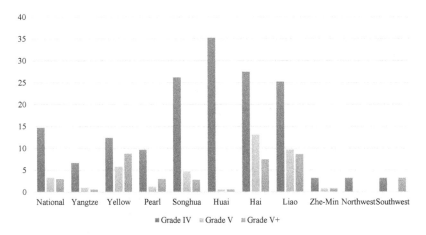

Fig. 3.9 Poor water quality of major river basins (Grade IV, V, and V+) in China, 2019 (Unit: %) (*Source* MEE [2019]; *Remarks* Zhe-Min indicates rivers in Zhejiang and Fujian Provinces)

the water quality of the combined coastal area, 57,000 km^2, was rated Grade IV or V (Han et al. 2016).

It is important to note that more attention should be paid to small tributaries of major river systems in China, since the water quality of them is often worse than that of the main water courses. This phenomenon comes into being thanks to less attenuation capacity of small streams compared with major rivers and lax regulations of industrial wastewater discharge to small streams that can often be observed in suburban and rural areas, especially toward Township Village Enterprises (TVEs). This situation posits that the overall assessment of surface water quality based on the national-level data might mislead the public, because the result only reflects severe surface water pollution in mainstreams, not small tributaries, which can be primary arteries of vast amounts of point and non-point source pollutions (Han et al. 2016).

According to the 2015 National Land and Resources Bulletin, 202 municipal and equivalent local governments conducted the investigation of the quality of groundwater with 5118 monitoring stations. A total of 466 monitoring stations, 9.1%, reached Grade I, and 1278 stations were rated Grade II, 25%. 236 stations, 4.6%, demonstrated Grade III, and 2174 stations were rated Grade IV, 42.5%. The rest of them, 964 stations, 18.8% reached Grade V (MLR 2015).

In 2015, 80% of groundwater samples from over 2000 shallow groundwater monitoring wells in the northern river basins were rated Grade IV and V. Water at or worse than Grade IV is not recommended for domestic or agricultural uses according to the National Groundwater Quality Standard (GB/T 14848-93). The primary pollutants for groundwater in China include the three types of nitrogen (NO_3-N, NO_2-N, NH_4-H), phenol, heavy metals and COD (Han et al. 2016; Liu and Zheng 2016).

A worrying trend is a gradual increase of groundwater pollution. The ratio of Grade IV and V surged from 55.1% in 2011 to 61.3% in 2015. The most recent data on shallow groundwater quality in 2018 indicates that Grade I–III aquifers account for 23.9%, Grade IV, 29.2%, and Grade V, 46.9%. This means that Grade IV and V aquifers in China amount to 76.1% (Han et al. 2016; MWR 2019).

This occurs because of the growing number of contaminants and more complicated mixture of pollutants discharged into groundwater resources. Currently, there has been no detailed information on individual data points and chemical concentrations so that these factors should seriously be considered for the enhancement of water quality monitoring and regulation for groundwater resources. An emerging concern is an extensive pollution of deep groundwater with the depths of more than 100 m. These deep aquifers are usually isolated from the rapid and surficial water cycle and need lengthy periods of time for being naturally recharged. The other dimension is that shallow and deep groundwater pollution is rampant at the national level, and the most seriously hit areas in the country are the North China Plain in which Grade I–III groundwater accounted for 22.1% of shallow groundwater and 26.4% of deep groundwater. The northwestern part of China, where much less people are around, is also subject to serious shallow and deep groundwater pollution (Han et al. 2016).

The most up-to-date information on groundwater quality are found in the 2019 Ecology and Environment Situation Bulletin. Out of 10,168 national-level groundwater monitoring stations, 14.4% of the stations reported Grade I–III, 66.9%, Grade IV, and 18.8%, Grade V. In addition, the 2019 bulletin discloses the situation of 2830 shallow groundwater wells, and Grade I–III wells occupied 23.7%, Grade IV 30.0%, and Grade V, 46.2%. The monitoring results above indicate that groundwater quality at the national level has continued to deteriorate, and more attention should be paid to detailed, localized, and specific regulatory frameworks and instruments for groundwater pollution control from the central and local governments (MEE 2019).

The water quality of large lakes in China has continued to deteriorate due to the sluggish circulation of water and the high concentration of pollutants in nearby areas. In 2019, Chinese authorities assessed the water quality of 110 major lakes at the national level. The lakes with the water quality of Grade I–III amounted to 69.1%, which was better than in 2018 by 2.4%, and those with the water quality of worse than Grade V accounted for 7.3%, which decreased by 0.8% compared with in 2018. Similar with the case of major river basins, primary indicators for assessing the degree of water pollution in lakes are Total Phosphorus (TP), Chemical Oxygen Demand (COD), and Permanganate (MEE 2019).

An imperative issue related to lake pollution is linked to eutrophication or oligotrophication. Eutrophication occurs when water bodies, often freshwater lakes, are full of minerals and nutrients, which triggers excessive growth of algae, thereby depleting dissolved oxygen of the water bodies. On the contrary, oligotrophication takes place when water bodies face nutrient depletion or reduction in rates of nutrient cycling. This phenomenon takes place as a consequence of acidification, typically the result of pollution, particularly air pollution and acid precipitation. In 2019, out of 107 monitored freshwater lakes in China, the lakes confronted with oligotrophication accounted for 9.3%, those with increasing levels of excessive minerals and nutrients, 62.6%, those with eutrophication, 28% (light degree 22.4%, medium degree, 5.6%) (MEE 2019). The data imply that although the eutrophication-hit lakes look less than 30%, more than 60% of the lakes may be highly subject to eutrophication if appropriate policy and technical measures are not introduced soon.

For instance, the water quality of the three major freshwater lakes in China, namely Tai, Dianchi, and Chao Lakes, are classified as Grade IV in 2019. Although the outcomes are still regarded as 'poor' (Grade IV, V and worse than V), the situation demonstrates slight improvement compared with the overall water quality of the lakes, Grade V, in 2017. Little progress was made regardless of tremendous amounts of investment for improving the water quality of the three lakes from 2000 to 2010 (MEE 2019; NSB and MEE 2019; Parton 2018; Yang et al. 2012).

Major Polluting Sources

Anthropogenic impacts on freshwater bodies originate from large amounts of wastewater discharge. The data from 2000 to 2015 unveil the trend of continuous surge of wastewater discharge, particularly with a sharp increase of household wastewater discharge. The upsurge of household wastewater discharge is attributed to wastewater from urban centers and semi-urban or rural areas nearby major urban areas. Although these macro-data do not necessarily show details, the data on household wastewater discharge involve the amounts of wastewater classified as Non-Point Source (NPS) pollution. Compared with wastewater from point sources, such as polluting factories and industrial complexes, pollution from non-point sources scattered in rural and semi-urban areas is difficult to detect and regulate.

Figure 3.10 visualizes the gradual decline of industrial wastewater discharge from 2007 to 2015, partially thanks to the levels of wastewater price and discharge fee being modified. Local environmental protection bureaus (now ecology and environment bureaus since 2018) seem to have been committed to controlling point source pollution, and their regulatory works have been paid off. However, the total amounts of wastewater discharge in China had continued to grow since 2005, and this trend appears to sustain because of rising income levels and consumption pattern changes of the wealthier Chinese (NSB and MEE 2019; Shen and Wu 2017).

Wastewater is primarily discharged from household and industrial sectors together with an increase of NPS pollutants from the agricultural

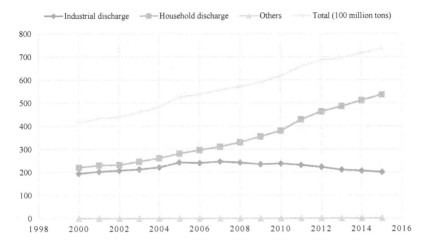

Fig. 3.10 Annual increase of wastewater discharge in China from 2000 to 2015 (*Source* NSB and MEE [2019])

sector, i.e., organic and chemical fertilizers, pesticides, and processing residuals. The total volume of untreated wastewater in the country had grown twice, amounting to 77.1 billion tons per annum over the past 15 years (MWR 2014). The official statistics of wastewater discharge embrace only household and industrial wastewater discharge. Since the mid-1990s, the amounts of wastewater discharge of households have overtaken those of industrial units (see Fig. 3.10).

These phenomena are closely associated with the dearth of sewer networks and wastewater treatment facilities in small- and medium-sized cities and the less clear economic gains of the wastewater treatment business. Many local governments have been reluctant to prioritize the construction and operation and maintenance of wastewater treatment plants for household wastewater, which has often been discharged to local streams and lakes without adequate treatment. In 2010, the untreated rate of household wastewater in small cities of China was as high as 90% (OECD 2011; Yang et al. 2012).

Environmental authorities endeavored to control industrial wastewater in the 1980s and the 1990s, which resulted in reducing large amounts of industrial wastewater discharge, mostly in urban areas and satisfying environmental standards, reaching 92% in 2010. Nevertheless, the overall water quality of major surface water bodies in the country has not enhanced remarkably, because the statistics of industrial wastewater discharge are applied to business units at and above the county level, which exclude units at the township and village levels. In addition, such a high rate of wastewater treatment in local areas would not necessarily demonstrate real situations with a suspicion of under-reporting by local governments to the central government. Almost no information on agricultural water pollution is available due to difficulty monitoring and measuring NPS pollution from agriculture, and a considerable number of TVEs are responsible for industrial pollution in rural areas, which are inadequately monitored by local environmental protection bureaus (Yang et al. 2012).

An aggregate of agricultural wastewater pollution has rapidly increased, and some water pollution indicators such as nitrogen (nitrate, nitrite and ammonia) and phosphorus demonstrate more loads of agricultural pollution than those of domestic and industrial pollution. In 2014, the major point source pollution originated from urban household wastewater, which amounted to 8.6 million tons of Chemical Oxygen Demand (COD), equal to 55% of total COD, and 1.4 million tons of NH_3-N (ammonia nitrogen), equivalent to 69% of total NH_3-N. Most of these effluents were dumped into water bodies without proper treatment, 77% of COD and 69% of NH_3-N. Rural areas were primarily responsible for NPS effluents, which were as large as 68.4 million tons of COD in 2014. Livestock and poultry production sectors contributed to 77% of total NPS effluents (ADB 2018).

There are different types of major pollutants depending upon regions, the intensity of sectors, and how effective local water quality control policies are. For instance, the northwestern part of China and the North China Plain have been the home of intensive irrigation farming for many years in which a heavy dependence on fertilizers and pesticides has resulted in serious nitrate contamination. An average per unit area application of fertilizer was 2.8 times the world average, and pesticide, three times the world average in 2011. The country is regarded as the largest consumer of agricultural chemicals, accounting for more than 30% of global fertilizers and pesticides with only 9% of the world's arable land (Han et al. 2016; Hanink 2018).

For instance, the applications of fertilizers and pesticides in Jiangxi Province, one of the traditional agricultural provinces in China, soared from 1.032 million to 426 million tons, and from 39,600 to 93,900 tons, from 1993 to 2015, respectively, and such amounts exceed the global guidelines for the safe use of chemical fertilizers and pesticides (Liu and Xie 2019). The overuse of fertilizers and pesticides has spawned severe groundwater pollution, and the Chinese government introduced ambitious action plans for reducing the exploitation of fertilizers and pesticides in 2015. These are the Action Plan for Zero Growth of Fertilizer Use by 2020 and the Action Plan for Zero Growth of Pesticide Use by 2020. Although the outcomes of the action plans should need more time, the implementation of the action plans does not seem to be smooth due to various pre-determined barriers, such as uncertainty of baseline data related to fertilizers and pesticides, data inconsistency, and complicated factors that influence the use of agricultural chemicals, such as farmers' perceptions, suppliers' interests, climate and market uncertainties (Jin and Zhou 2018; Liu and Xie 2019).

Side effects of intense and pervasive NPS pollution in rural areas have culminated in the emergence of 'Cancer Villages' in the North China Plain since the 1990s. Cancer villages are the rural settlements in which the occurrence of cancer morbidity is much higher than national averages. Since the first documentation of the case in 1954, the number of those villages was counted 459 by 2013, including 186 cases from 2000 to 2009, which is called 'decade of cancer' and exist in every province except for Qinghai and Tibet (Cheng and Nathanial 2019; Gong and Zhang 2013; Han et al. 2016). Cheng and Nathanial (2019) investigated the case of Jiangsu Province pertinent to a correlation between water pollution and Chinese cancer villages and provided the evidence that the villages close to polluted rivers show more cases of cancer outbreaks than those away from polluted rivers.

There are several polluting industries that are mainly responsible for producing large amounts of wastewater in China, which are the mining, textiles, paper, and chemical industries. Numerous TVEs in suburban or rural areas are often engaged in these fields so that untreated wastewater discharged from these industries are dumped into freshwater bodies without proper treatment, which keeps local freshwater bodies running dark brown or black. Such a grim situation of water quality in surface and groundwater resources is compounded by water shortages that decrease the assimilative capacity of freshwater bodies for diluting the level of water pollution (Yang et al. 2012).

Flood and Drought

Flood

Water-related disasters, particularly various kinds of flood (river basin flood, flash flood, urban flood, typhoon, ice flood, and dam-break flood), usually account for more than 90% of natural disasters in the world. In 2018, there were 315 climate-related and geophysical disaster events, affecting 68 million people with the death toll of 11,804. Out of the 315 disaster events, 273 events, about 87%, were associated with water-related disasters, which encompass floods (127 events), storms (95 events), droughts and extreme temperatures (41 events), and wildfires (10 events). This is the reason why primary attention should be paid to water-related disasters in natural disaster risk management policies at the global level (CRED 2019a, b).

In the past 100 years, out of the ten largest floods of the world, China has experienced seven of them, which are five in the Yangtze River, in 1911, 1931, 1935, 1954, and 1998, and two in the Yellow River in 1887 and 1938. The Chinese have struggled with harnessing water-related disasters for thousands of years, especially flood. The most devastating water-related disaster of China in the twentieth century was the 1931 floods in the Yangtze River and the Huai River, which caused 140,000 people to be drowned in addition to at least 3.7 million deaths in the following nine months (Biswas and Tortajada 2020; Wong 2020).

In China, flood-prone areas are scattered in its territory but primarily concentrated in the middle and lower reaches of seven major rivers and coastal plains. These areas occupy 8% of the total landmass of the country, reaching 800,000 km^2 in which almost half of the total population of China live, a third of farmland are located, and three quarters of total GDP are generated. Recently, increasing flash flood-prone areas are as large as 4.6 million km², almost half of the national territory where 2058 counties (72% of all counties) and a third of farmland are situated with 560 million people (about 40% of the total population) (Ding 2016).

China has had different scales of flood events every year, and for instance, in the period between 1980 and 2017, the annual average flood affected areas are 12.4 million ha, the annual average flood damaged areas, 6.8 million ha, and the annual mean flood damaged areas ratio, 53.7%. In addition, the annual mean flood-caused death toll is as large as 2165 persons (1994–2017), the annual mean direct economic losses, RMB 150.5 billion (US$ 23.1 billion), and the annual mean economic losses of water facilities, RMB 26.9 billion (US$ 4.14 billion) from 1998 to 2017 (MWR 2018).

The country has succeeded in undertaking flood prevention policies through both structural and non-structural measures, thereby achieving the downturn of human losses from 1990 to 2017. Structural flood protection measures embrace flood control levees, reservoirs (dams), floodways, and diversions of waterways. One of the most representative examples is the Three Gorges Dam, which has played a pivotal role in mitigating floods in the Yangtze River. From 2003 to 2019, the dam has been instrumental preventing and controlling floods 53 times.

Other examples of flood protection facilities are dikes which have been built more than 414,000 km along rivers and lakes, and reservoirs, amounting to 98,000 with the total storage capacity of 932 billion m³ at the national level. In addition to these, flood detention areas have been constructed, which occupy 33,700 km² with the total storage capacity of 107.4 billion m³. Focusing on urban areas, municipal governments have constructed urban flood control facilities that can withstand every 50-year to 200-year flood, and major residential areas along coastal lines have been protected with sea embankments whose design intends to control flood once every 50 to 100 years (Ding 2016).

In addition to gray infrastructures, the central government has encouraged local governments to opt for green solutions related to flood control, such as the Sponge City Initiative in 2014, artificial wetlands, rain gardens, and permeable pavements. Confronted with climate change-driven uncertainties, municipal governments have introduced green regulations and standards that encourage new buildings to be equipped with storage areas for rainwater that can temporarily store water during floods. Such infrastructures can retain substantial amounts of rainwater which otherwise would inundate into streets or give too much pressure to drainage systems in cities (Biswas and Tortajada 2020).

Non-structural measures are also significant in flood protection policy. There are three types of them: (1) avoiding inadequate use of floodplains and reducing risks to flood damage, i.e., development policies; (2) improving preparedness and resilience to floods, such as education and training, flood insurance, and tax adjustments; and (3) measures to fill the gap of structural measures, i.e., scaling up vegetative cover to reduce erosion. For example, several catastrophe insurance pilot projects in Shenzhen, Ningbo, and Chongqing have been introduced since June 2014. The scheme is aimed at expanding the insurance scope from earthquakes and hurricanes to other natural disasters, i.e., floods, droughts, and snowstorms. A variety of laws and regulations have been introduced for flood prevention and control, such as Water Law, Flood Control Law, and Flood Prevention Regulation together with technical standards associated with flood prevention and control, planning, design, forecasting, division, prediction scheme, and disaster evaluation (Ding 2016; Guo et al. 2020; Tan 2016).

There has been an upward trend of total direct economic losses by floods from 1990 to 2017 as seen from Fig. 3.11. Major flood events frequently occur in small rivers and urban areas. The number of cities suffering from floods and waterlogging increase, and the annual average number of cities hit by floods and water logging grew from 120 between 2006 and 2009 to 189 between 2010 and 2014. Construction of drainage networks often lags behind in the growth of cities, which also contributes to an increase of flood risk in urban areas (ADB 2018).

An increase of economic losses by floods is attributed to industrial locations that are highly subject to future flooding. Substantial volumes of Foreign Direct Investment (FDI) have been flowed into three major sectors, namely electronics, machinery and manufacturing, and retail and wholesale. In 2013, 52% of China's industrial sites were subject to flooding. These fields of industries are highly vulnerable to flood,

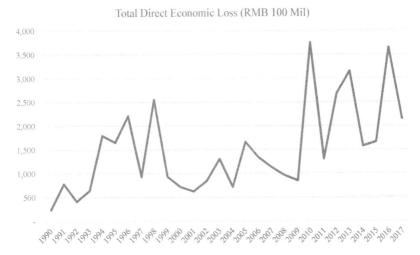

Fig. 3.11 Trends of total direct economic losses by flood in China from 1990 to 2017 (*Source* MWR [2018])

and insured losses for those industries in the Pearl River Delta could be equivalent to US$ 44 billion (Rij 2016).

Intense flood disasters wreaked havoc on the country in 2020. From 2 June 2020, 27 provinces in central (Yellow River Basin) and southern China (Yangtze River Basin) experienced severe rainfall for 41 consecutive days, which triggered calamities to numerous people. According to a report on 13 August 2020, at least 219 people were killed, 820,000 people required emergency assistance, and more than four million people had to be evacuated from flood-hit areas. A total of 64.5 million people were badly affected by the flood. More than three months' heavy rainfall in vast areas of central and southern parts of China decimated almost 400,000 houses and damaged five million ha of farmland (Guo et al. 2020; Myers 2020; Wong 2020).

The Ministry of Emergency Management announced that direct economic losses would exceed US$ 25 billion. The level of direct economic losses from the 2020 flood is beyond the annual mean direct economic losses from 1998 to 2017, US$ 22.1 billion, and can be worse than the 1998 floods which killed more than 4000 people, left 14 million homeless and affected approximately 239 million people alongside a total economic loss of US$ 30 billion. The scale of damage can swell more because the summer monsoon season usually goes on together with the possible arrival of a few more typhoons until the end of September (Guo et al. 2020; MWR 2018; Myers 2020; Rij 2016; Wong 2020).

Meteorologists and hydrologists in the world and China have incrementally been wary of unexpected patterns of weather change, primarily triggered by climate change. Climate change can intensify the risk of flood events over large regional and time scales. Among many conspicuous features caused by climate change, changes in precipitation and temperature are some of the most common features, which can accelerate the risk of flood events. Most of serious flood events in China since the 1950s have occurred due to extreme precipitation, including the 2020 flood. The average intensity, quantity and duration of precipitation in southern China in 2020 are some of the highest since 1961. Additional environmental elements, which are inseparable with the hydrological cycle, drainage basin conditions, and the soil status, are affected by both precipitation and temperature and closely associated with the risk of flood events. It is significant to recognize the ongoing impacts of El Nino Southern Oscillation, Pacific Decadal Oscillation, and atmospheric circulation patterns on the amount of precipitation and typhoons with heavy rainfall and strong wind, which can also entail serious flooding (Guo et al. 2020).

Various damages and impacts have been caused by flood events, and Chinese authorities have meticulously collected data on human and economic losses since the 1950s. Nevertheless, some aspects should have been reflected in flood- or drought-related data collection, such as health effects. Flood-related health effects include the direct contact of humans with flood waters, such as drowning and injuries from flood-related events (e.g., building collapse and damage, electrocution and fire).

In Fig. 3.12, the number of deaths by flood events had reduced from 1990 to 2017, and there is no record of indirect human health impacts due to flooding (MWR 2018). Flood events can escalate the risks of and deaths from cardiovascular disease and diabetes thanks to the suspension of health services or social support systems. Communicable disease outbreaks and infections can surge, particularly in places with poor hygiene and population displacement, including leptospirosis,

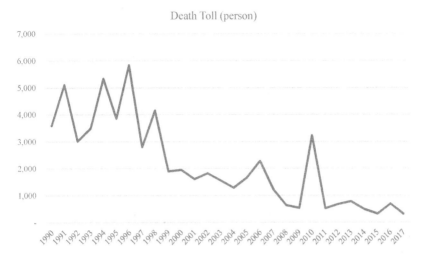

Fig. 3.12 Death toll caused by flood in China from 1990 to 2017 (*Source* MWR [2018])

hepatitis and gastrointestinal disease, and respiratory infections. More commonly, floodwaters are often full of contaminants, such as sewage, human and animal excreta, fertilizers, organic waste, and toxic chemicals, which can cause infections on the eye and skin and put people in jeopardy for being exposed to toxic and poisonous materials. Water-borne and vector-borne diseases can easily be transmitted through floodwaters, and Chinese authorities should keep a close eye on physical and mental health of the affected people (Guo et al. 2020).

Ding (2016) maintains that flood control benefits in China can be estimated at RMB 4600 billion (US$ 707.7 billion) with the size of reduced inundated farmland, 180 million ha and the amounts of reduced grain losses, 760 million tons. The collapse or damage of numerous properties have been avoided thanks to the strong commitment of governmental flood control efforts together with the substantial number of lives being saved.

Drought

Drought is defined as a long period of time featured with a lack of a region's water supply which results from constantly below average precipitation. Drought disasters are referred to as the phenomenon of water shortage due to unusually low levels of precipitation that affects livelihoods of local people, agricultural and industrial production and ecosystem services. China is historically portrayed as a drought-hit country, and historical documents of the country inform that the number of droughts recorded were 1074 from 1766 B.C. to A.D. 1937 with a frequency of once every three years and four months. The northern China is highly subject to drought (ADB 2018; Rij 2016; SFCDRH and MWR 2018).

In 2016, 2025 out of 2863 counties in China were affected by droughts, which is equivalent to 71% of the total number of counties and encompass 584 counties that were seriously damaged by droughts. In addition, 1400 counties suffered from drinking water shortage driven by droughts, and 1608 counties faced agricultural droughts (Ding 2016). In 2017, 26 out of 31 provinces and provincial level cities experienced droughts, drought-affected crop fields were as large as more than 18.2 million ha, and drought-affected areas were estimated at 9.95 million ha. More than 4.7 million people and more than 5.1 million livestock had limited access to drinking water due to droughts, and food losses reached 13.4 billion kg together with economic crop losses of RMB 11.6 billion (US$ 1.78 billion). Direct economic losses were estimated at RMB 43.7 billion (US$ 6.72 billion), which is equivalent to 0.05% of the country's total GDP (SFCDRH and MWR 2018).

Drought-prone areas in China are the southern, central, and eastern provinces. The deadliest drought of China in terms of the number of deaths took place in the southeastern part of the country from February and August 1991. 2000 people were killed with five million being affected by the drought. Two more severe droughts were recorded as some of the worst drought disasters in China until recently—1988 and 1994. The 1988 drought badly hit the eastern part of the country, and one-fifth of the arable land was severely damaged. The drought came together with extremely high temperatures of almost 40 degrees Celsius, making people have little access to drinking water and killing innumerable people, particularly the elderly. Economic losses of the 1988 drought reached US$ 942 million with 49 million people affected and the number of deaths, 1400. Long-spelled droughts in 1994 triggered huge amounts of economic losses, estimated at US$ 13.8 billion, and in particular, the agriculture sector was badly affected by the drought that impacted mainly on North China alongside some areas of the middle part of the country. Precipitation was recorded 40%–50% less than average in the affected areas (Rij 2016).

The trends of drought damage between 1980 and 2017 are shown in Fig. 3.13. Drought-affected and damaged areas in this period had gradually diminished although severe drought events occurred in 2000, 2001, 2007, and 2009 (MWR 2018). It can be argued that Chinese authorities have, to some extent, managed to abate drought-related negative impacts on crops, people, and economic assets in this period. However, the intensity and frequency of droughts would increasingly exacerbate thanks to adverse impacts of climate change and its associated phenomena, i.e.,

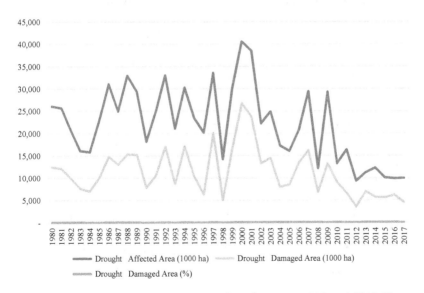

Fig. 3.13 Trends of drought damage in China between 1980 and 2017 (*Source* MWR [2018])

high temperatures and heatwaves, so that continuous commitments of the central and local governments should be required in due course. Regarding the agricultural sector, Chinese authorities should consider consequences of future droughts that would cause more substantial amounts of agricultural losses, thereby posing a serious threat to its food security. Relevant policies and programs against future drought disasters of China should be tailor-made for drought-prone areas, such as southwest, central, and northeast China (Rij 2016).

An important factor related to drought risk management is how to balance different water uses, water for energy or water against a drought. For instance, the Three Gorges Dam has been built in the middle reaches of the Yangtze River and has played a significant role in not only generating hydropower and facilitating inland navigation but also contributing to flood protection successfully. However, the dam was in an awkward position associated with the 2011 drought, because dam managers were forced to give up their future profits that can be gained from hydropower generation and instead, to release water for agricultural and domestic consumption. From January to May 2011, an unusual drought affected the middle reaches of the Yangtze River due to the lack of rainfall whereas the dam stored water for hydropower generation. As a result, little water was available downstream, which resulted in drying up freshwater lakes, killing fish stock and suspending inland navigation in the river. In the end, the dam had to release more water for relieving the downstream areas (Rij 2016; Wang 2011).

Structural measures against droughts have actively been implemented in China. There are more than two million irrigation districts, which occupy approximately 280 million mu, equivalent to 186,667 km^2. In 2016, there were 58.8 million rural water supply projects, which benefited 812 million farmers for agricultural and household water uses.

As non-structural measures, the Chinese government has prepared a cascade of laws and regulations related to drought prevention and control, including Drought Relief Regulation. In addition to that, diverse technical standards have been devised for protecting people and economic assets, i.e., drought prevention and control, planning, design, forecasting, prediction scheme, and disaster evaluation.

Unceasing efforts have been made by Chinese authorities for relieving negative impacts caused by droughts for many decades. From 1991 to 2015, 231.9 million people benefited from improved access to drinking water, irrigation areas of 430 million mu (286,667 km^2) became safe from droughts, and 38.02 million tons of grain were out of drought risks (Ding 2016).

Climate Change Impacts

According to the Intergovernmental Panel on Climate Change (IPCC)'s Fifth Assessment Report in October 2014, flood risks in the world will rise in the coming years. Among various future scenarios, a high-end climate scenario alongside high population growth and Green House Gas emissions indicates that floods would give far-reaching impacts on an average of additional 2.4 million people and trigger extra EURO 3 billion (US$ 3.65 billion) per annum. In the same scenario, the largest flood risks would be observed in the US, Asia, and Europe with a mean global temperature surge of 4 degrees Celsius, and flooding would affect China, resulting in damaging 40 million people with total economic losses estimated at EURO 110 billion (US$ 134 billion) per annum by the end of this century (Alfieri et al. 2017; Guo et al. 2020; IPCC 2014).

China may be confronted with three major water-related consequences caused by climate change: (1) an increase in temperature; (2) a geographical and seasonal change in water distribution; and (3) sea level rise. Climate change can exacerbate water scarcity in North China, and from the 1990s and the 2010s, mainstream water flows plummeted in the Yellow (Huang), the Huai, and the Hai River Basins, 15%, 15%, and 41%, respectively. In 2012, the amount of water shortage in the region was estimated at approximately 30–40 km^3 per annum, and the figure is expected to increase to 56.5 km^3 by 2050. The probability of basin-wide floods in the Huai River has increased due to the unpredictability of climate. From 1981 to 2015, extreme flood events in 23 sections of the Pearl River Basin soared significantly (MEE 2018b).

Glacier melting can be attributed to a temperature rise, which would seriously impact upon river runoff and flooding in China. Glaciers reduced by 10% between the 1970s and the early 2000s and are expected to shrink more in the coming decades. Global warming can be another factor to trigger glacial runoff, particularly in the spring and early summer (ADB 2018; Li et al. 2012; MST 2015; Sandalow 2018).

One of the distinguished phenomena related to climate change is a rise of sea level in coastal areas in China. From 1980 to 2005, the rising rate of sea level was 3.0 mm per annum, which was higher than the global average and would continue to increase in the coming years. Such a situation implies that coastal cities of China would be overwhelmed by large quantities of seawater inflow, which would not easily be tackled by the currently installed pumps and drainage facilities, and typhoons and tsunamis would threaten the safety of people and economic assets in coastal areas.

More than 550 million people live in the coastal provinces, and some of major Chinese cities are located in low-lying areas such as Shanghai, Qingdao, and Xiamen. China's Third National Assessment Report on Climate Change in 2015 reported that sea levels could rise by 40–60 cm beyond the twentieth century averages by the end of the twenty-first century, and an increase of one centimeter could cause the coastline to recede by more than 10 m in parts of China. A 100-year flood event in Shanghai could take place 40 times more likely with one meter of sea level rise (MST 2015; Sandalow 2018). In the next 30 years, the sea level of Bohai Sea will be expected to rise between 65 and 155 mm, that of Yellow Sea, between 75 and 165 mm, that of East China Sea, between 70 and 160 mm, and that of South China Sea, between 75 and 165 mm (Ding 2016).

As a different side of a coin, climate change can bring good news for certain sectors in the country, such as the irrigation sector, at least in the short term. China may undergo increased precipitation with the average growth rate, 2–5% by the end of the twenty-first century. Interestingly, semi-arid and drought-stricken northern parts of China can expect a surge of precipitation more than the average precipitation growth rate, 5–15%. The drought frequency will reduce in the future. In the arid regions of the country, more rainfall would be expected although runoff may decrease in the late summer and autumn (MST 2015).

But in the long term, water shortage will become more serious if a large fraction of glacier melts. The frequency of high temperature and extreme droughts can grow at the same time so that there will be the high risk of droughts in certain areas, such as North, East, central parts of northeast, and Southwest China. Climate change could extend growing seasons for some crops in North China but may bring less rainfall and less reliable soils that hold less water, the wide spread of dangerous pests and shorter growing seasons for many crops. From 1980 to 2015, droughts with the grade of severe or worse hit China every two years on average. Drought conditions in humid and semi-humid regions have intensified and expanded in recent decades. Interrupted rainfall patterns would give a pressure on reservoirs, triggering dam safety challenge including the Three Gorges Dam (ADB 2018; Ding 2016; MEE 2018b; Sandalow 2018).

Water-related disaster risk has increasingly risen thanks to climate variability and extreme weather events. Since 2000, extreme weather events have entailed natural disasters costing China RMB 4.5 trillion (US$ 692 billion). In 2016, more than 190 million people in China were affected by natural disasters, and related economic losses were estimated at RMB 503 billion (US$ 77.4 billion). Floods and geological disasters (i.e., landslides) contributed 62% of the total economic losses, RMB 313 billion (US$ 48 billion). To cope with such challenges, the government had to triple its budgets on flood prevention from RMB 69 billion (US$ 10.6 billion) in 2010 to RMB 208 billion (US$ 32 billion) in 2016. A recent phenomenon is the intensification of urban floods, especially in the Yangtze River Basin. Heavy rain days generally increased in major cities and suburbs in the river basin in the period from 1981 to 2010 compared with the period from 1961 to 1980 (Hu and Tan 2018).

Urgent and systematic polices and measures are necessary for reducing and managing such risks with the opportunity to beef up integrated disaster risk management in the country. Many types of water-related disasters are likely to occur owing to changes in temperature, precipitation, and other climate variables. Deforestation and irresponsible development of mountainous areas have triggered the loss of vegetation cover, which often multiplies the risk of landslides and flood disasters downstream. A rise of population density, the expansion of the built environment in urban areas, and infrastructure development all may shoot up vulnerability in hazardous areas.

Impacts of water-related disasters are direct, i.e., damages to buildings, crops, and infrastructure as well as losses of people and property, indirect, i.e., losses in productivity and livelihoods, greater investment risk, indebtedness, and impacts on public health. China's unprecedented depth and speed of socioeconomic development in the last few decades have amplified exposure to hazard-prone areas, which has culminated in putting people and economic assets in jeopardy. Such risks are compounded by climate change in recent years (ADB 2018; MEE 2018b).

China's response to climate change and its adverse impacts was entrenched in the 12th FYP (2011–2015) and the 13th FYP (2016–2020) coupled with the National Climate Change Adaptation Strategy (2013–2020). Possible impacts triggered by climate change were incorporated into socioeconomic development plans and programs for China's modernization, i.e., urban–rural development planning, infrastructure development, and distribution of industries between regions. The government will establish and modify technical standards in various sectors adapting to climate change and put more investments in research on climate change for enhancing forecasting and providing early warnings. Particularly, substantial improvement will be expected for increasing adaptation capacity and establishing regional spatial and functional planning for adaptation.

Relevant policies and actions include the publication of a series of policy documents about climate change, i.e., the 13th FYP for Construction of Water Saving Society, the Water Function Regionalization of Nationally Important Rivers and Lakes (2011–2030), and the Opinions of the Implementation of the Strictest Water Resources Management System.

In the 12th FYP (2011–2015), there were 100 national water saving society construction pilot projects together with 200 provincial pilot projects. Several water saving actions were introduced, including agricultural water saving measures with productivity increase, industrial water saving with efficiency increases, and urban water saving with consumption reduction. In addition, comprehensive control and protection of water resources were implemented, hygienic inspection and monitoring of drinking water were improved, and flood risk management and geological disaster prevention measures were undertaken (MEE 2018b).

A more efficient use of water resources was given a priority in the establishment of the 13th FYP (2016–2020), and the plan specified the outline of water security projects which would contribute to comprehensive flood control and mitigation systems. Simultaneously, the government pledged to focus on water resources protection and water pollution control through the implementation of IWRM.

Water resources management is a significant area to pay special attention to for climate change adaptation, which is also closely associated with the effective achievement of Nationally Determined Contributions (NDCs) of China, which was introduced in 2015 from the Paris Agreement. Climate resilience of the country should be improved based on the optimized allocation of water resources in tandem with the implementation of the strictest water resources management system. The strictest water resources management system divides the country into a four-level River Chief System (i.e., province, city, county, and township) and encourages the protection of water resources, river and lake shoreline management, water pollution prevention and control, water ecology restoration and supervision of law enforcement. As a part of the National Adaptation Strategy, an array of conservation, ecological restoration and use strategies are widely undertaken for the adaptation of the water sector whereas concerned authorities are committed to meeting diversified demands from various users and regions.

Among progress and achievements against adverse impacts of climate change, the South North Water Transfer Project has made substantial contribution to removing a large degree of uncertainty in water allocation. By October 2017, the Phase 1 of the middle route of the project had diverted a total of almost 11 billion m^3 of water to North China, including Beijing. Regarding flood risk management at the national level, 2058 county-level torrential flood monitoring and warning systems and public monitoring and prevention systems were established with an increase of flood reporting stations to 97,000. Eventually, the total number of death and missing people because of floods dropped to its lowest level since 1949 (MEE 2018a).

Unprecedented scales of challenges caused by climate change require innovative and novel approaches to financing climate change strategies, plans, and infrastructure development in China. According to a report in 2016, China's annual climate finance needs should be as large as RMB 2560 billion (US$ 394 billion) by 2020. China's climate-smart investment potential can amount to US$ 14.9 trillion between 2016 and 2030. The Belt and Road Initiative (BRI) accommodates the need to promote green and low-carbon infrastructure construction and operation management, which includes plans not only for trade and investment but also climate change and ecological conservation. China's new international financing vehicle, the Asian Infrastructure Development Bank (AIIB), recognizes the urgency to support and accelerate its members' transitions toward a low-carbon energy mix, which indirectly advocates climate change-concerned projects (Hu and Tan 2018).

Conclusion

The chapter has explored a critical set of information and data on water resources management and its association with socioeconomic and environmental situations in China. The total volumes of renewable water resources in the country are substantial, but the per capita water amount does not guarantee sufficient amounts of water supply to different water users who are in fierce competition, ranging from farmers in the agricultural sector, urbanites in households, industrial and commercial units to ecosystems. The uneven spatial distribution of water resources between North and South in the country is one of the most formidable challenges in terms of water allocation. North China occupies more than 64% of the total territory with 46% of the total population and about 60% of the total arable land whereas only less than 20% of the total water resources are available.

Whereas surface water resources are the most significant water supply source in China, accounting for 82%, it is important to recognize the contribution of groundwater resources, about 16%, particularly for the

North China Plain. Groundwater resources have seriously been degraded by overdraft and pollution, and such a grim situation has engendered chain effects, i.e., land subsidence, aquifer salinization, and seawater intrusion in coastal areas. Chinese authorities have been wary of the rapid deterioration of lake's water quality, often affected by eutrophication and oligotrophication. There are numerous rivers in China, and seven major river basins have intensively been exploited for socioeconomic development in many years, and more attention should be paid to not only keeping water quality good but also restoration of water environments.

The trend of water use over the last three decades discloses a surge of household and industrial water use with a sluggish growth of agricultural water demand and a minimum allocation of water for the environment from 1980 to 2009. Such a trend has been altered from 2010 to 2019, and there is little change for households, industries, and agricultural users' water demand whereas more water was allocated for the environment. This is a good hint for revitalizing water and ecosystems in the era of 'ecological civilization' and 'Beautiful China.'

Numerous surface and groundwater bodies are regularly monitored and evaluated by environmental authorities at the center and the local level in China. Strenuous works for law enforcement as well as the establishment of many laws and regulations have led to the betterment of the water quality of rivers, lakes, and groundwater resources over the four decades since the late 1970s. But polluting sources are scattered in and out of urban areas in the form of Non-Point Source (NPS) pollution, not only discharged by households and large industrial units but also numerous small-scale livestock farms and Township Village Enterprises (TVEs) which are not easily regulated. The term, 'Cancer Villages,' emerged in the 1990s due to heavy water pollution and its health impacts on local villagers who drank polluted water in rural areas. Chinese authorities have been warned of such a dreadful consequence if water pollution is not adequately dealt with.

Innumerable Chinese people are badly affected by floods and droughts almost every year. Some of the largest and worst flood events in the world have occurred in China, and Chinese authorities have striven to harness floods in many parts of the country, particularly in the Yangtze and the Yellow River Basins. China's Three Gorges Dam is touted as an icon to protect people as well as economic assets from major flood events, such as the 2020 summer floods. Structural and non-structural measures have been introduced for preventing flood, but climate change-driven uncertainties incapacitate such efforts.

Drought events often accompany long-spelled pains and devastation in food production, water shortage, and damage to public health through heat waves or high temperatures. For the last four decades, drought-damaged areas have diminished, which implies a successful abatement of drought-related negative impacts on crops, people, and economic assets. Nevertheless, climate change and its adverse impacts have made flood and drought prevention efforts sometimes ineffective. More resources should be mobilized and prepared for dealing with climate change, including financing and thorough scientific research and relevant technologies. Based on the lessons learned from the 13th FYP (2016–2020), more detailed and specific strategies and plans in China should be prepared for coping with new norms driven by climate change.

References

Alfieri, Lorenzo, Berny Bisselink, Francesco Dottori, Gustavo Naumann, Ad. Roo, Peter Salaman, Klaus Wyser, and Zuc Feyen. 2017. Global Projections of River Flood Risk in a Warmer World. *Earth's Future* 5: 171–182.
Asian Development Bank (ADB). 2018. *Managing Water Resources for Sustainable Socioeconomic Development: A Country Water Assessment for the People's Republic of China.* Manila: Asian Development Bank.
Biswas, Asit and Cecilia Tortajada. 2020. China Can Water Down Impact of Flood. *China Water Risk*, 18 August 2020.
Centre for Research on the Epidemiology of Disasters (CRED). 2019a. *Disasters 2018: Year in Review.* Brussels: CRED.
Centre for Research on the Epidemiology of Disasters (CRED). 2019b. *Natural Disasters 2018.* Brussels: CRED.
Cheng, Yuanyuan, and C. Nathanial. 2019. A Study of "Cancer Villages" in Jiangsu Province of China. *Environmental Science and Pollution Research* 26: 1932–1946.

Ding, Liuqian. 2016. *Flood and Drought Management in China*. Presented at China Water Resources Management Policy Seminar, June 22, 2016.

Global Water Partnership (GWP). 2015. *China's Water Resources Management Challenge: The 'Three Red Lines'*. Technical Focus Paper, Global Water Partnership.

Gong, Shengsheng, and Tao Zhang. 2013. Temporal-Spatial Distribution Changes of Cancer Villages in China (中国癌症村的空分布变迁研究). *China Population, Natural Resources and Environment (中国人口资源与环境)* 9: 156–164.

Guo, Yuming, Yao Wu, Bo Wen, Wenzhong Huang, Ke Ju, Yuan Gao, and Shanshan Li. 2020. Floods in China, COVID-19, and Climate Change. *Lancet Planet Health*. August 26, 2020, 1–2. Available Online: https://doi.org/10.1016/52542-5196(20)30203-5. Accessed 4 September 2020.

Han, Dongmei, Matthew Currell, and Guoliang Cao. 2016. Deep Challenges for China's War on Water Pollution. *Environmental Pollution* 218: 1222–1233.

Hanink, Nerissa. 2018. Overuse of Fertilizers and Pesticides in China Linked to Farm Size. *Stanford Earth*, June 17, 2018. Available Online: http://earth.stanford.edu. Accessed 14 August 2020.

Hu, Feng, and Debra Tan. 2018. *No Water, No Growth: Does Asia Have Enough Water to Develop?* Hong Kong: China Water Risk.

IPCC. 2014. *Climate Change 2014: Synthesis Report*. Geneva, Switzerland: IPCC.

Jin, Shuqin, and Fang Zhou. 2018. Zero Growth of Chemical Fertilizer and Pesticide Use: China's Objectives, Progress, Challenges. *Journal of Resources and Ecology* 9 (1): 50–58.

Li, Xiaokai, Graeme Turner, and Liping Jiang. 2012. *Grow in Concert with Nature: Sustaining East Asia's Water Resources Through Green Water Defense*. Washington, D.C.: World Bank.

Liu, Guiying, and Hualin Xie. 2019. Simulation of Regulation Policies for Fertilizer and Pesticide Reduction in Arable Land Based on Farmers' Behaviors—Using Jiangxi Province as an Example. *Sustainability* 11 (136): 1–22.

Liu, Jie and Chunmiao Zheng. 2016. Chapter 18: Towards Integrated Groundwater Management in China. In *Integrated Groundwater Management*, ed. Anthony Jakeman, Olivier Barreteau, Randall Hunt, Jean-Daniel Rimando, and Andrew Ross, 455–476. Springer Open.

Ma, Ronghua, Hongao Duan, Chuanmin Hu, Xuezhi Feng, Ainong Li, Weimin Ju, Jiahu Jiang, and Guishan Yang. 2010. A Half-Century Changes in China's Lakes: Global Warming or Human Influence? *Geophysical Research Letters* 37 (L24106): 1–6.

Myers, Steven. 2020. After COVID, China's Leaders Face New Challenges from Flooding. *The New York Times*, August 21, 2020.

Ministry of Ecology and Environment (MEE). 2015. *China Ecology and Environment Situation Bulletin 2015 (中国生态环境状况公报)*. Beijing: Ministry of Ecology and Environment of the People's Republic of China.

Ministry of Ecology and Environment (MEE). 2016. *China Ecology and Environment Situation Bulletin 2016 (中国生态环境状况公报)*. Beijing: Ministry of Ecology and Environment of the People's Republic of China.

Ministry of Ecology and Environment (MEE). 2017. *China Ecology and Environment Situation Bulletin 2017 (中国生态环境状况公报)*. Beijing: Ministry of Ecology and Environment of the People's Republic of China.

Ministry of Ecology and Environment (MEE). 2018a. *China Ecology and Environment Situation Bulletin 2018 (中国生态环境状况公报)*. Beijing: Ministry of Ecology and Environment of the People's Republic of China.

Ministry of Ecology and Environment (MEE). 2018b. *The People's Republic of China Third National Communication on Climate Change*. Beijing: Ministry of Ecology and Environment of the People's Republic of China.

Ministry of Ecology and Environment (MEE). 2019. *China Ecology and Environment Situation Bulletin 2019 (中国生态环境状况公报)*. Beijing: Ministry of Ecology and Environment of the People's Republic of China.

Ministry of Land and Resources (MLR). 2015. *2015 National Land Resources Bulletin (中国国土资源公报)*.

Ministry of Science and Technology (MST). 2015. *The Third National Assessment Report on Climate Change*. Beijing: Ministry of Science and Technology.

Ministry of Water Resources (MWR) and NBS (National Bureau of Statistics). 2014. *Bulletin of First National Census of Water*. Beijing: China Water Power Press.

Ministry of Water Resources (MWR). 2018. *China Water Statistical Yearbook 2018 (中国水利统计年鉴2018)*. Beijing: China Water Power Press.

Ministry of Water Resources (MWR). 2019. *China Water Resources Bulletin 2018 (中国国水资源公报2018)*. Beijing: China Water Power Press.

Ministry of Water Resources (MWR) (2020) China Water Resources Bulletin 2019 (中国国水资源公报2019). Beijing, China Water Power Press.

National Statistical Bureau (NSB). 2019. *China Statistical Yearbook 2019*. Beijing: China Statistics Press.

National Statistical Bureau (NSB) and Ministry of Ecology and Environment (MEE). 2019. *China Statistical Yearbook on Environment (中国环境统计年鉴)*. Beijing: China Statistics Press.
National Statistical Bureau (NSB) and Ministry of Ecology and Environment (MEE). 2019. *China Statistical Yearbook on Environment (中国环境统计年鉴)*. Beijing, China Statistics Press.
OECD. 2011. *Benefits of Investing in Water and Sanitation: An OECD Perspective*. Paris: OECD.
Parton, Charlie. 2018. China's Looking Water Crisis. *China Dialogue*. April 2018.
Rij, Emma Van. 2016. An Approach to the Disaster Profile of People's Republic of China 1980–2013. *Emergency and Disaster Reports* 3(4), 3–48. University of Oviedo – Department of Medicine, Unit for Research in Emergency and Disaster.
Sandalow, David. 2018. *Guide to Chinese Climate Policy 2018*. New York: Center on Global Energy Policy, Columbia University.
Shen, Dajun. 2015. Groundwater Management in China. *Water Policy* 17: 61–82.
Shen, Dajun, and Juan Wu. 2017. State of the Art Review: Water Pricing Reform in China. *International Journal of Water Resources Development* 33 (2): 198–232.
State Flood Control and Drought Relief Headquarters (SFCDRH) and Ministry of Water Resources (MWR). 2018. *China Flood and Drought Disaster Bulletin (中国水旱灾害公报) 2017*. Beijing: China Map Press.
Tan, Debra (2016) Beautiful China 2020: Water and the 13 Five Year Plan. *China Water Risk*. 16 March 2016.
Tao, Shengli, Jingyun Fang, Suhui Ma, Qiong Cai, Xinyu Xiong, Di Tian, Xia Zhao, Leqi Fang, Heng Zhang, Jiangling Zhu, and Shuqing Zhao. 2020. Changes in China's Lakes: Climate and Human Impacts. *National Science Review* 7: 132–140.
Wang, Qian. 2011. Worst Drought in 50 Years Along Yangtze. *China Daily*, 25 May 2020.
Wang, Xiaojun, Jian-yun Zhang, Juan Gao, Shamsuddin Shahid, Xing-hui Xia, Zhi Geng, and Li Tang. 2017. The Nw Concept of Water Resources Management in China: Ensuring Water Security in Changing Environment. *Environment, Development and Sustainability* 20 (4): 809–909.
Wong, Dennis. 2020. China's Worst Floods in Decades. *South China Morning Post*, July 27, 2020.

Xinhua. 2018. China's Temperature Rises Faster than Global Average: Blue Paper. *Xinhua*, 4 April 2018.

Yang, Hong, Zhuoying Zhang, and Minjun Shi. 2012. Chapter: The Impact of China's Economic Growth on Its Water Resources. In *Rebalancing and Sustaining Growth in China*, ed. Huw McKay and Ligang Song. Canberra: ANU Press.

4
Water Plan and Governance System

Introduction

This chapter assesses the commitment of the Chinese government to managing water resources with special reference to its plans, governing structure, and legal and regulatory settings, particularly focusing on the last ten years. Investigating the evolving nature of natural and anthropogenic challenges entrenched in water resources management, Chinese authorities decided to establish sustainable water resources management since the early 2010s, including the Three Red Lines in 2011 and the Water Ten Plan in 2015. These plans provide the evidence that top decision makers in the water sector have accommodated advice and wisdom extracted from decades of experiences since 1978 and are ready to upgrade China's water resources management for the middle of the twenty-first century.

There was the administrative reform in China, 2018, and one of the conspicuous changes related to the water sector was the reshuffling of water-related ministries, particularly the upgrade of the Ministry of Environmental Protection, which has become the Ministry of Ecology and the Environment (MEE). It seems that the MEE will be more influential in the water sector in due course although the Ministry of Water Resources (MWR) continues to be deeply involved in water resources planning, development and management issues. The Ministry of Natural Resources (MNR), which has been given water-related responsibilities from the precursor, the Ministry of Land and Resources, contributes to plans and projects in the context of the Integrated Water Resources Management (IWRM). The 2018 reform created a new ministry engaged in water-related disasters, flood and drought, the Ministry of Emergency Management, which is expected to play a leading role in conducting the works of response and recovery related to flood and drought events.

Such commitments of the government to the water sector have been supported with a constellation of water-related laws and regulations, including Environmental Protection Law, Water Law, Water Soil and Conservation Law, Water Prevention and Control Law, and Flood Control Law. A number of these institutions at the national as well as the local levels are impressive, but the gap between legal settings and law enforcement or policy implementation should be narrowed in the foreseeable future. This may depend upon the extent to which bargaining between the center and local governments would entail shared benefits based on mutual understanding, which reflects the subtle balance between economic growth and environmental sustainability.

The first part of the chapter focuses on the Three Red Lines and the Water Ten Plan. Discussions in the second part go to the governance structure of water resources management in China, primarily paying attention to major ministries involved in water issues, exploring the 2018 administrative reform and its implications. The third part is dedicated to evaluating a series of national-level water-related laws and regulations.

Plans

At the advent of the 2010s, there are two significant water resources management plans in China, the Three Red Lines in 2011 and the Water Ten Plan in 2015. The Three Red Lines was introduced in 2011 for strengthening the capacity of China's water resources management policy and plan with three imperative goals to be achieved by 2030. These 'red lines' are set for limiting the amount of water consumption, increasing water use efficiency in agricultural and industrial sectors, and improving water quality with carefully established criteria.

The Water Pollution and Control Action Plan or known as the Water Ten Plan in 2015 primarily focuses on reducing pollution loads to water and enhancing water quality by 2030. There are ten measure to achieve this plan, including industrial restructuring and the promotion of wastewater reuse and seawater desalination, the introduction of water saving practices and water use efficiency with a guarantee of ecological water flow.

Three Red Lines 2011

After the dramatically fast economic development in the three decades, the advent of the 2010s was a watershed for China's water resources management. The government was committed to resolving water shortage, enhancing water use efficiency, and addressing water pollution by issuing the No.1 Central Document 2011. The full name of the document is 'the Decision from the Central Party Committee and the State Council on Accelerating the Development of Water Resources Reform.' The dossier emphasized water resources development and proposed water conservancy as a top priority area for investment associated with economic development. This indicates the clear direction of national water policy development and acknowledges the magnitude of water infrastructure. The government intends to scale up and accelerate the construction of water infrastructure, such as water source projects, inter-connection projects, and water diversion projects (Jiang et al. 2020).

Following this, the central government issued the Opinions of the State Council on Implementing the Stringent Water Resources Management, which encompasses the 'Three Red Lines' with targets for water use, water use efficiency, and water quality at the provincial levels. The targets should be achieved according to the time schedule, the first phase by 2015, the second phase by 2020, and the third phase by 2030.

The first red line imposes the cap on the national annual water consumption at 700 billion m^3 by 2030, which is a drastic measure to restrict water use in every sector and delivers a strong message to various water users for using water carefully. This implies that an annual increase of water demand should be limited to less than 1%. A fundamental principle rooted in this scheme is to strike the balance between socioeconomic development and environmental sustainability pertinent to water consumption at all levels, river basin, province, city, and county (World Bank 2018).

According to the China Water Resources Bulletin in 2015, the total water demand of the country amounted to 635 billion m^3, which became the water use limit, and the demand can increase to 670 billion m^3 in 2020 and 700 billion m^3 in 2030. The demand for groundwater resources is projected to increase to 100 billion m^3 in the same period. To achieve such ambitious goals, multifaceted policy measures have been considered. An amount of permitted abstraction from river basins will be decided according to the available water resources, the master plan, and water allocation plans. Different levels of quotas will be introduced for regulating abstraction from rivers, lakes, and aquifers. Those quotas will serve as the baseline for the issuance and approval of water licenses that function as a regulatory tool to limit water use. In addition, the quotas at the provincial, city, and county levels will be determined once the quotas of river basins will be decided.

Water master plans at various administrative levels encompass water allocation schemes that control the total amount of water use. Maximum limits will be determined for regional water use and allocation plans for rivers, which gives the legal platform to decide the total permitted abstraction. Water managers have to undertake water inventories and to put restrictions on planned increases together with controlling water use via total water use indicators.

Under this Stringent Water Resources Management (SWRM) scheme, any new application for water abstraction will be limited and approved on an annual basis following the local water use agreement, and the water resources assessment based on water availability. Approval depends upon an investigation of water use efficiency, pollution control, the potential for water saving, and water rights trading. As for new applications for irrigation, stricter rules will be applied, and the only projects included in the National Plan of 50 billion tons of food grain can be allowed to have new access to irrigation water.

Room for an increase of new applications of industrial water abstraction will be little with water rights trading that are strongly encouraged. In the 12th FYP (2011–2015) for Water Resources Development, the government stressed the magnitude of dual tracks, infrastructure development as well as the implementation of national water rights trading (Jiang et al. 2020; NDRC et al. 2011). In a similar token, a new application for domestic water abstraction will be very limited and would only be allowed if cases were unavoidable. It is also important to note that no further water abstractions would be allowed if those were beyond the threshold (the red line) of total water use (GWP 2015; Wang et al. 2018). Water rights trading would be the only option left to those who or which require more water resources for their own socioeconomic and environmental activities. If effective, this policy would facilitate the achievement of sustainability in water resources management (Jiang et al. 2020).

In the second red line, specific targets are presented pertinent to industrial water use efficiency, thereby decreasing water use down to 40 m^3 per US\$1,600 (RBM 10,000) industrial added value. This is associated with seven high water-consuming industries, including thermal power generation, oil refining, steel and iron, textiles, paper making, chemistry, and food. Pertaining to agricultural water use efficiency, irrigation efficiency should increase by more than 60% in the same period. An increase of at least 3% should be achieved by 2015 for provinces that have not achieved irrigation efficiency at 52%.

The performance evaluation includes national- and self-evaluation, and responsibilities for monitoring and assessing water use and submitting the data to the national evaluation system are in the hands of chief officers in the provinces, and the Ministry of Water Resources oversees the whole process. The evaluation system gives rewards to the provinces that outperform their targets, which prioritize the provinces for annual water abstraction and project investment approval. Penalties are given to the provinces that do not meet their goals by restricting new water abstraction projects (GWP 2015; Moore and Yu 2020).

The third red line is linked to the enhancement of water quality in major water bodies. More specifically, 95% of primary 'Water Function Zones'[1] should meet water quality standards by 2030, and all sources of drinking water should comply with rural and urban standards by 2030. Several indicators will be used for measuring the circumstances of water quality in major freshwater bodies, including Chemical Oxygen Demand (COD) and ammonia nitrogen, which were included in the 11th FYP (1996–2010). Nutrient salt parameter has been included for freshwater lakes.

More performance indicators have been introduced, i.e., annual bearing capacity of main pollutants (COD and ammonia nitrogen for rivers, total nitrogen, and total phosphorus for lakes and reservoirs), and a total amount of allowed pollutant discharge into major river and lakes. Regarding drinking water sources of national significance, there are several indicators, such as the rate at which water resources are safeguarded, the proportion of drinking water sources reaching the required standard, and the safety of drinking water supply projects. Ecological water indicators embrace the degree of compliance between observed runoff at control nodes on major rivers and planned ecological flows, compliance of low water levels in major lakes with planned ecological water levels, and the general health of major rivers and lakes. Provincial governments take responsibility for decreasing total pollutant discharges, and the outcomes will give an influence on the assessment of performance of local government leaders (GWP 2015; World Bank 2018).

In order to facilitate the achievement of the targets of the plan, the government has introduced the five specific measures as follows (GWP 2015):

1. To bolster leadership for monitoring performance, specifying targets and establishing working systems

---

[1]This is an appropriate natural water area within a river basin for managing water resources. The main purpose of this is environmental quality management and water pollution control. This oversees the natural attributes of the water resource and its various uses in order to produce an appropriate and sustainable balance of economic, social, and environmental benefits (GWP 2015: 10).

2. To improve legislative settings for regulating water conservation
3. To enhance supervision and management between key water users
4. To undertake significant pilot and demonstration projects, including recycling industrial water, introducing and rehabilitating water saving technologies, and publicizing water saving devices in cities
5. To educate the public to regard water saving as a cultural norm.

Table 4.1 Summarizes specific targets of water planning according to the time schedule in three phases, 2015, 2020, and 2030.

An array of pilot projects were introduced for testing the effectiveness of such reforms in 2010, and the four provinces, Jiangsu, Shandong, Hebei, and Zhejiang, and three municipalities, such as Beijing, Tianjin, and Shanghai, participated in the projects. The Han River Basin,

Table 4.1 Future water planning targets—the Three Red Lines

Targets	2015	2020	2030
Total water consumption must not exceed	635 billion m^3	670 billion m^3	700 billion m^3
Industries will reduce their water use per US$ 1,600 (RMB 10,000) of industrial added value	30% below 2010 figures	65 m^3	40 m^3
Irrigation efficiency must exceed (%)	53	55	60
The number of water function zones complying with the water quality standard will be more than (%)	60	80	95
All sources of drinking water will meet standards for both rural and urban areas		Yes	Yes
All water function zones will comply with water quality standards			Yes

Source GWP (2015, p. 14)

Zhangjiagang City, and Yongkang City were also selected. As for Shandong Province, water abstraction is under strict management, and water resources assessment is necessary for all new construction projects within the pipe network. Water resources fees have been collected based on different pricing levels, and water quality targets for the water function zones of 17 cities are determined every year.

The Tianjin Municipality designated 10 industrial areas favoring recycled water and desalination as part of the management plan. A 'three-level water saving management network' was introduced, initiated by both government and businesses, and a chief officer was appointed for overseeing water resources management and the aquatic eco-environment and establishing a licensing system for groundwater abstraction. Regarding the Han River Basin, the Yangtze Water Resources Commission created a pilot project coordinating group, and each province within the river basin managed to draw up indicators for total water use (GWP 2015).

Criticism on the Three Red Lines exists, and there are six problematic issues. First, the 700 billion m^3 total water use cap has been set for restricting growth in water use in accordance with business-as-usual scenarios so that it might be ineffective in decreasing substantial amounts of water use in due course. Considering the amount of water use in 2011 reaching approximately 610 billion m^3, the 2030 cap can signify that there would be enough room for the increase of water use and supply.

Second, there are formidable challenges in monitoring and enforcing the caps established by water resources allocation plans at both the river basin and the regional levels. Against the government's intention, the caps would not serve as regulatory tools to limit water use but as hopeful yardstick or suggestions. Another noteworthy issue is linked to the water abstraction permit, and inconsistent water allocation plans at multiple levels are widely recognized together with the lack of clarity in implementing the water abstraction permit system (Jiang et al. 2020).

Third, the selection of indicators to assess performance might neglect other areas and intangibles reflecting silo approaches. Fourth, the system will depend upon indirect estimations and self-reporting by those who are evaluated for the reward or penalty system, leading to the creation of a conflict of interest. The fifth problem is related to the political nature of setting quotas. For instance, numerous provinces would like to be granted technical adjustments from the original plan figures. Whereas Guangdong Province, a water-abundant and highly industrialized area with less future water demand, agreed with lower caps, some water stressed places, the Tianjin Municipality and Shaanxi, Shanxi, Henan, Shandong, and Hebei Provinces, were allowed higher than average increases in their quotas (Nickum et al. 2017).

Sixth, there is the challenge of water data collection at the local level. The plan would strictly limit water consumption as well as rigorously regulate water rights in different localities. Such a nature of the plan might lead local authorities at various levels to provide incorrect water data for securing more water rights in the future. It is essential for the central government to take into careful consideration the loophole that might undermine the commitment of the country to sustainable water resources management (Long 2019).

Water Pollution Prevention and Control Action Plan 2015 (Water Ten Plan)

The Water Pollution and Control Action Plan or just called the 'Water Ten Plan' was announced in August 2015, aimed at decreasing water pollution and enhancing the water environment in China by 2030. The plan was introduced at the 5th Plenary Session of the 18th Central Committee of the Community Party of China in 2015 that stressed the government's commitment to economic restructuring, inclusive growth, and sustainable natural resources management.

With a projected total investment of RMB 2,000 billion (US$307.7 billion), the purpose of the plan is to tackle industrial and municipal point source as well as Non-Point Source (NPS) pollution from agriculture and stormwater. The short- and mid-term goals are that the water quality of more than 70% of seven major river basins, i.e., the Yangtze River, the Yellow River, and the Pearl River, should be as good as Grade III or above, and urban sewage control should be improved to a large degree. For these goals, the government would make efforts to decrease the portion of acutely polluted water bodies and ensure improving the quality of drinking water by 2020. The pervasive trend of groundwater over-exploitation should adequately be controlled and reduced by environmental authorities. Offshore and aquatic ecosystems of coastal areas, including the Beijing-Tianjin-Hebei Region, will receive more attention from the central and local governments by 2020. In the long term, by 2030, the government pledges to improve the ecological environment and revitalize ecosystems (ADB 2018; State Council 2015).

There are ten measures in the action plan. These measures embrace direct regulatory works for controlling pollutant discharge by industries, households, agriculture, rural areas, ships and ports whereas a particular emphasis is placed on industrial restructuring, and the promotion of wastewater reuse and seawater desalination. Demand management approaches are proposed, including water saving practices and water use efficiency together with a guarantee of ecological water flow. Scientific and technological breakthroughs are encouraged for water pollution control together with the promotion of environmental protection industries.

With an emphasis on market principles, the government would endeavor to accelerate the reform of water pricing, support diversified investment methods, and introduce more incentive mechanisms for increasing the capacity of wastewater treatment. Law enforcement against environmental violators will be strengthened based on the enhancement or enactment of environmental laws and regulations. Pollutants and environmental risks should rigorously be monitored and tested and discharging pollutants should be authorized. The overall aquatic ecosystems will adequately be protected through relevant regulatory works, such as protecting drinking water sources, tackling severe pollution of groundwater and major river basins and lakes, and safeguarding the oceanic environment.

Different levels of local governments will be given more responsibility for protecting the water environment, and industries should be self-regulated by not discharging emission beyond thresholds. Public participation will be improved, and there will be a regular publication of water quality situations on cities and provinces that show the best and worst performance on the quality of water environment (State Council 2015) (see Table 4.2).

Table 4.2 Water Ten Plan measures

Item	Contents
1	Direct regulatory works for controlling pollutant discharge by industries, households, agriculture, rural areas, ships and ports
2	Industrial restructuring, and the promotion of wastewater reuse and seawater desalination
3	Water saving practices and water use efficiency together with a guarantee of ecological water flow
4	Scientific and technological breakthroughs for water pollution control and the promotion of environmental protection industries
5	Acceleration of the reform of water pricing, support of diversified investment methods, and introduction of more incentive mechanisms for improving the capacity of wastewater treatment
6	Law enforcement against environmental violators strengthened based on the enhancement or enactment of environmental laws and regulations
7	Strict control of the amount of pollutants and environmental risks and discharging pollutants being authorized
8	The overall aquatic ecosystems protected through relevant regulatory works, such as protecting drinking water sources, tackling severe pollution of groundwater and major river basins and lakes, and safeguarding the oceanic environment
9	Different levels of local governments given more responsibility for protecting the water environment, and industries being self-regulated by not discharging emission beyond thresholds
10	Improvement of public participation and a regular publication of cities and provinces that show the best and worst performance on the quality of water environment

Source State Council (2015)

Alongside the plan, there were numerous policy initiatives for water pollution control. As for the reduction of water pollution, the flagship policy change is the revision of the Environmental Protection Law, the training of courts in environmental law, and the establishment of environmental targets for party cadres' career evaluation. Furthermore, the River (Lake) Chief System will be promoted by the central and local governments at all levels, which appoint an official who is responsible for managing rivers or lakes for a certain period and would be assessed partly based on the performance of such works. The system is officially embedded in the 2018 Water Pollution Prevention and Control Law and discussed further in Chapter 6.

A myriad of distinguishable approaches is introduced in the plan. First, the plan puts an emphasis on the philosophy of total water cycle health, such as surface water, groundwater, and marine water with the recognition of interrelatedness between them. In this sense, the plan directly mentions seaport pollution control, which has not been spotlighted in past environmental policies. Second, particular attention is paid to the scale-up of an extensive environmental protection industry and environmental protection services sector, and the government pushes forward relevant policies and investments for the sector. The service sector is expected to grow and contribute to a share of GDP by 2.3%.

Third, the clear indication of relevant entities in charge of particular jobs is given related to monitoring and compliance responsibilities. In the past, the lack of such an assignment by the central government often caused inefficiency and confusion in environmental regulation works. This new plan clearly assigns which agencies are in charge with responsibilities and should be accountable based on performance assessments (Han et al. 2016) (see Table 4.3). However, such a division of labor for implementing the plan should be adjusted thanks to the administrative reform in early 2018. The Ministry of Ecology and Environment (MEE) has been taking over all the responsibilities of the Ministry of Environmental Protection. More responsibilities have been transferred from several ministries, such as the Ministry of Water Resources, to the MEE, and other responsibilities have also been redistributed to newly established or reorganized ministries, which will be discussed later in this chapter.

Governance Structure

China's water resources have been administered based on a complexity of institutional arrangements. Since the late 1970s, the highly centralized system of the country has been transformed into a system of decentralized economies and a federal financing system. More leverage has been given to local governments, which becomes the financial and political backbone of the CCP-ruled China. This is called 'asymmetric decentralization,' which indicates economic decentralization without political

Table 4.3 Roles of government agencies in China involved in water pollution control under the Water Ten Plan 2015

Name of Ministry	Primary responsibilities	Leading role(s) under the Water Ten Plan
Ministry of Housing and Urban and Rural Development	Municipal water supply, draining, construction of sewage treatment plants	1 and 3 of the plan Strengthen urban pollution control, strengthen the construction of supporting pipe networks, promote sludge disposal and recycled water use, strengthen urban water saving
Ministry of Agriculture	Management of agricultural pollution (non-point source); fisheries and wildlife conservation	1, 3 and 8 of the plan Promote agricultural rural pollution prevention, control agricultural non-point source pollution, adjust the planting structure and layout, promote the ecological health of aquaculture, develop agricultural water-saving
Ministry of Water Resources	Protection of water resources; flood control and drought relief; water withdrawal licenses; transboundary water conflicts	3 of the plan Monitor and oversee the total quality of water, improve water use efficiency, protect water resources scientifically, strengthen the management of rivers and lakes, determine ecological flow limits, strict control on groundwater exploitation, pay special attention to industrial water-saving, develop water-saving agriculture

(continued)

Table 4.3 (continued)

Name of Ministry	Primary responsibilities	Leading role(s) under the Water Ten Plan
Ministry of Land and Resources (Ministry of Natural Resources since 2018)	Monitoring and protection of groundwater resources	3 of the plan Actively protect ecological space, strict control of groundwater exploitation
Ministry of Environmental Protection (Ministry of Ecology and Environment since 2018)	Pollution control, water quality monitoring; wastewater discharge monitoring; environment impact assessment	1, 2, 6, 7, 8, 9, and 10 of the plan Vigorously promote the prevention and control of industrial pollution, carry out a special program to control pollution from ten key industry types, centralized management of industrial agglomeration, speed up comprehensive improvement of the rural environment, strict environmental law enforcement, ensure the eco-hydrological environmental security, confirm and implement the responsibilities for all parties, strengthen public participation and social supervision
National Development and Reform Commission	Governance of major river basins	2, 4 and 5 of the plan

Name of Ministry	Primary responsibilities	Leading role(s) under the Water Ten Plan
		Promote sustainable seawater desalination, develop the environmental protection industry, set the price of taxes and fees, establish incentive mechanisms, speed up the development of environmental services, improve levy policies, examination of results of water pollution prevention and control for relevant capital allocation, promote pluralistic financing of improvement projects, lead optimization of land use and water network layout
State Oceanic Administration	Supervision, monitoring and control of marine pollution	8 of the plan Strengthen the near-shore environmental protection, protect marine ecological health
Ministry of Transport	Marine pollution control (shipping)	1 of the plan Strengthen port pollution control and prevention
Ministry of Industry and Information Technology	Prevention and control of industrial water pollution	2 of the plan Adjust the industrial structure, promote exit of pollute enterprises, pay special attention to the industrial water-saving, lead optimizing land-use zoning and urban growth layout

Source Han et al. (2016)

decentralization (Chien 2010). Regarding water resources, following the relationship between the center and local governments, negotiations have surged, supervision becomes weakened, and pollution control at localities is lenient. Provincial and local governments are only entitled to implement policies and programs related to water resources management, which are established at the central government. Therefore, water resources management in China are undertaken based on continuous bargaining between central and local authorities (Xie and Jia 2018).

Similar with many countries, numerous rivers, lakes, and groundwater resources are managed based on administrative jurisdictions in China. The unitary system of Chinese bureaucracy is complicated and fragmented through the Tiao-Kuai (条块) lines of authority. The term tiao means the vertical lines of authority over a variety of sectors reaching down from the ministries of the central government whereas the term kuai indicates the horizontal level of authority of the territorial government at the provincial or local level. In relation to the kuai relationship, various governmental units in the same jurisdiction have no mandate or power over others at the same level. This determines a traditional administrative boundary between government units that oversee two interrelated issues in water resources management, water resource development, and water quality control (Xie and Jia 2018).

Institutional arrangements for water resources management in China are described in Fig. 4.1. The State Council plays the major role in policymaking and is responsible for the formulation of water-related laws, policy on the planning and management of water resources, and makes decisions on water resource fees or taxes. Under the State Council, the two major ministries in water resources management are the Ministry of Water Resources (MWR), and the Ministry of Ecology and Environment (MEE) (Ministry of Environmental Protection, prior to the 2018 administrative reform). In addition to the MWR and the MEE, several other ministries are involved in water administration of China, such as the National Development and Reform Commission, the Ministry of Natural Resources (Ministry of Land and Resources prior to the 2018 administrative reform), the Ministry of Housing and Urban-Rural Development, the Ministry of Agriculture and Rural Affairs, the Ministry of Emergency Management (Xie and Jia 2018).

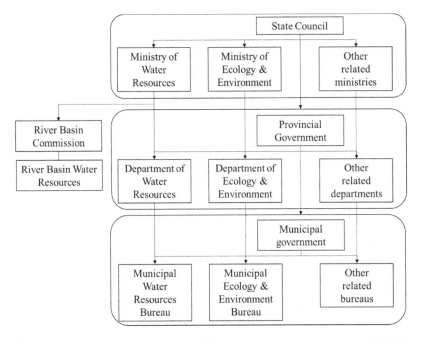

Fig. 4.1 Institutional arrangements of water resources management in China (*Source* Modified based on Xie and Jia [2018])

China's water governance structure underwent a large-scale transformation in March 2018. The institutional reform in 2018 was undertaken largely based on the promotion of ecological civilization that was neatly summarized in the Integrated Reform Plan for Promoting Ecological Civilization. The plan includes a list of principles and objectives coupled with key concepts such as the proper valuation and holistic management of natural resources (Dai 2019). The most conspicuous change in the Chinese water governance structure was the empowerment of environmental watchdog, the Ministry of Environmental Protection, in many aspects, including responsibilities, mandate, budget, and employees. The official name of the ministry was changed into the Ministry of Ecology and Environment (MEE), which implies the accommodation of ecological aspects in the range of responsibilities of the ministry and the seriousness of the central government related to ecosystem protection and rehabilitation.

Such an organizational reshuffling epitomizes China's unceasing efforts to bolster the capacity of environmental protection in various fields, including water resources management. Chan and Xu (2018) illustrate fragmented systems in water and environmental management prior to the reform in March 2018, by taking an example of a frog that swims in a river, administered by the Ministry of Water Resources, jumps on the riverbank, supervised by the Ministry of Land and Resources and the State Forest Administration, and enter a farm, managed by the Ministry of Agriculture. This vividly describes the extent to which water and environmental management has been ineffective due to fragmented roles and overlapped responsibilities between several ministries in China, which is described as 'nine dragons managing water.'

While the newly established Ministry of Ecology and Environment appears to be a large ministry with around 500 staff members in overseeing environmental and ecological issues, the implications of such a reshuffling of ministries and bureaus in 2018 are unclear for the overall water sector. The influence of the Ministry of Water Resources (MWR) can decrease due to the administrative reform in 2018, because the MEE has a profound influence on various water resources management areas in addition to water quality control and ecological protection and rehabilitation. Prior to the reform in 2018, the responsibility range of the Ministry of Environmental Protection was limited to environmental pollution prevention and control, including water pollution prevention and control activities. However, the incorporation of the term, 'ecology' into the title of the newly established ministry implies a broadening of the ministry's responsibilities that embrace ecological dimensions (Chan and Xu 2018; Chang and Li 2019; Voita 2018; Wang 2018).

There are the five distinctive features related to the creation of the Ministry of Ecology and Environment. The first noticeable feature of the reform is associated with the transfer of responsibilities of climate change and carbon emission reduction from the National Development and Reform Commission (NDRC) to the MEE. Carbon monoxide-caused air pollution and carbon dioxide related to climate change issues will be tackled by one ministry first time in China. Second, the responsibilities of groundwater pollution, which were born by the Ministry of Land and

Resources prior to 2018, have been transferred to the MEE. Synergistic effects for water quality enhancement would be expected thanks to the integration of responsibilities of surface and groundwater water pollution under one ministry.

Third, the newly transferred responsibilities of the MEE embrace the management of Non-Point Source (NPS) pollution from the Ministry of Agriculture and wastewater discharge outlets from the MWR, which indicates a step further toward the enhancement of river water quality. Fourth, managing NPS pollution means that the MEE would be able to tackle urban and rural water pollution in an integrated way, and fifth, the responsibility of marine environmental protection management has been transferred to the MEE from the State Oceanic Administration. This is iconic, because the change represents the integrated management of coastal and terrestrial ecologies, i.e., forests and wetlands, by one ministry, which is unprecedented in China's environmental management (Chan and Xu 2018).

Table 4.4 displays a list of departments involved in water resources management and policy under the State Council, and Fig. 4.2 delineates the path to the creation of the Ministry of Ecology and Environment (MEE) by taking responsibilities from the six ministries, the Ministry of Environmental Protection, the Ministry of Land and Resources, the Ministry of Agriculture, the State Ocean Administration, the NDRC, and the MWR, together with one office, the State Council Office of South North Water Transfer Project.

Attention is paid to the Ministry of Natural Resources (MNR). Prior to 2018, the primary foci of the Ministry of Land and Resources were on the exploitation and management of natural resources with little consideration of ecological protection and rehabilitation. Major works of the ministry were more closely related to the optimal use of natural resources, thereby leading to over-exploitation as well as environmental damage. The ministry will be responsible for managing grasslands, forests, wetlands, water and maritime resources, and urban and township planning, developing and protecting natural resources, and creating a system of paid use.

The significance of the creation of the MNS lies in the policy shift from natural exploitation for economic growth to the balance between

Table 4.4 Ministries involved in water resources management of China

Department	Water resources management responsibilities	Major responsibilities
Ministry of Water Resources	Surface and ground water management, flood control, water and soil conservation	Planning of water development and conservation, flood control, water and soil conservation, integrated water resources management
Ministry of Ecology and Environment (Ministry of Environmental Protection prior to 2018)	River basin management, water function area planning, wastewater discharge point management, ground water pollution control, agricultural non-point source pollution control, marine environmental pollution	Water pollution prevention and control, water environment function regionalization/zoning, establishment of water environmental quality standards and national pollutant discharge standards, climate change & carbon emission reduction, environmental protection project areas
Ministry of Housing and Urban–Rural Development	Urban and industrial water use, urban water supply and drainage	Planning, construction and management of water supply projects and drainage and sewage disposal projects
Ministry of Agriculture and Rural Affairs (Ministry of Agriculture prior to 2018)	Water uses for agriculture, and fishery and aquatic environment protection	Non-point source pollution control, protection of fishery water environment and aquatic environmental conservation
State Forest Administration	Water resources conservation	Forest protection and management for protecting watershed ecology and water resources
National Energy Administration	Hydropower development	Construction and management of large and mid-scale hydropower projects

Department	Water resources management responsibilities	Major responsibilities
National Development and Reform Commission	Participation in the planning of water resources development and ecosystem building	Planning of water resources development, allocation of production force and ecological environmental construction, coordinating the planning and policy of agriculture, forest and water resources development
Ministry of Transport	Pollution control related to navigation of ships on rivers	Pollution control and management of inland navigation
National Health Commission (National Health and Family Planning Commission prior to 2018)	Supervision and management of the drinking water standard	Supervision and management of environmental health
Ministry of Natural Resources (Ministry of Land and Resources prior to 2018)	Water resources rights registration management	Management of ferrous and non-ferrous metals, coal, critical raw materials, rare earths, surveying, mapping and geo-information, grasslands monitoring and registration rights management, marine affairs, forest and wetland survey and ownership registration, urban and rural planning, drafting of the plan of functional areas, grassland survey and ownership registration

(continued)

Table 4.4 (continued)

Department	Water resources management responsibilities	Major responsibilities
Ministry of Emergency Management (newly established in 2018)	Flood and drought management	Natural (i.e., fire & earthquake) and man-made disaster monitoring, management and reduction, work safety management, emergency rescue management

Source Modified based on Chan and Xu (2018), Chang and Li (2019), Long (2019), Voita (2018), and Wang (2018)

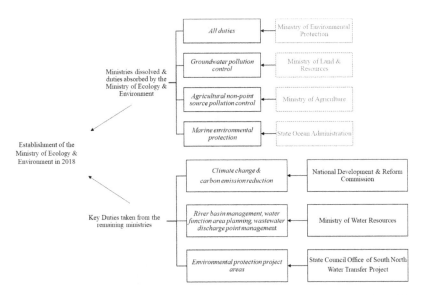

Fig. 4.2 Development of the Ministry of Ecology and Environment in China (*Source* Modified based on Chan and Xu [2018] and Chang and Li [2019])

economic development and ecosystems that embrace all the natural resources, including water. In that sense, the MNR is given the mandate to take all the responsibilities handed over by the Ministry of Land and Resources and be involved in monitoring and registration right management of water resources (ADB 2018; Chan and Xu 2018; Voita 2018; Wang 2018) (see Fig. 4.3).

The Ministry of Emergency Management (MEM) was newly established in the institutional reform in 2018. The primary purpose of the ministry is to put scattered responsibilities of diverse ministries regarding natural and human-made disasters into one ministry, which can entail an integrated management of catastrophic situations. This entity serves as a unique Chinese disaster management control tower to beef up the capacity of preventing and reducing disasters and rescuing people and protecting economic assets and to coordinate emergency works between the central and local governments.

A total of 13 ministries and bureaus have handed over their disaster and emergency management responsibilities to the MEM. These are:

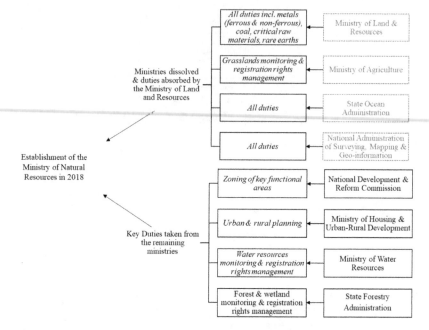

Fig. 4.3 Development of the Ministry of Natural Resources in China (*Source* Modified based on Chan and Xu [2018] and Voita [2018])

(1) all works of the State Safe Production Supervision Office; (2) emergency management works from the Administration Office of the State Council; (3) fire management works of the Ministry of Public Security; and (4) disaster relief activities of the Ministry of Civil Affairs. Additional responsibilities are: (5) geological disaster prevention works of the Ministry of Land and Resources; (6) water-related disaster prevention works of the Ministry of Water Resources; (7) grassland fire management works of the Ministry of Agriculture; (8) forest fire management-related works of the State Forest Bureau; and (9) emergency relief works related to earthquake disasters have been transferred to the MEM from the State Earthquake Office.

Major responsibilities from the four bureaus have also been transferred to the MEM, including the State Flood and Drought Control Command Center, the State Disaster Reduction Commission, the Earthquake Control and Disaster Relief Control Center of the State Council, and the State Forest Fire Control Center. Whereas local governments usually tackle small- and mid-scale disasters, the MEM is playing a coordinating role as a control tower. There will be an establishment of a special control center under the auspices of the MEM in case that large-scale or intense disasters take place and the launch of a 24-hour emergency response system to deal with natural disasters or major workplace accidents. Figure 4.4 illustrates the responsibilities of the MEM (ADB 2018; Ministry of Emergency Management 2020; Wang 2018).

Figure 4.5 illustrates the administrative structure of flood control and drought relief in China. This structure, however, does not necessarily reflect what has happened to natural disaster management since the 2018

Fig. 4.4 Responsibilities of the Ministry of Emergency Management in China (*Source* Liao [2019])

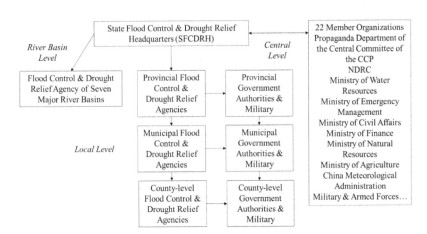

Fig. 4.5 The administrative structure of flood and disaster management in China. *Remarks* CCP means the Chinese Communist Party, and NDRC stands for the National Development and Reform Commission (*Source* Modified based on Ding [2016])

administrative reform. The newly established Ministry of Emergency Management is likely to take more responsibilities for flood and drought management in an integrated fashion, which implies that less ministries would be involved in natural disaster management.

The ministry is responsible for integrating emergency response forces and resources, promoting emergency management system and implementing emergency management in natural disasters and accidents. The Ministry of Water Resources is still the primary ministry in charge of river basin management together with overseeing hydraulic structures such as dams, levees, and embankments so that how to wisely adjust administrative duties on flood and disaster management between the two ministries would be the key to improving the capability of Chinese authorities to cope with water-related disasters.

China's institutional reform in 2018 provides food for thought in water resources management, particularly pertinent to departmental competitions between the ministries and bureaus involved in water issues. The outright winner appears to be the Ministry of Ecology and Environment, because the ministry has succeeded in absorbing a wide range of responsibilities in managing water resources, water quality, and climate change issues from various ministries and bureaus. One of major discourses in China's political economy in the Xi Jinping Era is ecological civilization, and therefore, more weight has been put on the emergence of environmentalism in ministerial reshuffling in 2018.

In addition, the development of super ministries in China has been ongoing, which represents the efforts of the government to streamline many ministries into a small number of them, thereby achieving efficiency and reducing administrative costs. The other cause for this trend is, more importantly, the attempt of the CCP to bolster its grip on governance systems at the center as well as the local levels, which eventually paves the way for the party to intervene and give more far-reaching influence on the governance system. Such an attempt is regarded as the revitalization of the Party-State China that would reverse the separation of the party and the state (Guo 2019).

The MEE has benefited from the trend and seems to win over the competition with the MWR in terms of expanding its mandate from water quality control to river basin management, water function area planning, and wastewater discharge point management. It is unclear about the extent to which the MEE will be able to implement relevant policies and projects in the responsible areas transferred from the MWR in a short period. Primary discourses in water resources management in China are gradually shifted toward water demand management, water quality enhancement, and the rehabilitation of water ecosystems from supply management-oriented plans and projects, i.e., infrastructure development such as dams and inter-basin water transfer projects. It is plausible to maintain that the MEE will expand its mandate much further and have more influence on water and environmental issues in due course thanks to not only the change of political economy landscape but also the emergence of environmentalism in society.

The MWR is still a powerful ministry in charge of water resources development including the construction, operation and maintenance of hydropower dams, water transfer projects, and other major water infrastructures. However, the ministry has lost in its competition with the MEE, the MNR, and the MEM. The MNR has been given the mandate to manage water resources monitoring and registration rights, which implies that some responsibilities of the MNR can be overlapped with those of the MWR. As for the MNR, these works would be regarded as part of the integrated land and resources management, however, the MWR may not be able to collect precise and correct data for water resources planning, modeling, and flood forecasting, which are even more significant than before in the era of climate change. Water-related disasters, such as flood and drought, will be managed by the MEM since 2018. The MWR may retain its responsibilities to operate and maintain dams for preventing and controlling floods and droughts but face an unclear division of labor with the MEM in terms of disaster management.

River Basin Management: Seven River Basin Commissions

The significance of the integrated approach to water resources management in China was recognized even around the establishment of the

People's Republic of China in 1949, and the first river basin commission was introduced in the Yellow River in 1950. Since then, the commissions have been established in the additional six major river or lake basins, namely the Hai River, the Huai River, the Song-Liao River, the Yangtze River, the Pearl River, and Tai Lake. Prior to 2018, the Seven River Basin Commissions were largely under the auspices of the MWR which had been responsible for providing funding, staff members, and guidelines on river basin management. The 2002 Water Law authorized the MWR to establish the river basin commissions for the unified management of and supervision over the water resources (Fry and Chong 2016).

The seven river basin commissions have several functions. For instance, the Yellow River Conservancy Commission oversees a constellation of issues within the river basin. The commission was given the mandate of water resources management in general by the MWR, particularly with the purpose of resolving conflicts or squabbles between different administrative entities although the Ministry of Ecology and Environment can take over the responsibility of river basin management since March 2018. The authority of the commission has been beefed up for developing basin-wide water resources development and water allocation plans through the introduction of 1997 Flood Control Law and the 2002 Water Law. At the same time, the commission is supposed to execute power over other administrative entities in certain fields, especially water pollution control (Moore 2018).

Securing more water resources is a sensitive and significant policy area in every local government in China, but this job is more seriously considered in semi-dry and arid areas like the Yellow River Basin. The river basin encompasses the large areas of the North China Plain, which receives much less precipitation compared with the other counterparts in South China. As the significance of the North China Plain is discussed in terms of food security in China, farmers in the river basin have no intention to compromise by giving way more water to other provinces, municipal or counties within the river basin. Aware of the fact that the Yellow River did not reach its own estuary for more than 100 days in the late 1990s, every local government in the river basin is endeavoring to have access to more water.

Some of the fundamental problems in river basin management of China are a fragmented administrative structure and the paucity of coordinating mechanism between entities with the mandate given by the central government and local authorities in charge of day-by-day water resources management.

These challenges are attributed to the unique combination of political centralization and fiscal and administrative decentralization in China since the late 1970s. Decision making is strongly top-down at the central government level, however, substantial administrative powers are delegated to local governments due to the sheer size and diversified territory of the country. China is perhaps one of the most decentralized countries in the world in terms of economic perspectives, and revenues and expenditure mandates are mostly controlled by local government officials. The institutional matrix, the Tiao-Kuai, determines the way local government officials behave, and for instance, the officials care for line control by various central government ministries and for territorial government leaders, such as mayors and provincial governors and to associated officials in the corresponding CCP structure. The distinctive governance structure of China has spawned the flexibility of local officials' responses to water issues and has sometimes resulted in prioritizing policies and projects for local interests ignoring instructions from the central government (Mertha 2008; Moore 2018).

Water allocation is one of the most challenging jobs for the commission, because this issue touches upon some of the fundamental problems that have been created thanks to the bureaucratic systems established since the late 1970s, including decentralization. Provincial, municipal, and county-level governments have been delegated to tackle their own water problems by introducing tailor-made policies, allocating water to various water users, regulating polluters, and planning and constructing infrastructures against flood.

The dramatic change of political economy landscape in water policy over the last four decades has engendered an imbalance of power between the center and local governments with regard to river basin management. For example, water managers in the Yellow River Conservancy Commission should have had the mandate to determine how much water resources are allocated into a number of local areas within its own jurisdiction. However, the commission has to undergo tough negotiations with different levels of local governments, from provincial, municipal, and county-level governments. The bargaining power of the commission is rather weak compared with that of provincial governments because those governments provide financial support to the commission. Consequently, there is little autonomy for the commission whose major decisions are highly subject to political influences of provincial governments (Xie and Jia 2018).

China's administrative hierarchy positions provincial governments as the same status as central ministries such as the MWR and the MEE, leaving confusion regarding authority. At the provincial level, a Water Affairs Department oversees water resources management and most of county-level governments have a branch in charge of water supply, conservation, flood control, and irrigation. There are special-purpose committees, i.e., dike maintenance committees, irrigation, and reservoir district committees, and water users' associations. Conflicts of interest come into being between these entities due to the unclear lines of responsibility, and tension is mounting between individual local officials as a whole. Disputes can only be resolved by the highest levels of government (Moore 2018). In the legal context, the 2002 Water Law does not explicitly stipulate which authority should be given to the river basin commissions, therefore leaving the system ineffective with little law enforcement power (Fry and Chong 2016; Xie 2009).

In addition, the legal settings for inter-provincial coordination are flimsy and inadequate to address current water resource quality and quantity problems. Water-related legislations stipulate only general principles for tackling inter-governmental disputes, which shall be settled through negotiation and consultation. Few environmental litigation cases, including water pollution cases, have been reported across administrative boundaries at the national level, and rulings of local courts are in effect only in their jurisdictions, not across administrative boundaries (Li et al. 2015; Moore 2018).

It is not uncommon to observe that some provincial governments openly oppose to the commission's decision on the volume of water allocation for the provinces and would be in dispute with the commission. Provincial water resources departments sometimes disregard plans and decisions made by river basin commissions in order to maximize provincial benefits. For example, an integrated water use plan that are set up by a river basin commission can be ignored by an upstream province, and the provincial government would overexploit water resources from rivers within the same river basin for boosting its local economies. Such a behavior might result in water shortage in downstream provinces, triggering a lack of drinking and agriculture water supply (Li et al. 2020).

Inter-jurisdictional cooperation between provinces has developed in several river basins since the 1990s, and one of the earliest efforts was the establishment and the operation of the Comprehensive Utilization Plan of the Yangtze River Basin as a water resources management mechanism in 1990. By analogy, Guangdong Province established its own regulations for local governments within its jurisdiction for water pollution-related conflict in 2006. These regulations require local governments to carefully monitor pollution sources near administrative boundaries, introduce administrative mechanisms for tackling disputes, and urge local government officials to be responsible for settling disputes. The Yellow River Conservancy Commission introduced the Yellow River Basin Emergency Pollution Incident Communication, Coordination, and Cooperation Mechanism for the nine riparian provinces in 2011 for pollution monitoring (Li et al. 2015; Moore 2018).

Nevertheless, these efforts do not seem to have been effective as expected, and the issues of inter-jurisdictional disputes have continued to be one of the major concerns for provincial governments. For instance, the National People's Congress and the Chinese People's Political Consultative Conference in 2013 recognized inter-provincial pollution disputes as an important issue. Similar news reports and government articles followed in 2015 and 2016, maintaining the need of the creation of an inter-provincial coordinating body chaired by provincial governors (China Democratic Alliance Guangdong Provincial Committee 2016; Li et al. 2015; Moore 2018).

Since the administrative reform in March 2018, the MEE has begun to give more influence on the River Protection Bureau in the commissions by providing more funding and guidelines. In fact, inter-provincial pollution compensation mechanisms were introduced in 2012 for enhancing inter-jurisdictional collective action in shared river basins

(Moore 2018). Although the commissions are still regarded as the entities under the heavy influence of the MWR, it seems that the MEE will possibly have more influence on the works of the commissions in the foreseeable future.

It is early to judge if the MEE would play a leading role in tackling a complexity of water challenges, aptly competing with the other conventionally powerful ministries in the sector, i.e., the Ministry of Water Resources (MWR), the Ministry of Agriculture and Rural Affairs (MOA), and the Ministry of Housing, Urban and Rural Development (MOHURD). According to the interviews with several Chinese water experts in September 2019, there would not be noticeable policy shifts or transfer of responsibilities from the MWR to the MEE, for instance, at least in the short term. A primary cause for this short-term stalemate would be linked to a continuous resistance from the MWR which will not give up their prolonged grip in some sectors, such river basin management, ongoing bargaining between the ministries in water resources management for the transfer of responsibilities, and the ill-preparation of implementing measures for these matters.

Although the 2018 administrative reform can bring about a new era in water resources management of China in terms of more power in a ministry, the fatal cleavage does not seem to be fixed, the lack of coordinating mechanism between ministries at the central level. Whether the 2018 new administrative system is successful for sustainable water resources management depends upon the extent to which different voices from diverse ministries and bureaus will be orchestrated, particularly led by the Ministry of Ecology and Environment.

Laws and Regulations

China's institutional frameworks on water consist of diverse levels of laws and regulations. It is necessary to divide swathes of laws and regulations depending upon major sectors, such as cross-cutting laws and regulations on water and the environment as a whole, water resources management and supply, water quality control, water-related disasters, and other water-related laws and regulations. Water-related laws are implemented alongside secondary legislation and regulations and are assisted by State Council decrees and sector administrative regulations. At the local level, local governments have adopted these laws and regulations as part of local laws and regulations depending upon their own socioeconomic and environmental circumstances (ADB 2018).

Primary attention is paid to the Environmental Protection Law 1989 and 2015 and the Environment Impact Assessment Law in 2002 and the Environment Protection Tax Law in 2018 as cross-cutting laws. Delving into the water sector, it is significant to explore the contents of Water Law in 1988 and 2002, which serves as a fundamental framework to oversee a variety of complex issues related to water resources management in China. The Water and Soil Conservation Law in 1991 and 2010 shows the close linkage between water and sediment. The Water Pollution and Prevention Control Law in 1984, 1996, 2008, and 2018 has played a key role in promoting water quality enhancement and the several revisions of the law implies an evolving policy emphasis toward water quality issues. In addition to these, this section sheds light on the Flood Control Law in 1997, and brief discussions are made on the Fisheries Law and Grassland Law that demonstrate their relatedness to water issues from the perspective of the Integrated Water Resources Management (IWRM).

Environmental Protection Law 1989 and 2015

The Environmental Protection Law was introduced in 1989. The law was the most significant foundation for environmental governance and policy of China in the early period of the open-door policy. Despite such an important position in environmental governance of the country, the law has been an epicenter of criticism associated with degraded and damaged environmental elements of China since 1989 (Lee 2006).

In almost three decades, the Chinese authorities decided to undertake a major revision of the law in 2014 and the law eventually became effective in 2015. A conspicuous feature of the 2015 law is the increase of articles from 47 in 1989 to 70 in 2015, which implies more sophisticated and detailed guidelines on addressing environmental pollution activities and misbehaviors. Several basic environmental protection rules have been updated, including environmental planning, standards, and monitoring.

The 2015 version has two different points compared with the 1989 version. First, the law has increased penalties on polluters by charging fines on a daily basis. For instance, if an environment polluter does not make a correction, ordered by the central or local government authority (usually Ecology and Environment Bureau), the authority has the mandate to impose a fine on a daily basis. This regulatory action is definitely much stricter than that in the 1989 version, because under the 1989 version of the law, only one-time penalty was imposed on each illegal entity.

Second, the law explicitly provides the process of public interest litigation against polluting acts. In many cases, the Chinese legal system discourages people from filing a lawsuit due to socio-cultural reasons, and therefore, alternatives, such as mediation, are often encouraged between concerned parties. However, the trend has increasingly been changing, and lawsuits appear to increase related to environment polluting activities. Reflecting such a way of thinking about lawsuits as well as the need to strengthen law enforcement, the government has shown its strong commitment to punishing environmental polluters severely in the 2015 law. (Lau 2014; Zhang 2014).

Environment Impact Assessment Law 2002

The Environment Impact Assessment (EIA) Law was introduced in 2002 and is one of the most significant laws in environmental management. The aim of the law is to prevent unfavorable influences of construction projects on the environment after the projects are implemented. Even though basic principles and rationales are presented in the Environmental Protection Law, the law encompasses technical guidelines for an environment impact assessment in surface and ground water resources, which gives detailed guidelines for environmental agencies to implement EIAs. The law includes principles, working procedures, working methodologies and requirements for a water quality assessment of possible impacts on surface water and ground water resources triggered by construction projects (Dai 2019; Lee 2006).

Environmental Protection Tax Law 2018

The introduction of the Environmental Protection Tax Law in 2018 epitomizes an unconventional approach of the Chinese government to tackling environmental pollution and symbolizes the launch of green taxation in practice. There are the four target fields in taxable emissions: (1) air pollution; (2) water pollution; (3) solid wastes; and (4) noise. The law retains the levied items as they are in the previous fee system, however, it will be local tax bureaus that collect the new environmental protection tax instead of ecology and environment bureaus.

Devolution has enabled local governments to decide the tax rate for the levied items within a threshold determined by the central government. The whole amount of the tax collected will be held as part of the local tax revenue, which will be of great help to local governments for enhancing their own environmental circumstances. The expected total scale of the tax may amount to RMB 50 billion (US$7.15 billion) per annum, and more than 260,000 companies are expected to pay the tax beginning on 1 April 2018. In the previous system, 10% of the emissions discharge fee had to be remitted to the central government (Chan 2018; Zhang 2018).

Water Law 1988 and 2002

Water Law of China was introduced in 1988, which was the first legal institution on water resources management in the country. The law, however, is regarded as a code in western terms, thanks to the lack of implementation provisions and advocates the introduction of the Integrated Water Resources Management. The collection of water resources fee and water engineering fee, and the subsequent creation of a water abstraction permit system have come from the law.

The revision came in 2002 in order to deal with innumerable pending issues for sustainable water resources management, ranging from water allocation, conservation, and protection. The revision partly occurred due to the two major events in the water sector, the Yellow River's not flowing to the sea in 1997 and the large-scale flood in 1998 in the Yangtze River. These events unveiled problematic management and regulatory capacities of water authorities as well as the poor quality of infrastructures in the water sector (Nickum et al. 2017).

The 2002 revision advocated the right to safe water and put more emphasis on conservation and pollution control as the objectives of water policy. A legal base was given for the river and lake basin commissions which had already been established under the MWR. In spite of such a positive development of water policy, China's legal framework in the water sector appears to have merely provided a myriad of principles without specific provisions for enforcement and with the opaque demarcation of authority between different sectors. In addition, a heavy reliance on local governments for law enforcement leads to less compliance in water regulatory works, and there is little transparency in decision making and less development of public participation (Nickum et al. 2017; Xie 2009).

Following the 2002 Water Law, a stellar of rules and regulations have been introduced, including the Rules for the Regulation of the Yellow River Water Resources and the Management Regulation of Water Abstraction Licenses and Water Resources Fee Collection. Ministerial regulations encompass the Management Methods for Construction Projects for Water Resources and the Management Methods for Pollution Discharge Outlets to Rivers. Relevant regulations at various local levels have followed. Since 2002, more attention has been placed on water conservation and water resources protection, which implies a policy shift from resource exploitation and use to conservation and rehabilitation in water and ecosystem management as well as more emphasis on law enforcement on water and the environment (GWP 2015; Wang et al. 2018).

Together with the 2002 law, the State Council introduced the Guidelines for the Implementation of Institutional Reform in the Water Project Management System for promoting institutional reform. Pilot projects for water institutional reforms were conducted in numerous cities prior to the extension of the scheme to the whole country. As part of this scheme, 1,740 (70%) of the total administrative regions at and above the county level (provinces, cities, or counties) created Water Affairs Authorities which was intended to integrate all water-related management activities in both urban and rural areas into one organization. These new entities were committed to integrating river basins with administrative regions, urban and rural areas, and water development and use with conservation and protection.

Diverse public authorities in China are given the mandate to deal with various roles in water resources management in accordance with the 2002 law, including local government agencies, water administration departments, and river basin authorities. In addition, communities and individuals are regarded as major stakeholders in water resources management and the roles and obligations of them for conserving water resources and protect water infrastructures are stipulated. Yet, law enforcement has not been effective related to the 2002 law (GWP 2015).

Water and Soil Conservation Law 1991 and 2010

The law aims to prevent and control water and soil loss, protect and reasonably utilize water and soil resources, reduce flood and drought disasters, and sandstorms. Rivers move not only water but also sediment, which is crucial to sustain flora and fauna in riverine areas, particularly downstream delta areas, because large amounts of sediment flowed from upstream contain different kinds of essential nutrient for living creatures. Such a significant role of sediment is often disrupted due to hydraulic infrastructures from upstream to downstream in many Chinese rivers. The law provides adequate rules on how to properly conserve water and soil against anthropogenic as well as natural disasters such as floods and droughts (Dai 2019).

Water Pollution Prevention and Control Law 1984, 1996, 2008, and 2018

The law was enacted in 1984. In the early 1980s, the burgeoning manufacturing industries already took a tremendous toll at surface water bodies such as river and lakes. The first revision occurred in 1996 that included three major purposes as follows. First, the 1996 law intended to introduce integrated plans on river basins and, second, to promote a system for controlling the total discharge of major pollutants for water bodies, the Total Pollution Load Control (TPLC), which is effective in regulating the situation that pollutant discharge meets discharge standards, but water quality does not satisfy national standards. Third, the law required urban sewage to be centrally treated and advocated the construction of local wastewater treatment plants for local governments.

Another revision of the law came into being in 2008. The situation of water pollution became grave, and numerous water pollution accidents occurred across the country even in the new millennium. Compared with the 1996 law, the primary differences of the 2008 version are fourfold. First, the 2008 law intended to solidify environmental responsibilities of local governments, asking local governments at or above the county level to put water protection into national economic and social planning. Specific targets were introduced for improving water quality in freshwater bodies, i.e., with the reduction of Chemical Oxygen Demand (COD) by 10% in the 11th Five Year Plan (2006–2010).

Second, public participation in environmental protection was encouraged by providing easy access to information on national water quality in a unified way. Third, much stricter fines are introduced against violators and novel penalties. Another feature was the adoption of double penalty for violators. Manufacturing units or companies which trigger severe water pollution incidents should be fined, and if the incident turned out to be severe, the person in charge of the unit should also be fined not more than 50% of the income earned from the entry in the previous years.

Fourth, the 2008 law encompasses the improvement of three existing schemes. The 2008 law intended to enhance the system specifying responsibilities of provincial, municipal, or county governments for reducing total loads of pollutants in their jurisdictions. If the total discharge of major water pollutants were beyond a threshold, environmental protection bureaus (now ecology and environment bureaus since 2018) would stop the examination of construction documents that may cause an increase of discharge of water pollutants.

The water pollution discharge permit system was strengthened by adding articles to make entities which discharge industrial or medical wastes or wastewater compulsory to seek for discharge permits. The 2008 law improved the protection of drinking water sources by providing more detailed guidelines and enhancing management systems of water source areas. For example, the law banned the setup of pollution outlets in a drinking water source area (water quality of Grade I and II) (Li and Liu 2009).

The third revision was conducted in 2018, and the revised law provides a legal platform to support the Three Red Lines which encompasses the element of strengthening agricultural and industrial water pollution control (Chan 2018). The 2018 law embraces four major elements. First, the River (Lake) Chief System is officially included in the law which was first piloted in Wuxi, Jiangsu Province, 2007 for addressing severe water pollution situations triggered by blue-green algae in Tai Lake (Zhang 2018). Second, the regulation of agricultural water pollution is particularly emphasized. The law specifies that quality and utilization standards for fertilizers should meet the standards of water environmental protection. Third, the government sheds light on drinking water sources and quality. Emergency and supplementary water resources should be in preparation in cities which rely on a single water resource. In terms of information disclosure, local governments at the county level and above should enable the public to have access to the information on water quality at least once a quarter. Larger fines will be levied to those which install sewage outlets or pipelines in a drinking water source area.

Fourth, against the previous criticism of low fines, the new law stipulates more fines imposed on violators which will be up to RMB one million (US$150,000). Those who engender serious water pollution accidents can face a closure of their business units or criminal punishment (China Water Risk 2017; Dai 2019).

Flood Control Law 1997

The law was introduced in 1997 and is aimed at preventing flood, minimizing damage, ensuring the safety of human life and property. Prevention plans for major rivers and lakes are decided by the central government and should be established by the MWR in collaboration with the Water Resources Bureau of related provinces, autonomous regions, and municipalities.

Local water bureaus are responsible for establishing the prevention plan for other rivers and lakes, and the plan should be approved by the higher-level government authority. Provisions are included for the implementation of flood prevention and control measures, the management of flood control facilities and legal responsibilities (National People's Congress 2007).

Fisheries Law 1986 and Grassland Law 1985

Other water-related laws are the Fisheries Law, which intends to enhance the protection, increase, development and reasonable utilization of fishery resources in rivers and lakes, and the Grassland Law which intends to protect, develop, and make rational use of grasslands. These laws are indirectly associated with water resources management from the perspective of the Integrated Water Resources Management (IWRM), river basin management, and the nexus between water, land, and other environment elements. Table 4.5 summarizes a series of major laws and regulations on water resources management in China.

Major progress has been made in water-related legislation of China for a few decades discussed above. Nevertheless, good practices in water governance from the international context inform that conventional water governance in China has depended too much on government and administrative measures instead of legislation and law enforcement. A viable legal framework for water resources management in China will win prominence in the foreseeable future together with policies, regulations, accountability, transparency, public participation, and law enforcement (ADB 2018).

4 Water Plan and Governance System   145

Table 4.5 Major state-level laws and regulations, and rules on water resources management in China

Category	Name	Promulgating organization	Provisions
Laws	Environment Protection Law	NPC (1989, 2014)	Guidelines for water control
	Environment Impact Assessment Law	NPC (2002)	Appraisal of water impacts
	Environmental Tax Law	NPC (2018)	Guidelines for levying water pollution taxes
	Water Law	NPC (1988, 2002)	The rational use, protection and management of water
	Water and Soil Conservation Law	NPC (2010)	Soil erosion control and conservation
	Water Pollution Prevention and Control Law	NPC (1984, 1996, 2008, 2018)	The prevention and control of water pollution, and protection of aquatic ecology
	Flood Control Law	NPC (1998)	Flood control and water safety
	Fisheries Law	NPC (1986, 2000, 2004, 2009, 2013)	Protection, development and reasonable utilization of fishery resources
	Grassland Law	NPC (1985, 2002, 2009, 2013)	Protection, development and rational use of grasslands
Regulations	Regulations on the Administration of the License for Water Abstraction and the Levy of Water Resources Fees	State Council Decree No.460 (2006)	The application for and justification, approval, inspection, and supervision of water abstraction permit

(continued)

Table 4.5 (continued)

Category	Name	Promulgating organization	Provisions
	Detailed Rules of Implementation of Water Pollution Prevention and Control Law	State Council Decree No. 284 (2000)	Responsibilities, supervision, and rewards for water pollution prevention and control, and penalties for water pollution
	River Channel Management Regulation	NPC (1988)	River channel use control, construction, protection, and obstacle clearing
	Interim Regulation of Water Pollution Prevention and Control of the Huai River Basin	State Council Decree No. 183 (1995)	Pollution prevention and control of the Huai River Basin
	Management Regulation of the Tai Lake Basin	State Council Decree No.604 (2011)	Protection of water resources in the Tai Lake Basin
	Regulation for the Prevention and Control of Geological Disasters	State Council Decree No. 394 (2003)	Prevention and control of disasters resulting from overexploitation of groundwater
	Opinions on Practicing the Stringent Water Resources Management System	State Council Decree No.3 (2012)	The identification of 'Three Red Lines', implementation of the strictest water resources management, and definition of administrative responsibilities

4 Water Plan and Governance System   147

Category	Name	Promulgating organization	Provisions
Rules and Regulatory Documents	Management Procedures for Water Resources Justification of Construction Projects	MWR & NDRC (2002)	Requirements for the procedure, contents and methodology of water resources justification
	Administrative Measures on the Water Abstraction Permit	MWR (2008)	Policies issued by water administrations for the management of water abstraction permit
	Administrative Measures on the Levy and Use of Water Resources Fees	Ministry of Finance, NDRC, and MWR (2008)	Procedures for paying water resources fees by organizations and individuals directly drawing water resources from rivers, lakes, or groundwater bodies, except for cases where water permits are exempted as per Article 4 of the Regulation
	Management Regulations for the Prevention and Control of Pollution in Water Source Protection Zones of Drinking Water	MEE (revised in 2010)	Regulations on zoning and management of water source protection zones
	Opinion on the Enforcement of Groundwater Management in Overexploited Areas	MWR (2003)	Objectives governing the overexploitation of groundwater and requirements for actions

Remarks NPC means the National People's Congress, MWR, the Ministry of Water Resources, NDRC, the National Development and Reform Commission, MEE, the Ministry of Ecology and Environment
Source Compiled based on ADB (2018), Dai (2019), Lee (2006), and World Bank (2013)

Conclusion

The chapter has examined the commitment of the Chinese government to water resources management, paying the first attention to the two major plans, the Three Red Lines in 2011 and the Water Ten Plan in 2015. These plans explicitly disclose the evolving trend of China's water policy in which the magnitude of water policy has been transferred from supply-centered to demand- and ecosystem-centered approaches. This also implies that top decision makers in Beijing become more serious about the rehabilitation of ecosystems with regard to water resources, and therefore, practical and actual plans on how to fill the gap between socioeconomic development and environmental sustainability are required at the national level as well as local level in which relevant policies and projects are being implemented on a daily basis.

The proposed three major challenges identified in the two plans, i.e., water shortage, low water use efficiency and water pollution and ecosystem degradation, have been, and are still difficult to tackle. The deadline of 2030 will come less than a decade as of 2021, and therefore, related policies and projects in various ministries and bureaus at the central and local levels will follow in accordance with guidelines imposed by the central government for achieving ecological civilization.

Special attention has been placed on more responsibilities given to the Ministry of Ecology and Environment, particularly regarding river basin management. Whereas the Ministry of Water Resources is still influential in the overall aspects of water resources development and management nationwide, the key to success of the new regime of water resources management system in China since 2018 depends on the extent to which the complexity of water issues can be resolved based on effective coordination between the ministries involved in water resources management.

The discussions on a variety of water-related laws and regulations unfold that country may have a substantial number of laws and regulations on water and the environment. However, the effective implementation of the institutions on water resources management is an essential element to realize sustainable water resources management. Such a result will be subject to the political willingness of the central and local governments and the enhancement of administrative structure which has been influenced by the traditional bureaucratic system as well as the CCP-dominated governing system. Another fundamental issue to be tackled is to narrow the gap between the central and local levels in terms of policy implementation, legal and regulatory settings and law enforcement.

References

Asian Development Bank. 2018. *Managing Water Resources for Sustainable Socioeconomic Development – a country assessment for the People's Republic of China*. Manila: Asian Development Bank.
Chan, Woody. 2018. Key Water Policies 2017–2018. *China Water Risk*, March 16.
Chan, Woody, and Yuanchao Xu. 2018. Ministry Reform: 9 Dragons to 2. *China Water Risk*, April 18.
Chang, Yen-Chiang, and Li Xinhua. 2019. The disappearance of the State Oceanic Administration in China? – Current developments. *Marine Policy* 107 (103588): 1–3.
Chien, Shiuh-Shen. 2010. Economic Freedom and Political Control in Post-Mao China: A Perspective of Upward Accountability and Asymmetric Decentralization. *Asian Journal of Political Science* 18 (1): 69–89.
China Democratic Alliance Guangdong Provincial Committee. 2016. Recommendations on Speeding Up the Establishment of an Inter-provincial Water Pollution Prevention and Cooperation Mechanism (关于加快推进珠江流域跨省 (区) 水资源保护与水污染防治协作机制建设的建议). 22 January 2016. Available Online http://www.gdszx.gov.cn/zxhy/qthy/2016/wyta2016/201601/t20160122_67179.htm. Accessed 5 November 2020.
China Water Risk. 2017. Revised 'Water Pollution Prevention and Control Law' Approved. *China Water Risk*, June 28.
Dai, Liping. 2019. *Politics and Governance in Water Pollution Prevention in China*. Cham, Switzerland: Palgrave Macmillan.
Ding, Liuqian. 2016. Flood and Drought Management in China. Presented at China Water Resources Management Policy Seminar, June 22.

Fry, James, and Agnes Chong. 2016. International Water Law and China's Management of Its International Rivers. *Boston College International and Comparative Law Review* 39 (2): 227–266.

Global Water Partnership (GWP). 2015. China's Water Resources Management Challenge: the 'three red lines.' A Technical Focus Paper, Global Water Partnership.

Guo, Baogang. 2019. Revitalizing the Chinese Party-State: Institutional Reform in the Xi Era. *China Currents* 18 (1). https://www.chinacenter.net/2019/china_currents/18-1/revitalizing-the-chinese-party-state-institutional-reform-in-the-xi-era/.

Han, Dongmei, Matthew Currell, and Guoliang Cao. 2016. Deep Challenges for China's War on Water Pollution. *Environmental Pollution* 218: 1222–1233.

Jiang, Min, Michael Webber, Jon Barnett, Sarah Rogers, Ian Rutherfurd, and Mark Wang. 2020. Beyond Contradiction: The State and the Market in Contemporary Chinese Water Governance. *Geoforum* 108: 246–254.

Lau, Lynia. 2014. China's Newly Revised Environmental Protection Law. *Clyde & Co*. May, 16. Available Online: http://www.clydeco.com. Accessed 21 January 2020.

Lee, Seungho. 2006. *Water and Development in China*. Singapore: World Scientific.

Li, Chengsi, Kenan Tog, and Feilong Xing. 2015. Why Are Transboundary Water Pollution Disputes Intractable? (跨界水污染纠纷为何多年难解?) *China Environment Report* (中国环境报), March 12. Available Online: http://www.chinanews.com/ny/2015/03-12/7122747.shtml. Accessed 4 November 2020.

Li, Jiahong, Yu. Xiaohui Lei, Aiqing Kang Qiao, and Peiru Yan. 2020. The Water Status in China and an Adaptive Governance Frame for Water Management. *International Journal of Environmental Research and Public Health* 17: 1–19.

Li, Jingyuan, and Jingjing Liu. 2009. Quest for Clean Water: China's Newly Amended Water Pollution Control Law. *A China Environmental Health Project Research Brief*. China Environment Forum, Woodrow Wilson Center. January.

Liao, Yongfeng. 2019. Application of Beidou in Natural Disaster Emergency Management in China. Available Online: https://www.unoosa.org/documents/pdf/psa/activities/2019/UN_Fiji_2019/S2-11.pdf. Accessed 8 July 2020.

Long, Qui-bo. 2019. An Overview of China's Water Statistics. *Acta Scientific Agriculture* 3 (1): 108–115.

Mertha, Andrew. 2008. *China's Water Warriors: Citizen Action and Policy Change*. Ithaca and New York: Cornell University Press.

Ministry of Emergency Management Website. Available Online: http://www.mem.gov.cn. Accessed 7 July 2020.

Moore, Scott. 2018. China's Domestic Hydropolitics: An Assessment and Implications for International Transboundary Dynamics. *International Journal of Water Resources Development* 34 (5): 732–746.

Moore, Scott, and Winston Yu. 2020. Environmental Politics and Policy Adaptation in China: The Case of Water Sector Reform. *Water Policy* 22: 850–866.

National Development and Reform Commission (NDRC), Ministry of Water Resources (MWR), Ministry of Housing and Urban-Rural Development (MOHURD). 2011. The 12th Five Year Plan for Water Development (水利发展规划 2011–2015).

National Peoples' Congress. 2007. Database of Laws and Regulations: Economic Laws – Flood Control Law of the People's Republic of China. Available Online: http://www.npc.gov.cn/zgrdw/englishnpc/Law/2007-12/11/content_1383581.htm. Accessed 22 January 2021.

Nickum, James, Shaofeng Jia, and Scott Moore. 2017. The Three Red Lines and China's Water Resources Policy in the Twenty-First Century. In *Routledge Handbook of Environmental Policy in China*, ed. Eva Sternfeld. Abingdon: Earthscan.

State Council. 2015. China Announces Action Plan to Tackle Water Pollution. english.gov.cn. The State Council of the People's Republic of China. 16 April 2015. Available Online: http://english.gov.cn. Accessed 5 March 2020.

Voita, Thibaud. 2018. Xi Jinping's Institutional Reforms: Environment over Energy? *Edito Energie*, IFRI, October 2. Available Online: https://www.ifri.org/en/publications/editoriaux-de-lifri/edito-energie/xi-jinpings-institutional-reforms-environment-over. Accessed 7 July 2020.

Wang, Yong. 2018. Explanation About China's Institutional Reform in 2018 (关于国务院机构改革方案的说明). *Xinhua Net*, March 14.

Wang, Xiao-jun, Jian-yun Zhang, Juan Gao, Shamsuddin Shahid, Xing-hui Xia, Zhi Geng, and Li Tang. 2018. The New Concept of Water Resources Management in China: Ensuring Water Security in Changing Environment. *Environment, Development and Sustainability* 20 (2): 897–909.

World Bank. 2013. China Water Resources Joint Strategy (中國國別水資源合作戰略) (2013–2020). World Bank and Ministry of Water Resources in China.

World Bank. 2018. *Watershed: A New Era of Water Governance in China—Synthesis Report*. Washington, DC: World Bank.

Xie, Jian. 2009. *Addressing China's Water Scarcity: Recommendations for Selected Water Resource Management Issues*. Washington, DC: World Bank.

Xie, Lei, and Shaofeng Jia. 2018. *China's International Transboundary Rivers*. Abingdon: Routledge.

Zhang, Laney. 2014. China: Environmental Protection Law Revised. *Library of Congress Law*. June 6. Available Online: http://loc.gov. Accessed 17 September 2020.

Zhang, Jiaqi. 2018. 2 New Environmental Laws to Go into Effect in 2018. China.org.cn, January 1. Available Online: https://china.org.cn. Accessed 29 February 2020.

5

Sustainable Water Use

Introduction

This chapter focuses on economic instruments for dealing with water shortage in China. Water shortage is particularly acute in North China. North China is endowed with less than 20% of water out of the total water resources but approximately 60% of the total farming areas of the country are in the region with about 40% of the total population living. South China boasts more water resources with less available land and less food production, which shows an undeniable contradiction between North and South China in terms of water availability and its relationship with arable land and food supply. North China's water shortage is serious because of the lack of surface water and groundwater resources, have been over-exploited as an alternative for quenching the thirst of North China. With the recognition of the severe water scarcity in North China, the central government has introduced several economic instruments for reducing water consumption, i.e., adequate levels of water tariffs in urban areas, water rights trading between different users and regions, and the consideration of the concept of virtual water and water footprint as sensible policy tools to abate water shortage.

An adequate level of pricing policy for water services in general has been regarded as a crucial tool to encourage the public to change their consumption behaviors in China. This will result in saving water and ameliorating the pressure for water shortage and guaranteeing sustainability in ecosystems by consuming less water resources, thereby leaving more room for the thriving of the environment.

In the 1980s, the level of urban water and wastewater service charges was very low so that there was no consideration in valuing water for the payback of all the costs involved for water and wastewater services. Some cities in arid and semi-arid areas of the country have incrementally introduced higher levels of water pricing over the past two decades, including Beijing and Tianjin, thanks to more serious water shortage and soaring demands of water by households and industrial units. But many urban centers keep a low level of water pricing based on egalitarianism and political concern, not necessarily reflecting socioeconomic aspects of water in Chinese society. The notion of rational water pricing is often heard in China's urban water and wastewater service sector, however, such a phrase seems to work well around the capital, Beijing, but give a limited influence on local governments which are more concerned about sociopolitical stability.

Water rights trading in China has been around more than a couple of decades. A recent push by the central and local governments has been possible in tandem with the launch of the Three Red Lines in 2011. This trading scheme has encouraged different water users in different areas to save water for economic gains and has made water users alter their consumption patterns based on economic incentives. Although it is still at the early stage, the scheme looks promising if a variety of institutional barriers are lifted for the upgrade of the water right trading in the future.

The virtual water and water footprint approach provides an innovative way of thinking about resolving China's water shortage challenge. China has been devoted to exporting diverse products to overseas, which implies that China has been exporting large quantities of water to other countries using domestic water resources. In order to keep more water for China's consumption, the government should seriously consider exporting less water-intensive products. In addition, water-intensive agricultural commodities of North China are advised not to be produced there, and instead, South China should take that responsibility based on its abundant water resources. Then, China may reach an equilibrium of water resources available in North and South.

The first part of the chapter is dedicated to discussing the overall situation of water shortage in China, especially the situations of North China. Second, attention is paid to the development of water pricing policies in China from the 1980s to the present, and discussions are made on the structure and the level of urban water and wastewater service charges. The third part is associated with the water rights trading scheme. The study explores the trend of water rights trading and evaluates achievements and challenges of the scheme. The last part focuses on virtual water and water footprint in China and sheds lights on China's virtual water export and import together with their implications. The contradiction between North and South China in terms of water, land, and population endowments will be investigated alongside potential solutions to alleviate water shortage in North China.

Water Shortage

Wen Jiabao, Ex-Premier of China, once stressed that the very survival of the Chinese nation would be threatened by severe water shortages (Parton 2018). It is useful to review the trend of water consumption of China in order to have a better understanding of how serious the situation of water shortage has been. The total water consumption in China reached 103.1 billion m^3 in 1949, 274.4 billion m^3 in 1965, and 443.7 billion m^3 in 1980. The abstracted amounts of groundwater resources were as large as 64.7 billion m^3 from 1949 to 1980. The seriousness of water shortage in China was evident in urban areas of North China, and more than 150 out of 191 cities in the region were regarded as the places running out of water in the 1970s (Shen and Wu 2017).

At the national level, the situations of water stress, which can be defined as the amount of per capita water consumption less than 1,700 m^3 directly or indirectly, are different, particularly in North or South China. As discussed in Chapter 3, the average per capita renewable water resources were estimated at 2,104 m^3 between 1998 and 2017, and the per capita water consumption reduced to 431 m^3 in 2019 from 456 m^3 in 2013 (MWR 2018, 2020). According to recent water resources assessments, the average annual rainfall in the Yellow River, the Huai River, the Hai River, and the Liao River had diminished by 6%, and the volume of surface water flow had decreased by 17%. In particular, there was a dramatic drop of the amount of water flow in the Hai River by 40%. The annual water shortages in the country would be estimated at 53.6 billion m^3, which badly affects agriculture, industries, daily living in households, and ecosystems. Intensifying droughts give seriously negative impacts on around 15 million ha of farmlands per annum (GWP 2015). Confronted with the severe water shortages, the government gave an urgent call for setting up the limit of water consumption at 670 billion m^3 in the 13th Five Year Plan (FYP) (2016–2020).

Despite the top-down restriction of water consumption per capita, more innate and problematic issue remains intact, namely the geographically uneven distribution of water resources in China. About 80% of the total amount of renewable water resources are available in South whereas North China has only around 20%. But North China includes approximately 60% of the arable land, and around 40% of the total population of China live in the area. The low level of water availability in North China has driven people to seek for alternative water sources except for rivers and lakes, including groundwater. No less than 50% of the total water consumption in Henan, Shanxi, and Inner Mongolia Provinces, and the Beijing Municipality stem from groundwater (GWP 2015; Wu and Edmonds 2017).

Groundwater levels have been falling in North China in recent few decades, which exacerbates the already serious water shortage circumstances caused by surface water overuse. The levels of groundwater tables in North China drop further on average at 1–3 m per annum. For instance, some areas of the Beijing Municipality subside at approximately 11 cm per annum. The rapid drop of groundwater table has entailed collapsing aquifers and depleting groundwater resources, and eventually local streams and rivers become dry, thereby severely damaging the viability of ecosystems. Other areas in North China, such as Hebei and Shanxi Provinces, have experienced growing demands of water resources from mining and industrial sectors, and local authorities in those provinces have striven to meet the demands by digging deeper for groundwater. In the 2010s, aquifers in the Central-North China were found 7.4 m below the surface on average, and groundwater seekers had to dig even much deeper than 10 m, and in some areas, down to 20 to 30 m (Parton 2018; Wu and Edmonds 2017).

A total of 12 provinces in 2017 suffered from water scarcity, including the Beijing, the Tianjin, and the Shanghai Municipalities, and Ningxia, Shandong, Hebei, Henan and Shanxi, Guangxi, Shaanxi, Liaoning and Jiangsu Provinces. To make things worse, such water shortage-stricken areas accounted for 35% of the total agricultural bases, 50% of power generation, 46% of industries, and 41% of the population of the country. The newly designated regional development area, the Beijing-Tianjin-Hebei Region, might be confronted with even more serious water stress considering the possible population of more than 120 million and relevant services, including power generation, public utility services, and housings. Tensions are also mounting in the Yangtze River Delta Region thanks to growing demands of water (Hao et al. 2019; Parton 2018). In the period between 1980 and 2017, the statistics disclose a gradual increase of water consumption in China. An annual average increase of the total water consumption was recorded at 1.23% from 1980 to 1993, 1.72% from 1993 to 1997, −0.75% from 1997 to 2003, 1.92% from 2003 to 2009, and 0.89% from 2009 to 2013 (Shen and Wu 2017).

The following section pays special attention to Beijing in order to explore the intensity of water shortage in North China. An average annual precipitation of the Beijing Municipality amounted to 585 mm in 2019. There are five rivers within the municipality coupled with 88 reservoirs, but these do not meet the growing demands of water in the city. More water resources are channeled through water transfer projects, such as the middle route of the South North Water Transfer Project (SNWT Project), the Yongding Water Diversion Canal, and the Jingmi Water Diversion Canal.

Beijing embraces a total of 72 water supply plants, 67 wastewater treatment plants which includes water recycle plants, and the network of water supply pipelines stretches over 16,000 km in tandem with a length of 21,000 km drainage network, which serves as critical infrastructures to protect the municipality from excessive run-off and urban floods. Despite well-equipped water supply and sewage networks, the city has suffered from a paucity of water resources for many decades. With the population of over 20 million in the metropolitan area, the current level of per capita water resource in the city was only 1/20 of the national average and 1/80 of the world average, 100 m^3 in 2011. In 2016, there was a warning from the Beijing Mayor that the ecological limit for the maximum population size of the city should be less than 23 million due to the serious scarcity of water resources (Pan and Wei 2019; Parton 2018; Wang et al. 2015).

Such a dire situation has been relieved thanks to diverted water resources through the middle route of the SNWT Project since 2014, which has helped augment water resources with a total volume of 5 billion m^3 from 2014 to 2019, thereby making per capita water resources, 150 m^3. In 2019, the Beijing Water Authority provided Beijing people with the tap water that comprises half from rivers and reservoirs around the city and half from the SNWT Project (Pan and Wei 2019).

According to the future network map of tap water supply in Beijing, water resources of the city will never be dried up, because the water veins of the city are planned to be connected from North, South, East, and West through transfer pipelines. It is still questionable if water shortage challenges in the city have been resolved thanks to the transferred water through the middle route of the SNWT Project, however, people in Beijing do not have to worry about running out of water. The prestigious position of the city in the country secures clean water any time because water can be pumped up or diverted by those who have power and money as Reisner (1993) maintained, 'water flows uphill to power and money,' discussing the development of Los Angeles equipped with political prowess and monetary power, fetching water from the central part of California.

Water Pricing

Water pricing is one of demand management options to improve the equity of water supply and ensures cost recovery together with sustainability of water resources. Demand management encompasses various measures that will result in making consumers use water sustainably. This type of management involves the adaptation and implementation of policies and initiatives for water managers to change the water demand and use of water for meeting economic efficiency, social development, social equity, environmental protection, sustainability of water supply and services, and political legitimacy.

Specific instruments in demand management are intermittent water supply, water loss reduction, i.e., leak detection and repair, comprehensive metering, changes in water pricing concepts, installation of water saving devices (retrofitting), wastewater reuse, institutional development, and public awareness and educational campaigns. The purposes of water pricing, however, are self-contradictory, because economic efficiency gained through water pricing does not automatically ensure water affordability for the general public and environmental sustainability (Stephenson 2005; Vairavamoorthy and Mansoor 2006).

It is imperative to explore such multifaceted nature in determining water pricing policies reflecting economic, social, and environmental concerns. The principle of 'CAFES' is useful as follows. First, the principle of 'Conserving' is that the water pricing should be structured for consumers to meet their own needs without wasting water. Second, the principle of 'Adequate' indicates that the level of water pricing should generate a sufficient level of revenues so that water service companies or entities can afford to make future investment. Third, the principle of 'Fair' means that water service companies or entities should achieve financial sustainability with adequate pricing. Revenue shall evenly and fairly be allocated between consumer groups for both the poor and the rich.

Fourth, the principle of 'Enforceable' is discussed. Water service entities should have the means to enforce the price through various sanctions, i.e., court action and disconnections. The fifth principle is 'Simple,' which means that the structure of water pricing should be simple enough for water service entities to manage and easy for customers to understand. It is common to notice that customers are willing to pay water bills when the bills are easy to understand (Vairamoorthy and Mansoor 2006).

With regard to the practices of Chinese water pricing, mainly in urban areas, the principle of CAFES is not necessarily neatly applied.

From 1949 to 1964, there was absolutely no charge for water supply services, and all the costs and expenditures were subsidized by governments. Little charge was levied for water services from 1965 to 1984, and between 1985 to 1993, attempts were made for applying the cost-recovery principle in water services although the level of service charges was insufficient. The complete form of the cost-recovery level of water pricing was introduced from 1994 to 2003 based on the Notice on the Implementation of Water Pricing Reform, Acceleration of Water Saving, and Protection of Water Resources. From January 2009 to June 2010, more than a third of 669 cities adjusted water prices (Shen et al. 2017).

Although there has been an upsurge of water prices in China over the last four decades since 1978, the level of water pricing in general is still low. Economic efficiency and environmental sustainability have not well been reflected whereas social equity has seemed to be over-emphasized, even buttressed by political ideology, such as egalitarianism embedded in the socialist market economy. However, the water pricing system in China has undergone eye-catching transformations that increasingly accommodate economic and environmental dimensions (Shen and Wu 2017).

Delving into the water pricing framework in China, there are four fees and charges: (1) water resources fee; (2) water supply engineering fee; (3) water supply fee; and (4) wastewater treatment fee. Prior to 2018, there was another fee, pollutant discharge fee, however, this has been converted into a pollution discharge tax thanks to the introduction of the Environmental Pollution Tax Law in 2018.

China's water tariff structure is described below. The total amount of water tariff for households is calculated following the formula according to the Water Law (1988 and 2002).

$$\text{Water Tariff} = \text{Water Resources Fee} + \text{Water Supply Engineering Fee} + \text{Water Supply Fee} + \text{Wastewater Treatment Fee}$$

This structure, however, has been modified in accordance with the launch of the Water Resources Tax Reform in 2016. The water resources fee has been replaced with the water resource tax since 2017 although the system requires more time to be fully applied to everywhere in China.

The first testing ground for this reform was Hebei Province in which a series of positive results were noticed, i.e., improved water resources management, a rise of tax revenue, better water use efficiency, and higher water saving awareness.

To begin with, take a close look at the water resources fee. This is a resource charge and is collected in accordance with the system of payment, which was introduced in the 1988 and 2002 Water Law. The primary purpose of the fee is to save, protect, manage, and develop water resources. Entities which directly abstract water from rivers, lakes, and aquifers are obliged to pay the fee based on the actual volume of water and should gain a permit from the provincial government. But water abstraction for domestic use, little amounts of water for livestock and emergency, is not charged. The fee is collected by water departments under the MWR (Shen and Wu 2017; Shen et al. 2017).

Second, the water supply engineering fee is a service charge for water supplied by hydraulic engineering to water users. Hydraulic engineering means barring, storage, diversion, pumping projects, and supplying water without treatment. This service charge is differently levied to agricultural and non-agricultural groups, and the level of charge for non-agricultural groups is higher. The charge for agricultural ones is set low because the level of charge reflects costs and operational and management expenditures with no profits or taxes. This is another example of sociopolitical acceptability being prioritized over economic and environmental concerns in the Chinese water pricing system.

Third, the water supply fee is a service charge that is levied by water supply entities and accommodates the principles of cost recovery, adequate profits, promotion of water saving and fair allocation. In addition, provincial-level governments should convene public hearings and notify the adjustment of water prices. These practices largely meet the principle of CAFES discussed above. Fourth, the wastewater treatment fee as a service charge is levied by a wastewater treatment entity for collection and treatment services. The 2014 Wastewater Treatment Tariff Collection and Use Management Methods were instrumental for improving collecting practices by allowing local governments to collect the fee together with the urban water supply fee (Shen and Wu 2017; Shen et al. 2017).

A total of nine provinces have been chosen for implementing the pilot projects for water resources tax, including Beijing, Tianjin, Shandong, Henan, Shanxi, Inner Mongolia, Shaanxi, Ningxia, and Sichuan. Such a reform imposes the minimum tax rates for the pilot provinces, which are RMB 0.1–0.6 (US$ 0.015–0.09)/m^3 for surface water and RMB 0.2–4 (US$ 0.03–0.61)/m^3 for groundwater. In the middle of 2018, the overall amounts of groundwater extraction in the pilot provinces dropped at 9.28% for just six months since the onset of the reform, and a thermal power plant dramatically reduced groundwater use at 64%. Although the reform limits the geographical boundary within the nine administrative areas, it seems that the government would push forward the reform in full throttle in a few years (MOF and MWR 2017; Yu and Hu 2018; Xu 2018).

The Circular on Accelerating the Reform of Water Price, Promoting Water Saving Practices, and Protecting Water Resources in 2004 served as a turning point to raise the level of water tariffs in China. Many local governments have begun to adjust water tariffs, and some cities decided to increase the tariffs three or four times between 2004 and 2006. The water tariff reform reaffirmed the continuous commitment of the government to making private sector participation in the water sector viable commercially and ensuring long-term profitability for private sector players (Qian et al. 2020).

In 2012–2013, the State Council and the NDRC issued several documents on water pricing, requesting ministries, provincial, municipal, and county governments to conducting three tasks: (1) to achieve a full cost recovery of water services by user charges; (2) to introduce progressive water prices in urban areas; and (3) to implement bulk water pricing based on a water scarcity principle. These documents have pushed forward the rise of water charges in recent few years (Spooner 2018).

Water tariffs are collected by a local water company based on consumer metered readings. The level of tariffs is decided by the local government which takes into account diverse factors, advice, and opinions, including public hearings. Then, the local water company collects the tariffs or is paid for the volume of water treated (m^3) by the local government.

Within the local government, it is the Price Department in the Development and Reform Bureau that makes final decisions on water tariffs. Although the bureau seems to have been reluctant to implement the principle of full cost recovery until recently, more political and social pressure favoring ecological civilization has forced them to raise the level of water tariffs (Spooner 2018).

Chinese authorities have introduced the increasing block tariffs for water saving since the issuance of the Guiding Opinion on the Acceleration and Completion of Urban Household Water Supply Increasing Block Tariff System in 2013 by the NDRC and MOHURD. As of April 2019, 35 out of 36 major cities in China adopted the increasing block water tariff system except for Xining that did not create relevant standards yet. This demonstrates the wide adoption of the increasing block tariff system in major cities of China (Water Resources Institute 2019).

The level of water tariffs in China is much lower than that in other parts of the world, including Europe. For instance, the tap water tariffs for household use in the Beijing Municipality are on the increasing block tariffs that were set in January 2018. Water resources fees are decided depending upon the source and the region, and the fees are higher in the area in which groundwater resources are over-exploited and subject to being depleted. In water shortage areas in the North China Plain, the level of water tariffs is much higher, since raw water is provided together with the water through long-distance transfer projects, such as the South North Water Transfer Project. The average water resources fee for surface water in Beijing is RMB 1.6 (US$ 0.24)/m^3 whereas that in Shanghai reaches only RMB 0.1 (US$ 0.015)/m^3. Beijing's tap water stems from its own reservoirs as well as the middle route of the South North Water Transfer Project.

Wastewater tariffs are charged differently depending upon which discharge standards are applied. More investment and cost are involved if an amount of wastewater is treated to meet the water quality standards from Grade II to Grade I, and therefore, upgrading wastewater treatment plants to meet higher standards would cost more (Spooner 2018; Water Resources Institute 2019). Table 5.1 describes the details of water tariffs

Table 5.1 Increasing block tariffs in Beijing since January 2018

Use	Water supply types	Stepped	Annual water consumption (m³)	Water tariff	Fee Structure (RMB/m³)		
					Water supply fee	Water resources fee	Wastewater treatment fee
Household	Tap water	1st	0–180	5	2.07	1.57	1.36
		2nd	181–260	7	4.07		
		3rd	260 above	9	6.07		
	Groundwater	1st	0–180	5	1.03		1.36
		2nd	181–260	7	3.03		
		3rd	260 above	9	5.03		
Non-household	Six districts in the city center	–	–	9.5	4.2		3
	Other areas in the city	–	–	9	2.2		
Special industries (i.e., Health spa)		–	–	160	4		

Source Beijing Water Authority http://swj.beijing.gov.cn (accessed 22 September 2020) in Beijing and Fig. 5.1 illustrates the water tariffs of 36 major cities in China in 2020. The process of deciding water and wastewater tariffs in localities is complex and political. No direct responsibility is given to the price bureau at a local government, and there are various elements to be considered prior to the finalization of the level of water and wastewater tariffs, i.e., economic inflation, water resources and environmental quality metrics, the balance between economy and the economy, and the assurance of keeping local water companies in business.

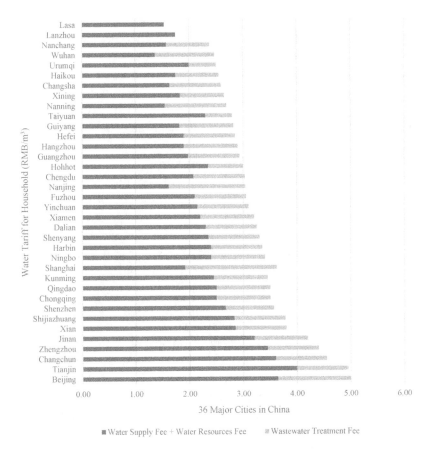

Fig. 5.1 Water tariff for household in 36 major cities in China, 2020 (*Source* H2O Website. http://h2o-china.com [accessed 25 September 2020])

As for a local water or wastewater company, if the tariffs paid to them are insufficient to recoup their investment and compensate operation and maintenance costs, they can have the opportunity to negotiate with the local government for adjusting payments. The negotiation processes are subtle and highly political, but the amounts of adjusting payment would be substantial. This elucidates the reason many Chinese water and wastewater companies are able to survive despite such a low level of water and wastewater tariffs. This sort of bargaining process may be more difficult for foreign companies of water and wastewater treatment services to receive the balancing payments (Spooner 2018).

One of the most water-scarce areas in China, Beijing, demonstrates the steep rise of urban water supply prices in the past few decades. The urban water supply prices had increased at dozen times from 1990 to 2017 with the adjustment of water supply pricing structure from a single rate to an increasing block tariff. The total sum of urban water supply and wastewater service prices in Beijing was as high as RMB 5 (US\$ 0.71)/m^3 in 2020, which is twice more expensive in Nanchang (see Fig. 5.1) (H2O China 2020; Shen and Wu 2017).

However, Beijing introduced different categories for urban and rural water users and imposed lower water tariffs on poor farmers, which implies the gravity of political implications in the capital region (Moore and Yu 2020). This level of water price in the city does not necessarily reach the level of water price to pay back all the necessary costs and reasonable profits for future investment because of social stability worries as the capital of the country. The share of water and wastewater service costs in relation to disposable income in the city ranged from 0.62 to 1.23% from 2003 to 2012, and the ratio of Beijing is similar with that of counterparts in the OECD member countries, which was less than 0.8% in 2008. The highest water tariff on special use is established at RMB 160 (US\$ 24.6)/m^3, which is regarded still low compared with the maximum water tariff in Singapore, US\$ 42/m^3 in 2020 (Moore and Yu 2020; Singapore Public Utilities Board 2020). It can be maintained that considering the sustainability of water resources, Beijing still has much room for higher water and wastewater service prices for the sustainability of water resources, because the international benchmarking threshold for water affordability in an urban household ranges from 2.5 to 3.0% (Shen et al. 2017; Shen and Wu 2017).

While carefully designed reform policies for water pricing have been introduced and implemented, particularly for urban water and wastewater services, such reforms demonstrate mixed results. From 2002 to 2009, water price increases for urban water users culminated in reducing water demands by 5% although there was almost no significant level of water tariff levied for agricultural water users. The cost of agricultural water supply was estimated at RMB 1.18 (US$ 0.18)/m^3 from a 2010 survey of 414 large-scale irrigation districts. Yet, the highest water tariff for water users reached only RMB 0.35 (US$ 0.053)/m^3. At the national level, bulk water supply prices for agricultural uses were charged at approximately RMB 0.0611 (US$ 0.009)/m^3 (Moore and Yu 2020; World Bank 2019).

Considering the rationale behind water pricing reforms in China, i.e., conservation of water resources, financing water infrastructures as well as encouragement of efficient water use, the overall assessment of the reforms is not necessarily successful because the level of water and wastewater tariffs has turned out to be still too low to achieve the full cost recovery or sustainable water resources management. This outcome is partly attributed to China's unique political economy system of water administration based on gradualism which means a step-by-step approach to policymaking and reform where diverse versions of policies are tested and observed. This principle has augured well in socioeconomic development of China but has failed to be applied to water pricing reforms (Moore and Yu 2020).

Water Rights Trading

A noticeable transformation in water policy and management has occurred since the late 1990s, which encapsulates the shift from state-dominated water administration to new management paradigms that advocate roles of markets in water allocation. Such a policy shift, however, does not necessarily mean more weight on water services in the hands of the private sector and a retreat of the public sector in water resources management. Instead, the interplay between the public and the private sectors paves the way for a new hybrid system, the state, and the market, to facilitate the balance between economic development, social equity, and environmental sustainability (Jiang et al. 2020).

China has attempted to establish a unique form of the hybridization of water governance between the state and the market. The government is still dominant in water governance and strives to secure sufficient amounts of water resources for every sector whereas water markets for allocation are being created by the central government and participated by local collectives and farms. Confronted with soaring demands of water and limited amounts of water resources, the country has endeavored to establish a 'Water Saving Society' since 2000, which encourages the development of water markets and tradable water rights. Interestingly, there is no water market in China according to western standards whereas the volume of water rights trading has surged in recent few years (Jiang et al. 2020).

It is worth delving into what a water saving society in China means. This initiative was proposed first by the central committee of the CCP for the 10th FYP (2001–2005) in 2000. According to the initiative, water resources management should take into account both increase in water supply and decrease in water demand. In order to provide more water for North China, the party recommended the South North Water Transfer Project as a solution and the water saving society concept for demand management by proposing water conservation measures and developing water saving agriculture, industry, and services. On the basis of this dual track, the government pushed forward a tremendous increase in investment for water projects, from RMB 46.8 billion (US$ 7.2 billion) in 1998 to RMB 94.5 billion (US$ 14.5 billion) in 2007 (Jiang et al. 2020; MWR 2013).

In October 2000, the then-Minister of Water Resources (MWR), Wang Shucheng, gave a speech on 'Water rights and water markets: economic measures for achieving optimal allocation of water resources' and stressed that all the works (construction, operation, and management) of the SNWT Project should be undertaken based on a 'quasi-market' principle (Jiang et al. 2020; Sheng and Webber 2019). Water recipient regions through the project can acquire a water use right for the project by contributing to the SNWT Project Construction Fund that would be created based on the collection of water resource fees. The project, which has been praised as a hydraulic mission-type commitment, can serve as a conduit for facilitating water rights trading based on a quasi-market together with the tight national control across provinces and municipalities (Jiang et al. 2020; Zhang et al. 2015).

The first ever water trading occurred in November 2000 between two counties in Dongyang and Yiwu, Zhejiang Province in which the water rights were transferred for 50 million m^3 per annum from Dongyang to Yiwu with the payment of RMB 200 million (US$ 30.7 million). The deal was unprecedented but controversial, because there was no legal foundation for supporting the transaction either in the Constitution or the Water Law 1988. These laws explicitly state that water resources belong to the state, and therefore, no legal rights were given to local governments to trade water rights. Against such legal loopholes, the MWR advocated the 'policy experimentation,' and similar projects were introduced in Zhangye, Gansu Province, water rights transfer between farmers, and in Inner Mongolia and Ningxia Provinces, regarding large-scale water transfers from the agriculture to industries (Jiang et al. 2020; Shen et al. 2017).

Water rights trading became officially legalized through the introduction of the revised Water Law in 2002, which encompasses the water rights strategy by stating 'building a water saving society' as a basic principle (Article 8) and the adoption of water abstraction rights (Article 48) although the 2002 law does not explicitly endorse water rights trading (Jiang 2018; Jiang et al. 2020). More institutional backups accompanied, including the Framework for Establishing a National Water Rights System comprising three essential elements: (1) water resources ownership; (2) water resources use rights; and (3) water rights transfer (JICA 2006; Shen et al. 2017).

In addition, the 2006 Regulation on Administration of Water Abstraction Permits and Water Resources Fee Collection relaxed restrictions on the transfer of water abstraction permits and paved the way for companies to make investments in the construction of irrigation rehabilitation projects for enhancing irrigation efficiency by exchanging abstraction rights with the saved water (JICA 2006; Shen et al. 2017). Water abstraction rights were stipulated in the 2007 Property Law in China as a type of water right, which distinguishes from the state ownership of water resources, and the law allows water abstract rights to be transferrable under special conditions. The institutional changes brought about a rise of water rights trading between 1998 and 2007, and this indicates the transformation of water abstraction permits from a tool of governmental control over water use to a foundation for water rights trading. The path to the development of water rights trading in China connotes the incorporation of market-oriented water allocation mechanisms into the conventional government-dominated regime embodied with infrastructure-centered and supply management focused policies (Jiang et al. 2020) (see Fig. 5.2).

Major developments in water rights trading have been observed since 2014. In July 2014, a nationwide water rights pilot scheme was introduced by the MWR in order to accelerate its previous local-level pilot programs. The ministry designated seven provinces and autonomous regions for experimenting a variety of forms of water rights trading, such as the verification and registration of water rights (Ningxia, Jiangxi, and

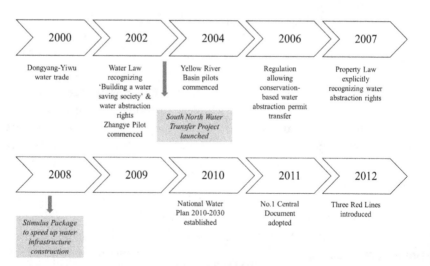

Fig. 5.2 Major water policy developments in China from 2000 to 2012 (*Source* Modified based on Jiang et al. [2020])

Hubei Provinces), and the test of water rights transfer models (Inner Mongolia, Henan, Gansu, and Guangdong Provinces).

The MWR issued the Provisional Measures on Administration of Water Rights Trading in April 2016 for leading water trading practices by suggesting three types of water rights trading. These are: (1) regional water rights trading (between county-and-above local governments); (2) water abstraction rights trading (from water abstraction permit holders, excluding urban public water supply companies); and (3) water rights trading between irrigators (water users' associations or individual users of irrigation water with well-defined water rights). The magnitude of this regulation lies in the inclusion of regional water rights trading first time, which legitimizes local governments above the county level to make transactions of their water rights (Jiang 2018; Jiang et al. 2020; MWR 2016).

In 2013, 500 water rights trading occurred in Gansu Province, and the volume of traded water resources amounted to 25 million m^3. The Shiyang River Basin Water Users Associations (WUAs) had actively encouraged farmers in the basin to be aware of the cap of the annual water use consumption related to irrigation fields, save water, and sell the amount of saved water resources in the water market. In this system, there are two types of water rights for a year, long-term water entitlement and seasonal water allocation (Zhao 2018).

Attention is paid to the pattern of transactions of the water rights trading in the basin. Most of the transactions took place at 5–10 km within the same county, because farmers would like to sell saved water to relatives, friends, and neighbors, not total strangers. This phenomenon can be something to do with the cost of delivery and conditions of irrigation pipelines (Zheng 2019).

Although it is a bit early to judge the extent to which successful this system is, the water rights trading scheme of the region shows several positive impacts. First, this institutional experiment will encourage farmers to change their consumption behaviors. Second, the inappropriate level of water pricing for the agricultural sector can be resolved. Farmers realize the need to save water resources by themselves through such a social learning. Third, the amount of saved water through the water rights trading can contribute to sustainability since less water in the agricultural sector implies more water for the other sectors, including ecosystems.

Fourth, significant development in water rights trading of China has culminated in the establishment of national water exchange in Beijing, 2016, the China Water Exchange, with an initial investment of RMB 600 million (US$ 92 million). Major investors and sponsors include enterprises owned by the MWR, seven river basin commissions, and the Beijing Municipal Government. This high-profile ownership structure has aided the smooth take-off of the institution. This institution helps accelerate water rights trading and provides services, such as trading consultation, technical evaluation, market information, intermediary services, and public services. In addition, the exchange has introduced numerous standards to promote water rights trading with the guidelines covering fees, trading protocols, awareness-raising, and risk management. Shandong Province has shown its interest in joining the China Water Exchange as a full shareholder. Several provincial governments have attempted to create their own water rights trading markets through the introduction of new policy, legal settings, and organizations, i.e., Inner Mongolia, Henan, and Guangdong Provinces, which have established their own water trading mechanisms (Moore and Yu 2020; Jiang et al. 2020; Wang and Yang 2018; World Bank 2018).

At the national level, 152 trades had been made in the period from 2013 to August 2019, which means around 2.8 billion m^3 of water volume traded worth RMB 1.7 billion (US$ 260 million) (Jiang et al. 2020). The data in early July 2020 show that in the period between June 2016 and December 2019, 89 trades were undertaken regarding regional and water abstraction rights trading with the volume of 2.8 billion m^3, and 258 trades occurred between irrigators from March 2017 to June 2020 with the volume of 145.3 billion m^3 (China Water Exchange 2020). The upward trend of water rights trading is projected in the future based on the strong emphasis of the central government on the state-market mixture in water resources management. Figure 5.3 illustrates China's water policy developments related to water rights trading between 2013 and 2017.

Despite the progress, there are various challenges in China's water rights trading. First, a fundamental problem is the low level of local

Fig. 5.3 Major water policy developments in China between 2013 and 2017 (*Source* Modified based on Jiang et al. [2020])

organizations' understanding of water rights trading. There is no public consensus over the aim and the roles of government and market in the creation of water rights trading system. It is imperative to clarify water rights from the perspective of property right, which will pave the way to reduce relevant costs as well as facilitate an increase of operational efficiency in the water rights trading market. Recognizing good progress in water rights trading between local governments and between public entities and private users, transactions between private users are relatively less (Shen et al. 2017; Wang and Yang 2018).

Second, it is difficult to undertake the price measurement of water resources, which is closely related to difficulty metering and monitoring. This phenomenon is particularly true in most of the rural areas in the country, which can entail the lack of creditworthiness of the system within farmers and rural collectives in the long term. For instance, the Pearl River Delta, Guangdong Province has adopted tidal irrigation, and the measurement of agricultural water is primarily conducted through installed meters at the start point of tidal canals. This measurement, however, may lead to inaccurate data due to the fluctuated levels of tide, which can affect water rights determination and the metering and monitoring of trading water volumes (Wang and Yang 2018).

Third, the current system only allows farmers to engage in water rights trading within the concerned year so that some remaining portion of saved water in the year one cannot be transferred to the year two. Fourth, the system has been established recently and requires more detailed regulations and guidelines that do not necessarily function properly faced with a variety of situations. For instance, a buyer decides to buy a certain amount of water rights because of the lack of water for his or her irrigation fields but what if it rains sufficiently for his or her own fields prior to the transaction, there is a question if the authorities allow the buyer to resell the water rights to the seller.

Fifth, an infrastructure-related problem is the capacity of local areas to store extra amounts of water resources, which can facilitate more active water rights trading. Some local areas can renovate existing agricultural reservoirs or develop new ones where other areas cannot afford to have those options.

Sixth, water rights trading has been taking place only within the same province, county, or village. Cross-provincial or any local-level water rights trade has rarely occurred due to the reluctance of local governments which would not give up their entitled shares of water resources and administrative complexities they might face. On top of that, water shortage may come into being any time and no one knows how long the phenomenon lasts. Local authorities are extremely careful sharing such invaluable resources with their neighbors and willing to keep their water resources as much as possible for future socioeconomic development (Moore and Yu 2020; Yang et al. 2012; Zheng 2019).

China's water rights trading scheme needs improvement. First, the government should play a leading role in promoting the scheme through the establishment of a clear accountability system, the use of all media, such as newspaper, television, broadcast and the internet, and the engagement of the public. Second, administrative incentives and private sector participation should be encouraged for more financing in the scheme. More investments should be allocated for the overall implementation of water rights reform, i.e., the metering and monitoring of water resources, the creation of information platforms, efficient water conservation, and project constructions. More accurate data shall be available based on the introduction of smart monitoring systems, and reward and punishment systems are considered for localities to support the water rights trading scheme.

Third, the water rights trading platforms should be refined and enhanced. A better form of the platform can develop through two phases. In the first phase, the China Water Exchange and water trading platforms for pilot provinces are allowed to be in operation together with trades within each province through local trading platforms. As the trade's size grows, the China Water Exchange will absorb all local platforms and serve as the only official platform to provide services for access to trading information and complete trading database (Wang and Yang 2018).

Chinese authorities have adopted the dual mechanisms in terms of water supply management by putting an emphasis on the continuous construction of water infrastructure such as the South North Water Transfer Project, and the creation of quasi-water market through water rights trading. The two approaches work in the unique Chinese context, and these are necessary and useful foundations to achieve sustainable water resources management.

Virtual Water and Water Footprint

The nexus between water, food, and trade is neatly elucidated by the virtual water and water footprint approach. Since China has become a major contributor to the global supply chain, the country has been regarded as the world factory that manufactures and exports all kinds of goods and products, which serves as a powerful engine to make its economy grow rapidly over the past decades. China joined the World Trade Organization (WTO) in 2001, which has accelerated industrialization and economic growth, entailing the transformation of the country into the largest manufacturer and trader in the world. In addition, the continuous commitment of the CCP over the 'Go Out Strategy' from the eras of Hu Jintao and Xi Jinping has promoted the outward investment of Chinese companies. These phenomena have triggered an export of pollution to other countries, especially developing countries (Wu and Edmonds 2017). Such a trend of globalization of ecological destruction from China to other parts of the world has increasingly exacerbated thanks to China's recent global strategy, the Belt and Road Initiative.

The impressive economic development of China has been possible at the expense of the environment, including water resources. Growing demands of water resources within China, particularly related to food security, have led the country to turn its attention to the global food market, which can deteriorate global water shortage in the long term if there would be no adequate policy shift for water use efficiency and consumption pattern change. In this sense, it is necessary to investigate viable policy options for alleviating water stress in the context of virtual water and water footprint, whose approaches are useful in tackling challenges of China's water stress as well as global water shortage.

Virtual water means the volume of water needed to produce agricultural, livestock, and industrial commodities (Allan 2001). Whereas virtual water has been touted as a useful tool to explore the nexus between water, food, and trade, water footprint is instrumental in quantifying water resource requirements for products and services used by individuals, industries, cities, or countries (Chapagain and Hoekstra 2002; Hoekstra et al. 2011; Tian et al. 2018).

It is interesting to ponder over ramifications of the virtual water and water footprint approach for China's water resources management. In 2007, the total amount of water resources available in the country were 2,812 billion m^3 per annum, and the total volume of virtual water export reached 68.2 billion m^3 per annum, which is equivalent to approximately 2.4% of its total water resources. Although the percentage is relatively little, this amount of water is as vast as the total volume of water delivered per annum through the three routes at the final stage in the South North Water Transfer Project. Virtual water export implies that invaluable domestic water resources within China have been sent to overseas countries, and water-scarce regions in the country, i.e., Northwest and Northeast China, have no access to massive amounts of water embedded in export products. It is maintained that China's economic well-being has been achieved based on manufacturing industries that induced extremely high costs for water resources and the environment, especially for water-scarce regions (Yang et al. 2012).

According to the World Trade Organization (WTO) (2018), China was the largest merchandize exporter in the world, totaling more than

US$ 2,263 billion from 2016 to 2017. Such a powerful status in international trade is also confirmed in the virtual water trade, and the annual virtual water export of China amounted to 39.04 billion m^3 in 2002 and 68.18 billion m^3 in 2007, which indicates a growth of 74% in five years (Yang et al. 2012). In 2009, the figure increased more to 153.3 billion m^3 with virtual water import, 99.8 billion m^3, producing a net outflow of 53.5 billion m^3 (Hou et al. 2018). Among various industrial sectors, main industries for virtual water export include textiles and clothing, electrical equipment and telecommunication equipment, wholesale and retail trade, passenger transport, and metal smelting and products. Various industrial sectors are virtual water net importers, including agriculture, petroleum processing, machinery and equipment, and utilities (Yang et al. 2012).

China appears to sustain its status of net virtual water exporter at the global level considering the architecture of global supply chain. It is significant to recognize that more virtual water export means more loss of domestic water resources and gives more water stress to China. Realistic options for saving water resources are to enhance industrial water use efficiency in the industrial sectors through the introduction of water saving technologies, and incentive mechanisms, and related institutional settings should be established at the local and national levels, including water rights trading between regions and sectors (Hou et al. 2018).

The fundamental contributions of the virtual water and water footprint approach to China's water resources policy are several. First, the approach reminds the magnitude of the role of international food trade in resolving water shortage in the country. Tremendous political, financial, and technical challenges can exist considering large-scale water resources development projects, such as dams, aqueducts, and water transfer projects. Food import from other countries would ensure a stable supply of various food items, and therefore, there would be no need to mobilize mega-scale water infrastructure projects. Massive amounts of virtual water import in the form of food products should further be considered as a significant factor in establishing water supply plans, particularly for water-scarce regions in China, and this can lead to questioning the long-term viability of the South North Water Transfer Project.

For instance, North China exported 52 billion m^3 of virtual water to South China in 2006, which surpassed the planned amount of transferred water through the South North Water Transfer Project, from 38 billion m^3 to 43 billion m^3 per annum. Furthermore, the 3H Region (Yellow/Huang, Huai, and Hai Rivers), water recipient areas of the east and middle route, exported 26 billion m^3 of virtual water to South China. The maximum amounts of transferred water through the two routes of the project will reach 28 billion m^3, including the water supply to other provinces in different sub-regions, e.g., the Beijing and the Tianjin Municipalities, and Jiangsu Province. The actual amounts of water diverted to the 3H Region, however, will be much less than its virtual water export excluding the water diverted to other areas in the routes (Ma et al. 2006).

Second, those who are aware of the principles of virtual water and water footprint are encouraged to change their consumption behaviors for saving water in their daily lives. The people intuitively understand their true consumption of water resources, not only from their direct use of water resources in their communities but also from indirect use of virtual water from other areas. In this sense, recognizing China's role in the global economic system, a major manufacturing powerhouse, the global communities would appreciate China's contribution to water supply for manufacturing and exporting products, goods, and services overseas. China is a net virtual water exporter, which means that water resources consumed for export goods have been larger than those spared through import products (Yang et al. 2012; Tian et al. 2018).

The third contribution is associated with the issue of water use efficiency. According to Allan (2001), water managers, who aptly recognize the value of efficiency in water allocation, tend to relocate more water from less GDP contributing sectors, such as the agricultural sector, to more value-added industries in manufacturing or service sectors. Such a policy shift can occur based on the assumption that industrial and service sectors could realize a better degree of water use efficiency. Whereas water use efficiency in non-agricultural sectors is generally higher than that in the agricultural sectors, it is worth having a close investigation into China's circumstances.

The industrial restructuring for a tertiary-centered economy has been one of the most urgent policy priorities among top leaders in China, and the percentage of the first and second industries contributing to the annual GDP growth has decreased whereas the contribution of service industries has surged in recent few decades. For instance, the first industries contributed to 50.5% of the annual GDP growth of China in 1952 whereas the second industries, 20.8%, and the tertiary industries, 28.7%. In 2016, the profiles of each sector were changed dramatically, the first, 8.1%, the second, 39.6%, and the tertiary 52.4%. The tertiary industries' contribution to the annual GDP growth of the country has grown to 54.5% in 2020 whereas the first, 7.7% and the second, 37.8% (NBS 2021; Xinhua 2019). However, water use efficiency in industries does not show an impressive progress yet. Chinese industries consume two or three more times of water per unit of value produced compared with the average of world's developed economies, and China's water productivity was as low as US$5.3/$m^3$, which would be dwarfed by developed economies' US$35/$m^3$ on average in 2010 (Parton 2018).

Whereas large attention has been placed to China's export and import of virtual water at the global level, it is also significant to have a close look at inter-regional trade of virtual water within China. As discussed in the previous chapters, various regions of China have faced serious water scarcity triggered by a mismatch between available water resources and water demands in addition to its physical imbalance of water resources distribution as well as a wide margin of seasonal variability of precipitation. For instance, the arid North China has striven to meet the water demands of almost half of China's population with only less than 20% of regionally available water resources. More than 20 water diversion projects, including the South North Water Transfer Project, have been under development by the government (Cai et al. 2019).

However, water diversion projects accompany adverse impacts on the natural environment, such as degradation on ecosystems and negative impacts on local livelihoods and social well-being in water donor regions although a series of socioeconomic and environmental benefits are expected for water recipient regions. These are ecological rehabilitation as well as an increase of agricultural and industrial production and better livelihoods. In addition, China's national development programs to revitalize marginalized regions have seemed to have far-reaching effects on virtual water trade since the new millennium, e.g., the Great Western Development in 2001, the Revitalization of the Northeast in 2004, and the Rise of Central China in 2003, considering the close linkage between water use and socioeconomic development. For example, chemical industries and thermal power plants have been relocated from coastal areas to inland areas according to the national development programs. These industries consume vast amounts of water, and therefore, this relocation has spawned changes in inter-provincial virtual water flows between the water-rich southeast and the water-scarce northwest and has exacerbated water shortage in water donor regions (Cai et al. 2019).

Virtual water and water footprint proponents maintain that considering a variety of challenges entrenched in large-scale water diversion projects, a new water allocation policy based on the virtual water and water footprint approach would help rebalance the spatial mismatch between available water resources and water demands in China. A probable hypothesis is that the water-abundant southern regions should export water-intensive products to the water poor northern regions instead of North producing themselves. Also, more water allocated for southern regions, which would replace imports from North China, would be better in terms of water use efficiency and ecological protection. But this might entail the loss of comparative advantages for industrial production in South China (Feng et al. 2014; Cai et al. 2019; Ma et al. 2006).

Inter-provincial grain trade in China can play a significant role in saving water resources through the optimization of virtual water flow. For instance, in 2015, the total volume of virtual water flow in China's inter-provincial grain trade amounted to 79.2 billion m^3, occupying 20.5% of the total agricultural water consumption. Eventually, 31.4 billion m^3 of water resources were not used for grain production thanks to virtual water flows, which accounted for 39.7% of the total amount of virtual water flows. Despite less water availability, the reason why North China exports virtual water to South China is that the productivity of land in North is higher than that in South. Therefore, virtual water flows from arable land-rich and water-scarce regions to arable land poor and water-rich regions in China. It is worth pointing out that such a saving of water resources was possible thanks to water transfer from water poor regions to water-rich regions, which would put the northern regions in more jeopardy in terms of water resources availability (Wang et al. 2019).

In 2002, the total volume of virtual water within China reached 110.3 billion m^3/year, which accounted for 20.1% of the total water withdrawal in China. This level of virtual water flows increased to 209.5 billion m^3/year, 36% in 2007 and further grew to 248.1 billion m^3/year, 40.5% in 2012. In the period between 2002 and 2012, virtual water flows amplified by more than 124%, and the share of virtual water in relation to withdrawals increased. Besides, Northwest and Northeast China, less developed regions, scaled up virtual water export to East, North, and South coast regions. Especially, Xinjiang and Heilongjiang Provinces were the two major virtual water exporters, exporting 48.2% and 23.8% of the respective provincial water withdrawal in 2002. The shares surged to 64.6% and 78.3% in 2012, respectively. As major virtual water importers, Shandong, Zhejiang, and Guangdong Provinces were responsible for more than 27% of the total virtual water import in the same period, which replaced the Beijing Municipality and Gansu and Hainan Provinces in 2002 as major virtual water import regions.

Agriculture, electricity, and chemical sectors contributed mainly to virtual water flows in China in the same period. The agricultural sector was the largest contributor, accounting for about 70% of total virtual water flows, increasing from 68.2 billion m^3/year in 2002, 138.5 billion m^3/year in 2007, and 174.2 billion m^3/year in 2012. This expansion can be attributed to the increase in meat consumption. The electric sector of the major virtual water exporters contributed to more than 15% of the total virtual water export from the electricity sector, and virtual water export associated with the electricity sector was involved in Northwest and Northeast China, being destined for east and south coastal areas (Cai et al. 2019).

Virtual water flows have played an incrementally significant role in water consumption, socioeconomic development, and water allocation in China. Northwest and Northeast China became major virtual water exporters for the take-off of their economic development and meeting water demands in developed regions at the expense of invaluable water resources. Considering the growth of per capita GDP by more than

250% from 2002 to 2012, virtual water flows tremendously contributed to the scale-up of socioeconomic development in China. But such an outstanding result of China's modernization seems to be attributed to the sacrifice of virtual water exporting regions with scarce water resources (Cai et al. 2019).

Flows of scarce water footprint between regions in China are worth having attention to rather than virtual water flow only, which can demonstrate relative scarcity and environmental impacts of water flow. Scarce water footprint is referred to as scarce water consumption per unit of economic output for all economic sectors in all regions. In 2007, the water poor regions of Northwest China, including Xinjiang, Inner Mongolia, and Ningxia Provinces, showed the highest scores in terms of scarce water footprint. Guangdong, Fujian, Hunan, and Jiangxi Provinces generally had large virtual water flows but relatively small scarce water footprint thanks to their abundant water resources and compensating imports from water poor regions (Feng et al. 2014).

Increasing virtual water flows between regions would relocate the problem of water scarcity from the coastal to inland regions instead of alleviating water shortage in North China. Such a paradox in the virtual water flow from north to south manifests itself in some of highly developed areas with less water availability, such as Beijing, which imports almost half of virtual water from other provinces (Feng et al. 2014; Wang et al. 2019). For instance, the total amount of water footprint in Beijing had increased from 1990, 2000 to 2010 from 3.254 billion m^3, 6.392 billion m^3 to 7.707 billion m^3. The per capita water footprint per m^3 showed an upward trend from 483.32 m^3 in 1990 to 651.03 m^3 in 2000 and then a downward trend to 525.26 m^3. This occurs because of the development of water saving technologies in irrigation as well as water saving equipment in households. The annual mean total water resources for Beijing from 1956 to 2000 was 3.7 billion m^3, and based on the water footprint volume in 2000, 6.392 billion m^3, the capital depended upon approximately 42% of virtual water import from other regions (Hou et al. 2018).

A prerequisite for using virtual water and water footprint as a policy tool is to investigate virtual water flows between water poor and water-rich areas associated with water scarcity. Then, tailor-made water policy options for localities should be considered in tandem with the exploration of virtual water flows on a macro level within and out of China, such as minimizing the export of water-intensive products, e.g., agricultural products and processed food. In addition, water pricing for agricultural water consumption should consider the degree of water scarcity and the value of water in grain production alongside provincial disparities in water scarcity. The regions with large outflows of virtual water, such as Inner Mongolia, Henan, and Shaanxi Provinces, should take into serious consideration specific measures for enhancing water productivity and water use efficiency, including water saving technology and production practices. There is a big gap of irrigation water use efficiency between China and developed countries, since the irrigation water use coefficients of developed countries is 40–60% higher than China's. Simultaneously, it is wise to consider more allocation of limited water resources to secondary and tertiary industries for their socioeconomic development. At the national level, regional compensation mechanisms should be under serious consideration for balanced development as well as sustainable development (Cai et al. 2019; Wang et al. 2019).

Without serious policies for enhancing water use efficiency and water allocation, northern regions of China would be confronted with ecological disasters owing to soaring demands of virtual water from coastal and developed regions. Water and environmental policies at the national level should focus on the reduction of water consumption in water poor regions and prevent water rich from becoming water poor by keeping their water withdrawal to a sustainable level (Feng et al. 2014).

Conclusion

This chapter has evaluated a variety of means to sustainably use water resources with special reference to institutional methods, water pricing, water rights trading, and virtual water and water footprint. Prior to delving into the economic instruments to resolve water shortage, the study has examined the situation of water shortage, particularly focusing on North China. Discussions on water scarcity in North China confirm the severe water shortage in surface water resources and groundwater overexploitation. Another significant dimension in water shortage discourses is the unsustainable water exploitation of North China for food production favoring South China despite the serious water shortage. Together with many water infrastructures projects, including the South North Water Transfer Project, the central government has encouraged local governments to introduce various institutional methods for a sustainable use of water even prior to the introduction of the Three Red Lines in 2011.

Pertinent to water tariffs for urban water and wastewater services, an adequate level of pricing policy for water and wastewater services has incrementally been accepted for encouraging the general public to change their consumption behaviors. The 1988 and 2002 Water Law has contributed to clarifying what local governments can charge for urban water and wastewater services, and a recent policy change for water resource fee will further the water pricing reform. Beijing's example demonstrates that the severe water shortage can lead a local authority to raise urban water and wastewater service levels which are often not favored by the general public. Nevertheless, it still needs some time for other local governments to adequately levy water and wastewater service charges to urban users for considering the sustainability of water resources.

Whereas water rights trading was proposed more than a decade in China, the major development began in the early 2010s, and based on several pilot projects in the northwestern parts of China, the establishment of the China Water Exchange in 2016 became a watershed for the take-off of the water rights trading system in the country. Water rights trading has increasingly been active, particularly between agricultural water users in the same area although cross-boundary transactions would lead to more sustainable water use for agricultural, household, and industrial water users. There is an array of institutional barriers to hinder further development of water rights trading that should draw more attention of both central and local governments.

Discourses on virtual water and water footprint have unveiled the global dimension of China's water challenges. China has made substantial contributions to the global economy via manufacturing and exporting various products over the past few decades, and such a contribution has been possible at the exploitation of domestic water resources of the country. China's water managers have to consider the impacts of virtual water trade for water resources management policy, especially regarding virtual water export, and less water-intensive items are encouraged to be produced as an important step to save domestic water.

Inter-provincial virtual water transactions between North and South China have been explored, and North China seems to sacrifice itself for the socioeconomic well-being of South China by excessively using its invaluable water resources for producing and exporting food to people in South, which is endowed with more water. One of the long-term policy options for China's water managers would be to replace North China as major farming areas with South China which have better conditions for cropping including water availability. Such a policy shift may not be feasible, and more realistic approach should actively be considered and implemented, including a gradual transformation of cropping patterns and crop varieties that would be more resilient in less water conditions. This policy requires not only the development of resilient crop varieties through science and technology, but also relevant institutional support based on laws and regulations at the national and local levels. An equilibrium of water resources between North and South China would then be possible through the dual track of physical and non-physical infrastructure development.

References

Allan, John. 2001. *The Middle East Water Question.* London: I.B. Tauris.
Beijing Water Authority. Available Online: http://swj.beijing.gov.cn. Accessed 22 September 2020.
Cai, Beiming, Wei Zhang, Klaus Hubacek, Kuishuang Feng, Zhengliang Li, Yawen Liu, and Yu Liu. 2019. Drivers of Virtual Water Flows on Regional Water Scarcity in China. *Journal of Cleaner Production* 207: 1112–1122.
Chapagain, A., and A. Hoekstra. 2002. Virtual Water Trade: A Quantification of Virtual Water Flows Between Nations in Relation to International Crop Trade. Water Research Report Series. IHE, Delft.

China Water Exchange. 2020. Transaction Information (成交信息). Available Online: http://cwex.org.cn/lising/. Accessed 2 July 2020.

Global Water Partnership (GWP). 2015. China's Water Resources Management Challenge: The 'three red lines.' A Technical Focus Paper, Global Water Partnership.

Feng, Kuishuang, Klaus Hubacek, Stephan Pfister, Yang Yu, and Laixiang Sun. 2014. Virtual Scarce Water in China. *Environmental Science and Technology* 48: 7704–7713.

H2O China. 2020. Beijing's Water Price. Available Online: http://h2o-China.com. Accessed 25 September 2020.

Hao, Yu., Xinlei Hu, and Heyin Chen. 2019. On the Relationship Between Water Use and Economic Growth in China: New Evidence from Simultaneous Equation Model Analysis. *Journal of Cleaner Production* 235: 953–965.

Hoekstra, A., A. Chapagain, M. Aldaya, and M. Mekonnen. 2011. *The Water Footprint Assessment Manual: Setting the Global Standard*. London: Earthscan.

Hou, Siyu, Yu Liu, Xu Zhao, Martin Tilotson, Wei Guo, and Yiping Li. 2018. Blue and Green Water Footprint Assessment for China – A Multi-Region Input-Output Approach. *Sustainability* 10 (2822): 1–15.

Japan International Development Agency (JICA). 2006. *The Final Report on Water Rights Trading System Study Project in China: Volume 4* (水权制度建设研究项目 最终报告书 第4卷).

Jiang, Min. 2018. *Towards Tradable Water Rights: Water Law and Policy Reform in China*. Cham, Switzerland: Springer.

Jiang, Min, Michael Webber, Jon Barnett, Sarah Rogers, Ian Rutherfurd, and Mark Wang. 2020. Beyond Contradiction: The State and the Market in Contemporary Chinese Water Governance. *Geoforum* 108: 246–254.

Ma, Jing, Arjen Hoekstra, Hao Wang, Ashok Chapagain, and Dangxian Wang. 2006. Virtual Versus Real Water Transfers within China. *Philosophical Transactions of The Royal Society B Biological Sciences* 361: 835–842.

Ministry of Finance (MOF) and Ministry of Water Resources (MWR). 2017. Implementation Measure for the Expansion of Water Resources Tax Reform Pilot Projects (扩大水资源税改革试点实施办法), November 24.

Ministry of Water Resources (MWR). 2013. 中国水利统计年鉴 2013 *[China Water Statistical Yearbook 2013]*. Beijing: China Water Conservancy and Hydropower Press.

Ministry of Water Resources (MWR). 2016. The Provisional Measures on Administration of Water Rights Trading (水权交易管理暂行办法).

Ministry of Water Resources (MWR). 2018. 中国水利统计年鉴 *2018 [China Water Statistical Yearbook 2018]*. Beijing: China Water Power Press.

Ministry of Water Resources (MWR). 2020. 中国水资源公报 *2019* [China Water Resources Bulletin]. Beijing: China Water Power Press.

Moore, Scott, and Winston Yu. 2020. Environmental Politics and Policy Adaptation in China: The case of water sector reform. *Water Policy* 22: 850–866.

National Bureau of Statistics (NBS). 2021. Statistical Communique of the People's Republic of China on the 2020 National Economic and Social Development, February 28.

Pan, Anjun, and Duan Wei. 2019. Firmly Establish the Concept of Ecological Civilization and Accelerate the Construction of Harmonious Relationship Between Human-Water and Beautiful Beijing. Presented at the 2019 International Specialty Conference of the American Water Resources Association & Center for Water Resources Research, Chinese Academy of Science. Beijing, September 16.

Parton, Charlie. 2018. China's Looming Water Crisis. *Chinadialogue*, April.

Qian, Neng, Schuyler House, Alfred Wu, and Xun Wu. 2020. Public-Private Partnerships in the Water Sector in China: A Comparative Analysis. *International Journal of Water Resources Development* 36 (4): 631–650.

Reisner, Marc. 1993. *Cadillac Desert: The American West and Its Disappearing Water*. New York: Penguin Books.

Shen, Dajun, and Juan Wu. 2017. State of the Art Review: Water Pricing Reform in China. *International Journal of Water Resources Development* 33 (2): 198–232.

Shen, Manhong, Huiming Xie, and Yuwen Li. 2017. 中国水制度研究 [Chinese Water Policy Research], vols. 1 and 2. Beijing: Renmin Press.

Sheng, Jichuan, and Michael Webber. 2019. Governance Rescaling and Neoliberalisation of China's Water Governance: The Case of China's South-North Water Transfer Project. *EPA: Economy and Space* 51 (8): 1644–1664.

Singapore Public Utilities Board. 2020. Water Price. Available Online: https://www.pub.gov.sg/watersupply/waterprice. Accessed 1 February 2021.

Spooner, Simon. 2018. The Water Sector in China: Market Opportunities and Challenges for European Companies. The EU Horizon 2020 Report.

Stephenson, David. 2005. *Water Services Management*. London: IWA Publishing.

Tian, Xu, Joseph Sarkis, Yong Geng, Yiying Qian, Cuixia Gao, Raimund Bleischwitz, and Yue Xu. 2018. Evolution of China's Water Footprint and Virtual

Water Trade: A Global Trade Assessment. *Environment International* 121: 178–188.

Vairavamoorthy, Kalanithy, and Mohamed Mansoor. 2006. Demand Management in Developing Countries. In *Water Demand Management*, ed. David Butler and Fayyaz Ali Memon. London: IWA Publishing.

Wang, Jianhua, Yizi Shanga, Hao Wang, Yong Zhao, and Yin Yin. 2015. Beijing's Water Resources: Challenges and Solutions. *Journal of American Water Resources Association* 51 (3): 614–623.

Wang, Xin, and Shizhong Yang. 2018. Research on Status Quo of Water Rights Trading in China. *IOP Conference Series: Earth & Environmental Science* 171: 1–7.

Wang, Zongzhi, Lingling Zhang, Xueli Ding, and Zhifu Mi. 2019. Virtual Water Flow Pattern of Grain Trade and Its Benefits in China. *Journal of Cleaner Production* 223: 445–455.

Water Resources Institute. 2019. 基于水压力的水价, 水资源税与绿色发展 [Water-Stress Based Water Tariffs, Water Resources Tax and Green Development]. Beijing: Water Resources Institute (世界资源研究所).

World Bank. 2018. *Watershed: A New Era of Water Governance in China – Synthesis Report*. Washington, DC: World Bank.

World Bank. 2019. *Watershed: A New Ear of Water Governance in China – Thematic Report*. Washington, DC: World Bank.

World Trade Organization (WTO). 2018. *World Trade Statistical Review* 2018. World Trade Organization.

Wu, Fengshi, and Richard Edmonds. 2017. Chapter 7: Environmental Degradation. In *Critical Issues in Contemporary China*, 2nd ed., ed. Czeslaw Tubilewicz, 105–119. London: Routledge.

Xinhua. 2019. China's Industrial Structure Optimized in Past Seven Decades. *Xinhua*, July 2.

Xu, Yuanchao. 2018. China's Water Resource Tax Expansion. *China Water Risk*, February 14.

Yang, Hong, Zhuoying Zhang, and Minjun Shi. 2012. The Impact of China's Economic Growth on its Water Resources. In *Rebalancing and Sustaining Growth in China*, ed. Huw McKay and Ligang Song. Canberra: ANU Press.

Yu, Jialing, and Jian Wu. 2018. The Sustainability of Agricultural Development in China: The Agriculture-Environment Nexus. *Sustainability* 10 (1776): 1–17.

Yu, Jiangyuan, and Lu Hu. 2018. Reduction of Groundwater Resource Extraction at 9.28% in the 9 Pilot Project Provinces Related to the Water Resources Tax Reform during the First Half of 2018 (上半年九个水资

源税改革扩大试点省区超采区取用地下水量同比下降9.28%). *Xinhua News Net*, August 30.

Zhang, Hong, Gui Jin, and Yan Yu. 2018. Review of River Basin Water Resource Management in China. *Water* 10 (425): 1–14.

Zhang, Yu, Qingshan Yang, and Donghui Lu. 2015. A Case Study on a Quasi-Market Mechanism for Water Resources Allocation Using Laboratory Experiments: The South-to-North Water Transfer Project, China. *Water Policy* 17: 409–422.

Zhao, Jianshi. 2018. Water Trading: Shiyang River Basin Case. *China Water Risk*, September 18.

Zheng, Han. 2019. Water Rights Trade Platform. Presented at the 2019 International Specialty Conference of the American Water Resources Association & Center for Water Resources Research, Chinese Academy of Science. Beijing, September 17.

6

Water Quality Management

Introduction

The chapter discusses a myriad of water quality management issues in major rivers, lakes, and aquifers in China. Institutional dimensions are examined, focusing on the recent policy initiative to reduce water pollution, the Ecological Red Lines in 2016, and a list of relevant laws and regulations. China's surface and groundwater bodies have severely been contaminated over the past four decades. More than 200 million people in China use unsafe water sources, and since 1995, 11,000 water quality-related emergencies have occurred, including the instance of the temporary shutdown of the city water supply in Lanzhou due to contamination by benzene from a petrochemical facility. Public protests and unrest cases are reported regarding water pollution, mainly in rural communities (Han et al. 2016).

The conundrum of policy focus on water quality enhancement of China has been shifted from industrial wastewater control between the 1980s and the 1990s to household and agricultural sewage control since the beginning of the new millennium. Particular attention has been paid to an increase of agricultural wastewater discharge, Non-Point Source

(NPS) pollution, because it is often difficult to detect wastewater outlets in rural areas.

Based on the previous discussions on governance issues pertinent to water quality control in Chapter 4, this chapter pays closer attention to water quality enhancement policy in recent years. Premier Li Keqiang declared a 'war on pollution' in March 2014, and the central government of China has produced several policies, including the Ecological Red Lines in 2016 as part of the 13th FYP (2016–2020).

The chapter evaluates practices and experiences of the Environment Impact Assessment (EIA), the Three Synchronizations, and the Pollution Discharge Fee System that were implemented by Environmental Protection Bureaus (EPBs, now Ecology and Environment Bureaus, EEBs) at the provincial, municipal/prefectural, county, and township levels. Since the 'war on pollution' policy kicked off in 2014, the policy ambience on water and environmental issues has been changed, and putting environmental matters first, including the enhancement of water quality, gains more attention in Chinese society. Institutional rearrangements have been made in the field of water quality enhancement. In 2018, the Pollution Discharge Fee System was demolished and replaced with the Environmental Pollution Tax System based on the enactment of the Environmental Pollution Tax Law. This initiative signifies China's new policy direction to introduce green taxation schemes in environmental regulatory fields, including water pollution regulation.

Particular attention is placed on the recent development of the River (Lake) Chief System that was introduced in 2007. The newly revised Water Pollution Prevention and Control Law in 2018 officially recognizes the system, which has helped disseminate the successful experiences of Wuxi to other localities in China for combating water pollution. The River (Lake) Chief System demonstrates the central government's renewed commitment to law enforcement for non-compliance entities pertaining to water quality enhancement.

The Water Pollution Trading Scheme is one of the new economic instruments that is aimed at reducing water pollution in China. Although the first trading was made in Shanghai in the 1980s, it took more than two decades to recognize active transactions of water pollution trading between various entities. Nevertheless, the scheme is not necessarily working well compared with the water rights trading system due to several setbacks entrenched in the institutional and socioeconomic settings.

The blue-green algae outbreak of Tai Lake in 2007 is thoroughly examined for policy implications regarding water quality enhancement. Such an emblematic event has played an important role in reviewing water quality control policies and strategies not only in the Take Lake Basin as well as other rivers and lakes of the country. Several lessons learned from the case have been reflected in the government's policies in water quality control, including the improvement of Water Pollution Trading.

The chapter consists of four parts. The first section of the chapter sketches the general framework of water quality management with a special focus on the Ecological Red Lines in 2016 and the conventional three approaches, the EIA, the Three Synchronizations, and the Pollution Discharge Fee System. Second, the study sheds light on the River (Lake) Chief System, evaluating the development, achievements, and challenges of the system. Third, attention is paid to the Water Pollution Trading System, and the effectiveness of the scheme is discussed investigating the background and development, achievements and challenges. The last part is dedicated to the analysis of the Tai Lake blue-green algae outbreak in 2007 with its policy implications.

General Framework

Water pollution control policy of China is based on an array of national-level laws and subsequent local-level regulations. The Environmental Protection Law that was enacted in 1989 first and revised in 2015 provides the basic foundation for regulatory frameworks for the environment, including the water sector, particularly water quality improvement. The most conspicuous feature in the 2015 version is the prioritization of the environment rather than putting the environment in parallel with socioeconomic development which is clearly stated in the 1989 version. Premier Li Keqiang's pledge to wage war on pollution in March 2014 was introduced a month before the final draft of the 2015 version, which represents the government's firm commitment to environment and ecological issues onwards (Lau 2014).

The 1989 version included 47 articles, and the 2015 version has been expanded into 70 articles which updates environmental planning, standards, and monitoring. Highlights of the new version in 2015 encompass the increased penalties or fines on a daily basis, the improvement of pollution permission management system, the enhancement of the Environment Impact Assessment (EIA) system, and the clear rules on public interest litigation against polluting activities, indicating which social groups can file lawsuits with courts against environmentally damaging acts.

In addition to these, the revised law encompasses public information disclosure and transparency on information, clean production and the recycling of resources, and the integrated prevention and control. The integrated prevention and control system includes the establishment of key areas across various administrative regions and prevents air and water pollution and ecological disruption and carries out unified planning, standards, monitoring, and prevention measures (Dai 2019; Lau 2014).

Relevant national-level laws include the Water Pollution Prevention and Control Law, which was introduced first time in 1984 and revised in 1996, 2008, and 2014, and the Water Law that was enacted first in 1988 and revised in 2002. There are other laws and administrative regulations pertaining to water pollution at the central level (discussed in more details in Chapter 4).

At the local level, innumerable local laws, regulations, and standards for water pollution prevention and control have been enacted, particularly in accordance with the Water Pollution Prevention and Control Law. For instance, the Jiangsu Provincial Government created the Regulation on Tai Lake Water Pollution Prevention and Control in Jiangsu Province in 1996 and amended it in 2007, which is regarded as the most significant legal dossier tackling water pollution control of Tai Lake in the province. This regulation explicitly stipulates the obligations of provincial and municipal governments to control water pollution in the Tai Lake Basin based on relevant plans and to establish Total Pollution Load Control (TPLC) plans at different levels. By doing so, these plans should be in harmony with Tai Lake water pollution control targets of the Jiangsu Provincial Government (Zhang et al. 2012).

Together with the legal framework, it is significant to have a firm and systematic planning of water pollution prevention and control in China. Alongside the socioeconomic Five Year Plans, the Chinese central government has introduced the Five Year Environmental Plans (FYEPs). These plans are divided into sector-specific FYEPs, which embrace water resources management in key rivers and lakes, air pollution reduction in particularly designated regions, hazardous waste management, and nature conservation. Similar to any target-oriented policies imposed by the central government in China, the FYEPs encompass a top-down and hierarchical establishment of environmental protection targets and related works in every five years and determine the allocation of tasks to provincial, municipal, and prefectural governments. Local FYEPs are drawn up reflecting local circumstances down to the county levels (Zhang et al. 2012). The following section delves into the latest FYEP, the 13th FYEP, which is dubbed as 'Ecological Red Lines'.

Ecological Red Lines 2016

China's commitment to achieving the Sustainable Development Goals has been reflected in the 13th Five Year Plan (2016–2020), which is aimed at promoting a cleaner and greener economy with a special emphasis on environmental management and protection, ecological protection and security, and the development of green industries. The 13th FYP for Ecological and Environmental Protection, also called, Ecological Red Lines, aims to achieve the three main goals. First, the plan includes the vital ecosystem service areas, i.e., eco-function hotspots and the provision of basic water and environmental services including clean and safe drinking water, improved water storage, and carbon sequestration. In addition, ecological well-being should adequately be maintained for supporting national and local socioeconomic development.

Second, the government will safeguard ecologically fragile areas including eco-fragile hotspots, administer land degradation, desertification, and soil erosion, and take good care of the safety of human habitats with special reference to living environments at the local level. Third, the government will protect living environments and habitats, particularly for important species and maintain biodiversity on both national and global scales.

The plan has its full name, Ecological Function Protection Red Lines, and defines the bottom red lines for ecological safety, public safety, and sustainable development. In 2000, the government rolled out the National Outline for Ecology and Environment focusing on the significance of ecological functional zones, resource development districts, and good ecological situation districts and emphasized the need to establish the bottom red lines for ecology and environment. The State Council introduced the Opinion on Strengthening Important Works of Environmental Protection in 2011 and the concept of 'Ecological Red Lines' stressing the salience of ecological functional zones, terrestrial and marine ecology, and environment-sensitive zones, and environmentally damaged zones for related projects. More developed and mature policies and strategies were suggested in the 13th FYP (Shen et al. 2017).

A total of 15 provincial Ecological Red Lines were approved by the central government in 2018, including the Beijing and Tianjin Municipalities, and Hebei Province, Ningxia Hui Autonomous Region, and the 11 regions along the Yangtze River Economic Belt (YREB). These regions have a combined area of 610,000 km^2 and accommodate diverse nature reserves, scenic areas, forest parks, geological parks, and wetlands. Whist the red line zones allow human activities, those activities will be subject to strict regulation (ADB 2018).

There are the three core environmental sectors in the plan, air, water, and soil. The approach of green development is emphasized as a practical tool to pursue sustainable development by creating a resource-conserving, environmentally friendly society and introducing a brand-new type of modernization. Amongst environmental elements, water is a key element of the plan, and there are the six goals for enhancing China's water quality from 2016 to 2020.

(1) To establish unit-based management of water quality.
(2) To establish holistic and basin-wide strategies to tackle pollution.
(3) To prioritize protection of good quality water bodies.

(4) To tackle groundwater pollution with holistic strategies.
(5) To strongly enhance polluted urban water bodies.
(6) To improve water quality of river mouth and nearshore areas.

More specifically, more than 70% of China's surface water must reach Grade III or equivalent by 2020. In 2015, 66% of China's surface water met Grade III. In addition, the government set the target that the proportion of groundwater in the 'very bad' category is to drop from 15.7% in 2013 to approximately 15% by 2020. More detailed targets for water quality enhancement are shown in Table 6.1 (China Water Risk 2016; Dai 2019; State Council 2016).

In May 2020, Huang Runqui, the Minister for Ecology and Environment informed that ecological and environmental goals for the 13th FYP had neatly been achieved, and seven out of nine binding targets for assessing the protection work had been accomplished, including major

Table 6.1 Water quality enhancement targets for the 13th Five Year Plan (FYP) (2016–2020)

Sector	Indicator	12th FYP Targets 2010–2015	13th FYP Targets 2015	13th FYP Targets 2020
Water	% of surface water at Grade II or above[a]	55 to >60	66	>70
	% of surface water at Grade V	17.7 to <15	9.7	<5
	Water quality compliance rate of key water body function areas	N/A	70.8	>80
	% of groundwater in 'very bad' category	N/A	15.7[b]	About 15
	% of nearshore water areas of good quality (Grade I or II)	N/A	70.5	About 70

Remarks [a]State-controlled sections only (increased from 972 to 1940 during the 12th FYP), [b]indicates 2013 data
Source China Water Risk (2016) and State Council (2016)

rivers' water quality. He stressed that the proportion of major rivers showing a good water quality surged by 4.8% in 2019 (Xinhua 2020).

Conventional Programs

The Chinese environmental authorities are equipped with several policy instruments for water pollution prevention and control in various water bodies, including rivers, freshwater lakes, and groundwater resources. The two key environmental legislations, the Environmental Protection Law and the Water Pollution Prevention and Control Law, have been the legal foundation to introduce a myriad of water pollution control policy schemes: (1) emission, discharge and environmental quality standards; (2) Environmental Impact Assessment (EIA); (3) the Three Synchronizations; (4) Pollution Discharge Fee; and (5) the Pollutant Discharge Permit System (OECD 2006; Zhang et al. 2012).

Various environmental standards serve as a fundamental component with which environmental regulators monitor and assess environmental performance of potential water polluting units, which are particularly crucial for local Ecology and Environment Bureaus (EEBs) officials addressing water pollution challenges on a daily basis. An EIA exercise should be undertaken for a new project or the expansion of an existing facility which can give adverse impacts on an aquatic environment. Fundamental and basic guidelines and rationales of the law were presented in the Environment Protection Law in 1989, which was modified for enhancing implementing capacities of environmental agencies in 2015 (Dai 2019).

Another essential instrument of local EEBs is the Three Synchronizations in an EIA process, which indicates that the design, construction, and operation of a new industrial units (or an existing manufacturing unit that is expanding or altering its production processes) should be synchronized with the design, construction, and operation of adequate waste (water) treatment facilities. Between 1996 and 2009, approximately 90% of 766,368 projects were given the permit of the Three Synchronizations, and the implementation rate of projects EIA grew from 61% in 1991 to 99.8% in 2009. These two schemes demonstrate preventive measures of Chinese environmental authorities against water pollution (He et al. 2012; Ma and Ortolano 2000; Lee 2006; Zhang et al. 2012).

Nevertheless, the effectiveness of the Three Synchronizations has been under question, and there was a joke about the system, which may stand for 'eating, drinking and singing' in restaurants and karaoke bars. Lax law enforcement in water quality control of China is a critically weak point, which originates from the paucity of adequate legal and political accountability (Tilt 2015).

The schemes of Pollution Discharge Fee and the Pollution Discharge Permit System advocate the polluter-pays principle in which polluters should be responsible for any damage or losses caused by their own polluting behaviors. Water polluting units should pay pollution discharge fees for their wastewater discharge, and these discharges need to report their water pollutant discharge status to concerned EEBs for verifying the status and obtaining the discharge permits.

These units' discharge amount should be less than the threshold of authorized permits, and the concentration of sewage in emissions should comply with discharge standards. The system was first experimented in Shanghai and other 16 cities involving water as well as air pollutants from 1987 to 1998 and practiced nationwide. The Pollution Discharge Permit System is part of the government's attempt to introduce economic instruments or market-based mechanisms in addition to the command-and-control measures for environmental regulatory works. But the Pollution Discharge Fee System has been replaced with the Environmental Protection Tax Law in 2018 (Chan 2018; He et al. 2012; Lee 2006; Zhang 2018; Zhang et al. 2012).

The enactment of the Environmental Protection Tax Law has paved the way for the country's 'war on pollution' to be implemented in more effective fashion. Alongside vast scales of investment for pollution abatement, RMB 255.5 billion (US$39.3 billion) in 2018, the Environmental Protection Tax Law was applied to approximately 260,000 enterprises at the national level. Regulatory law enforcement was strengthened for non-compliant commercial and manufacturing units, and as a result, in the first ten months of 2018, 30,000 violations were reported, about RMB

11.8 billion (US$1.8 billion) fines levied and 6,500 violators detained (CCICED 2019).

Despite good progress in water pollution control for more than three decades, He et al. (2012) specify a series of problems in environmental regulatory works, i.e., partial results, little capacity of the government for strategic planning, fragmented administrative works in policymaking and implementation rather than integrated approaches, high costs, no or little attention to NPS pollution, and implementation and supervision failures.

Environmental authorities in local governments increasingly recognize the magnitude of NPS pollution over the past decades, however, rapidly developing urban centers in China have become the home of foreign polluting or toxic industries, which represents the movement of environmental pollution from developed to developing countries, including China. For instance, the city of Guiyu in the Guangdong Province received massive amounts of electronic wastes and became a main center of electronic waste recycling in the mid-2000s. Such a phenomenon has triggered chain effects, causing water, soil and air pollution, and in particular, the quality of freshwater bodies has deteriorated. This situation poses a grave threat to public health due to contamination in drinking water (He et al. 2012; Wu and Edmonds 2017).

The conventional five regulatory programs are the typical examples of China's command-and-control environmental policies with a mixture of market-based approaches related to water quality enhancement. China's recent environmental regulatory frameworks have been changed, which encompass not only a new type of command-and-control program, i.e., the River (Lake) Chief System but also a market-based instrument, Water Pollution Trading. There have been more regulatory measures reflecting environmental economic principles since 2002, e.g., the eco-compensation program conducted in water function zoning and water basin environmental protection. The River (Lake) Chief System and the Water Pollution Trading Scheme better encapsulate recent efforts of the government to undertake surgical solutions for water quality enhancement via a conventional command-and-control approach (river chief system) and a market principle-based instrument (Water Pollution Trading Scheme).

River (Lake) Chief System

The River (Lake) Chief System is the recent manifestation of the central government of China for putting more emphasis on 'law enforcement' in environmental management for non-compliant entities. The system is a byproduct of local governments' efforts on how to enhance the water quality of local rivers and lakes amid socioeconomic development. Local party leaders are given more incentives to properly undertake water resources management in their jurisdiction and to strike the balance between economic development and environmental protection.

The system came into being in the course of addressing the cyanobacteria (blue-green algae) pollution incident in Tai Lake, 2007, which threatened the quality of drinking water for five million people in Wuxi City, Jiangsu Province. The Wuxi City Government devised a brand-new system which should guarantee a good degree of the implementation of water quality protection guidelines and standards in 2007, the River (Lake) Chief System. A total of 64 river chiefs were appointed for enhancing the water quality of 64 river sections within the city. The monitoring results of the water quality of the river sections under the system in one year demonstrated remarkable improvement so that the Wuxi City Government decided to adopt the river chief system for the entire river and lake areas of the city in 2008. At the same time, the Jiangsu Provincial Government recognized the effectiveness of the river chief system and became determined to adopt the system for the whole province. Furthermore, the functional field of the river chief system was expanded from water quality to flood prevention and protection, and ecosystem protection in September 2012 (Fang et al. 2018; Smith 2018; Xu 2017).

An important backdrop of the system is something to do with the recognition of blurry accountability embedded in the institutional framework, the Tiao-Kuai, between the center and water-related local bureaus, and between water-related local bureaus at different levels, which should be resolved by a competent high-ranking official that can play a leading role in tackling river and freshwater lake management issues in an integrated way.

In addition, the revised Water Pollution Control Law in 2008 emphasized the establishment of clear responsibilities and assessment at the local level, which called for more direct commitment of local governments and their leaders to water and environmental issues. The river chief system can be regarded as a new platform to consolidate local governments-led water quality regulatory policies and actions (Hao 2017). Whilst the system can be applicable to all the fields of water resources management, an emphasis is placed on water quality improvement.

On 11 December 2016, the central government of China issued the document to implement the River (Lake) Chief System at the national level, called 'the Opinions on Fully Promoting the River Chief Mechanism'. The system has officially been included in Article 5, the revised Water Pollution Control Law in June 2017. Article 5 reads that river (lake) chiefs at the provincial, municipal/prefectural, county, and township levels should be in charge of water resources protection, waterfront and bank areas management, water pollution control, and water environment management (Fang et al. 2018; Hao 2017; Xu 2017; Zhan 2019).

Since the introduction of the River (Lake) Chief System in 2007, Zhejiang, Kunming, Liaoning, Shanghai, and other local areas have adopted the system. A pilot section of a heavily polluted river in Nanjing City was adequately treated after the river chief system was introduced, and river inspections were undertaken three times a week in Qingdao City that asks river chiefs to take full responsibility for their own jurisdictional areas and would lose their jobs if river chiefs failed their missions. In September 2019, more than a million river and lake chiefs have been appointed and active (Liu 2019; Xu 2017).

The system encapsulates a strict, authoritative, and top-down approach that calls for the direct and serious engagement of top-ranking party cadres or government officials at the local level in controlling water pollution activities, especially by chemical and manufacturing industries (Fang et al. 2018). River (lake) chiefs are appointed among the CCP top leaders, vice mayor, or governor of localities so that their involvement can facilitate the implementation of relevant laws and regulations. The most important element of the system is the assessment exercise, which is regarded as a key aspect of the mechanism. River (lake) chiefs assume lifetime accountability for environmental performance in their jurisdiction.

This is called the Target Responsibility System in which the head office of the CCP in Beijing sets its overall targets to provincial governments, which reallocate their quotas to municipal/prefectural, county, and township levels. Such opportunities would be given to CCP leaders at local levels who excel in performance of various conventional areas, including economic development, birth control, and tax collection, and a new field has been added, environmental pollution, since the 12th FYP (2011–2015) (Dai 2015; Minzner 2011). The River (Lake) Chief System is well-positioned at an appropriate time for helping raise the profile of water pollution abatement issues on top of policy priorities.

As a specific example, the Huishan District of Wuxi City requires each river chief to place a deposit of RMB 3,000 (about US$430) at the beginning of each year and evaluates their performance at three levels based on diverse assessment criteria: (1) water quality improved; (2) water quality the same as before; and (3) water quality deteriorated. The deposit can be returned to those who have shown good performance with additional 100% financial incentive, to those who have shown a decent level of performance with a return of the deposit, and to those who have poorly performed without the return of the deposit as a penalty (Dai 2015; Fang et al. 2018).

Dai (2015) stresses that such an incentive policy does not necessarily encourage river (lake) chiefs to work harder for meeting the targets in water quality enhancement. Considering the annual per capita disposable income of Wuxi City in 2019, RMB 54,847 (US$7,834), the amount of RMB 3,000 (US$430) does not seem to be an attractive incentive for the chiefs to be engaged in arduous works related to water pollution control (CEIC 2020). More important motivation can be possible opportunities CCP leaders can attain, such as promotion in rank or position, additional wages or bonus payments, or other benefits (free transport, entertainment, training and travel, subsidized housing, and health care).

The conventional assessment of local government officials based on only GDP growth has been replaced with the assessment result of those through their performance of environmental regulation (Liu 2019; Xu 2017; Zhan 2019). There are four levels in the river chief system following the hierarchy of the administrative system of China, from top to bottom: (1) province; (2) municipality/prefecture; (3) county; and (4) township (Fang et al. 2018).

The Lake Chief System is similar to the river chief system. For instance, Hubei Province boasts innumerable freshwater lakes with its nickname of 'the province of 1,000 freshwater lakes', and in fact, there are 755 lakes occupying the total area of 2,706 km^2. Wuhan City, Jiangsu Province in 2012 began to implement the Lake Chief System whereas the provincial government introduced relevant laws and regulations. In 2014, the system was officially adopted as an important regulatory instrument of the province to enhance the water quality of freshwater lakes. The central government officially recognized the significance of the Lake Chief System by issuing the Guiding Opinion of Implementing the Lake Chief System in Freshwater Lakes in January 2018 (Fang et al. 2018).

Figure 6.1 delineates the hierarchy of the River (Lake) Chief System at the provincial, municipal/prefectural, county, and township levels. At the top of the hierarchy, the General River Chief is in charge of the

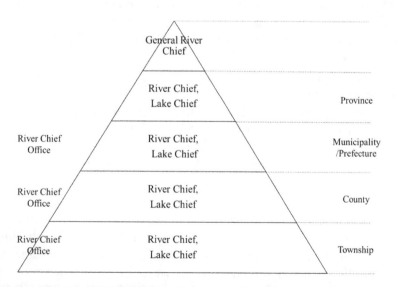

Fig. 6.1 Organization of the river and lake system (*Source* Modified based on Liu [2019])

overall management of rivers and lakes and is appointed between CCP leaders or high-ranking government officials at the provincial, municipal/prefectural, or county levels. River (Lake) Chiefs are appointed among CCP leaders or local government officials with the direct responsibility of managing local rivers and lakes. The River (Lake) Chief Office consists of water resources bureaus and other relevant bureaus tackling water quality management issues (Liu 2019; Xu 2017).

Merits of the River (Lake) Chief System are threefold. First, in 2015, the State Council issued the Water Pollution Control Action Plan or 'the Water Ten Plan', which specified a variety of duties and responsibilities for water pollution regulatory authorities at the local level, and the River (Lake) Chief System furthered the implementation of the plan. The system appears to have contributed to promoting and accelerating the Water Ten Plan in 2015 for abating water pollution in localities and can be regarded as a new instrument of environmental regulation (Fang et al. 2018).

Second, the system requires top-ranking party cadres to be directly involved in water resources protection, water pollution abatement, and water environment enhancement. Such a practice demonstrates the CCP's tactic of using the party hierarchy as a leverage to enhance coordination between various government bureaus and agencies (Smith 2018). The Wuxi City's deposit mechanism is noted.

Furthermore, the system clearly defines the roles and responsibilities of river (lake) chiefs, which guarantees what they have to do, a good degree of policy implementation, and positive outcomes of their performance. Consequently, local water environments would further be enhanced thanks to such an administrative experiment (Hao 2017).

Third, there has been no major adjustment of responsibility or mandate between ministries or bureaus related to water resources management in the introduction of the river chief system. Yet, top-ranking party cadres in localities are in charge of river (lake) management for water pollution abatement in the system, which has naturally paved the way for the facilitation of an integrated approach to water pollution control policies and projects. The River (Lake) Chief System serves as a brand-new platform to put water-related bureaus and institutions together and to make relevant policies and projects implemented in an integrated way.

The ineffectiveness of water resources management at the local level has been ameliorated thanks to the introduction of the river chief system. The system has culminated in putting water-related bureaus and agencies together under the strong political leadership of the CCP. For instance, Zhejiang Province succeeded in establishing a new coordinating mechanism under the river chief system to tackle water resources, water ecosystems, and water safety issues in 2013. Whereas the environmental protection bureau was leading for the implementation of the river chief system in the fields of water pollution abatement, the water bureau was in charge of flood prevention and protection, and water leakage and water saving issues were dealt with by the construction bureau. The water management leading small group office played a mediating role in coordinating water affairs between the bureaus. By analogy, there are 23 related bureaus and agencies involved for the river chief system at the provincial level in Jiangxi Province (Fang et al. 2018; Hao 2017).

Fourth, the system has facilitated the protection and management of river and lake front areas where diverse socioeconomic activities are taking place and ecosystems are often vulnerable to such activities, requiring more attention of environmental regulatory authorities. For instance, the Ministry of Water Resources, the Ministry of Transport and Logistics, and the Ministry of Land and Resources jointly issued the Comprehensive Plan for the Protection and Development of the River Front Areas in the Yangtze River. Better river or lake front area planning, and law enforcement would be expected since the introduction of the River (Lake) Chief System because of the integrated nature as well as strong emphasis of implementation capacity and law enforcement.

In the river chief system, local governments at different levels have come to communicate and collaborate better, which often results in the improvement of overall planning, regulatory actions, and law enforcement against polluting units. The strongly hierarchical river chief system from a province to a township helps different levels of local water-related bureaus communicate and collaborate well, because each level's river chief, mainly local CCP leader, is politically powerful and influential at each local government. Therefore, integrated approaches to river or lake basin management can be introduced by breaking down the administrative barriers between local governments.

In addition, inter-provincial collaborations have occurred thanks to the introduction of the system. Such an activity has been supported by 'the Improvement of the Collaborative Mechanism in Water Resources Protection and Water Pollution Control in 2011' and 'the Important River Basin Water Pollution Control Plan (2011–2015)'. Through these policies, the central government has encouraged inter-provincial collaborative works, and an example is found in the Song-Liao River Basin Protection Leading Small Group Collaborative Mechanism, the Yellow River Basin Water Pollution Accident Communication Small Group Collaborative Mechanism, and the Middle Route of the South North Water Transfer Project Water Resources Protection and Water Pollution Control Collaborative Meeting Mechanism (Dai 2019; Hao 2017; Xu 2017).

Fifth, a total of nine ministries, spearheaded by the MWR, at the center showed the firm commitment to implementing the river (lake) chief system in the most effective way in the 12th Five Year Plan (2011–2015). The Opinions on Fully Promoting the River Chief Mechanism in 2016 pointed out that the range of mandate given to the system encompasses water resources protection, water pollution control, and water environment management. This document explicitly shows the linkage with the Three Red Lines in 2011, which strongly emphasizes the urgent need to optimize water use for future needs confronted with anthropogenic as well as natural risks and uncertainties related to water supply and demand (Fang et al. 2018).

Sixth, numerous local governments have encouraged civil associations to take an active part in promoting the system. At the central level, the revised Environmental Protection Law in 2015 added the section of information disclosure and public participation. The 2015 law specifies the way ordinary people, legal person, and other civil associations acquire environmental information and participate in environmental protection inspection (Zhan 2019).

The Jiangsu Provincial Government selected six rivers and asked six environmental protection civil organizations to monitor comprehensive environmental management works for the rivers. These civil organizations implemented site-visits, public survey, filming, public announcement, and environmental education. In August 2016, the Jiangsu Provincial Government appointed a civil river chief related to Caoqiao River, which demonstrates the local government's efforts to establish alliance with civil society for consolidating sociopolitical legitimacy in fighting against non-complying units for water pollution. Smith (2018) also informs that there was a case of a foreign river chief appointed in December 2017, Carlos Brito, the Brazilian CEO of Anheuser Busch InBev, the world's largest brewer of beer. AB InBev manages the inspection and monitors several rivers in China (Fang et al. 2018; Smith 2018; Zhan 2019).

There is criticism about the river (lake) system. First, the system heavily depends upon the strong hierarchical system of the CCP, which imposes incentives and punishment upon local party cadres according to their performance in river or lake basin management. This initiative embraces two edges of a sword. If a local party cadre were capable and dedicated to such works with enthusiasm, relevant responsibilities and duties would neatly be fulfilled. Such a rule of man rather than a rule of law would entail worsening results of water regulatory policies and activities if the person was not available for the job anymore. Another potential weak point of the system is that party cadres might be replaced in a short period of time because party cadres often stay in a jurisdiction for two to three years. This short period of time of duties can hamper long-term sustainable development of their assigned areas and jeopardize policy coordination and implementation (Smith 2018; Tan 2014).

Second, stringent assessment mechanisms in the system would encourage river (lake) chiefs to endeavor to produce better performance and would work well only if such assessment mechanisms were adequately designed reflecting various environmental, socioeconomic, and political conditions and circumstances at the local level. For instance, every locality has its own distinctive water environment features so that assessment criteria or perimeters should be different, and various demands in water environmental protection exist in water resources protection, tourist, agricultural and livestock farm, and urban districts. In addition, diverse river (lake) chiefs often have different levels of capacity to tackle water pollution or manage water ecosystems, and there is a gap between municipalities/prefectures, counties, and townships in terms of socioeconomic development, which requires river (lake) chiefs to adjust their regulatory policies and law enforcement activities (Zhan 2019).

Third, the assessment mechanism of the system can become a redundant administrative burden in the water environmental assessment system of local governments. Taking an example of the Tai Lake Basin Environment Protection Responsibility Assessment Mechanism. The mechanism requires several assessment exercises. These include the Urban Environmental Comprehensive Management Quantification Assessment, the Total Pollution Reduction Assessment, the Jiangsu Province Tai Lake Management Work Inspection Assessment, the Water Resources Protection and Water Pollution Management Assessment, the Jiangsu Province Basic Modernization Index System Assessment, the Jiangsu Province Ecological Civilization Construction Assessment, and the Strictest Water Resources Management Policy Assessment. The existent assessment criteria is often overlapped, and measuring or calculation methods related to various assessment exercises are inconsistent. Water environmental statistics can be incorrect and are not able to reflect real situations, which may result in providing inadequate regulatory activities, thereby leaving damaged ecosystems intact. This is the reason why opponents question the usefulness of the system (Zhan 2019).

Fourth, there is a gap between rich and poor local areas in terms of the success of the system, and for instance, the case of Qinghai Province illustrates that the effectiveness of the system is in question since the province in general lacks sufficient amounts of budget or enough staff members for extra works or responsibilities. Although the system sounds promising, a low degree of implementation capacity can undermine the viability of the system (Zhang 2019).

Fifth, little consideration is given to the promotion of public participation in the system. Dai (2019) evaluates that decent progress has been made in recent years, but the system shows limited participation. This consequence is attributed to the result-oriented responsibility system which is devised solely for governmental internal hierarchical control rather than public accountability. China's authoritarian system prevents the participation of non-governmental organizations in water resources management decision making by design, and major water users in Chinese society are collective agricultural units, state-owned enterprises, and municipally owned utilities so that there is little room left for public participation (Moore and Yu 2020; World Bank 2019). It seems that the top-down nature of China's water quality control remains firmly entrenched in the system.

Water Pollution Trading

China has endeavored to reduce its water pollution to an acceptable level since the 1980s, and there are two primary tools to tackle the problem: (1) environmental regulatory programs with a top-down nature, including the Environmental Impact Assessment (EIA), the Three Synchronizations, the Pollution Discharge Fee, and the Pollution Permit System; and (2) market-based instruments, such as Water Pollution Trading Scheme. The trading scheme is hailed as a useful market-based instrument for controlling the total amount of water pollution and for lowering emissions with lowest possible economic cost. In the 9th FYP (1996–2000), the Chinese government introduced the Total Pollution Load Control (TPLC) policy, which worked as a catalyst to draw attention of water and environmental communities in China for the possible establishment of water polluting trading schemes.

Shanghai's Minhang District was the first place to introduce the early type of Water Pollution Trading Scheme in 1985, and in 2001, Jiaxing City in Zhejiang Province established the first rules on the Water Pollution Trading Scheme, called the Provisional Measures on the Total Water Pollution Load Control and Water Pollution Trading. In 2007, the Ministry of Finance, the Ministry of Environment Protection, and the National Development and Reform Commission selected 11 local areas for WPT pilot projects, including the Tai Lake Basin. The government decided to apply the scheme at the national level in 2014 based on the Guiding Opinion on Further Piloting the Paid use of Trading Emission Permits. In 2016, more detailed policy measures were issued in the Implementation Scheme for Pollutant Emission Permit Control, which

6 Water Quality Management

permits the establishment of markets in tradable pollutant emission permits in the pilot areas.

The significance of the trading scheme lies in paving the way for regulated companies to sell and buy emission allowances, which makes the marginal costs of emission reductions equal between facilities. As a result, such transfers can entail an overall reduction of emissions from many companies and decrease the total costs of emissions control (Moore and Yu 2020; Shen et al. 2017; Zhang et al. 2012).

Figure 6.2 describes the workflow of the Water Pollution Trading Scheme of China, which does not necessarily reflect the current stage of emission trading. Whilst most of the trading has occurred between public sector organizations, i.e., water pollution right saving center, or between public sector organizations and companies so far, the workflow demonstrates the long-term development of the trading scheme, particularly between private players. In terms of the administrative hierarchy, Shen et al. (2017) maintain that China needs the three-layered market of Water Pollution Trading: (1) national level; (2) provincial level; and (3)

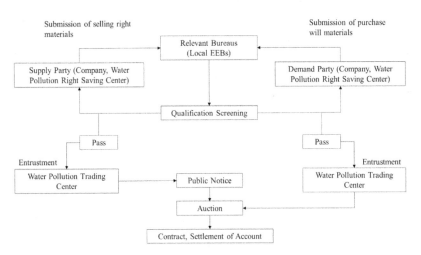

Fig. 6.2 Workflow of the water pollution trading scheme in China. *Remarks* EEBs mean Ecology and Environment Bureaus (*Source* Modified based on Shen et al. [2017])

city level. The Water Pollution Trading Center does not directly participate in the process but serves as a trading platform dealing with the enrollment of pollution right and settling account.

The WPT is a newly coined environmental policy and has been utilized in China on a limited scale owing to a myriad of institutional barriers. These are conflicts of different policies and contradictory issues between the WPT and other environmental regulations. It is necessary to identify regulatory loopholes or problematic issues that can be tackled through the introduction of the WPT. In order to avoid unnecessary administrative confusions or deficiencies, it is imperative to consider the WPT in the context of implementing existing environmental regulatory programs (Zhang et al. 2012).

The introduction of the WPT can drive manufacturing companies in China to be in an awkward position for keeping their existing polluting rights which are permitted by present method and regulating standards. OECD (2006) and Zhang et al. (2012) warned that several environmental regulatory instruments could trigger extra administrative costs than a single or simpler instrument.

Figure 6.3 describes the structure of the Water Pollution Control System in the Tai Lake Basin including the implementation of Water Pollution Trading program. Although it depends on local circumstances, the overall framework of water pollution control in China works in a similar fashion (Zhang et al. 2012).

One of the cases for Water Pollution Trading is found in the Tai Lake Basin back in late 2007. The large-scale blue-green algae incident in the lake in late May 2007 alarmed the Jiangsu Provincial Government as well as the central government, which was a turning point for environmental regulatory authorities to introduce a more innovative instrument for improving water pollution control. Regarding the WPT program, it is worthwhile to pay close attention to previous market-based instruments for water pollution control, e.g., the Total Pollution Load Control (TPLC). In 1994, the Chinese government introduced the TPLC and officially recognized the policy in the amended Water Pollution Prevention and Control Law in 1996. The total load of key water pollutants, i.e., COD pollutants and ammonia nitrogen, began to be determined by the central and local governments in the 10th FYP (2001–2005).

6 Water Quality Management

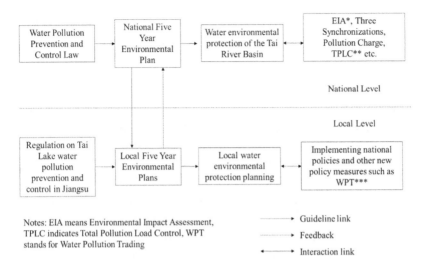

Fig. 6.3 Structure of the water pollution control system in the Tai Lake Basin (*Source* Modified based on Zhang et al. [2012])

The central government decides the TLPC quota (the maximum load of a specific water pollutant), which was distributed to all provinces according to a specific allocation principle. In a hierarchical fashion, provincial governments allocated quotas to their municipal/prefectural governments, and municipal/prefectural governments assigned quotas to county and township level governments. Eventually, the quotas were officially issued to polluting units as discharge permits (Zhang et al. 2012).

In 2007, the Water Pollution Trading Pilot Scheme was endorsed by the Ministry of Finance and the Ministry of Environmental Protection for the Tai Lake Basin, and the pilot project would issue and charge for COD discharge permits. The Jiangsu Provincial Government revised the Regulation on Tai Lake Water Pollution Prevention and Control and included a clause on the WPT. The environmental protection and finance bureaus of the province were scheduled to determine the COD discharge fee based on clean-up costs, environmental carrying capacity, and levels of economic development in various parts of the basin.

In the second phase, provincial bureaus would establish an open market for companies to make transactions of their COD discharge quotas, which would incentivize companies to decrease their organic pollution levels as the companies could sell their unused quotas. In addition, the scheme embraces several ambitious sub-programs, the automatic monitoring of water quality, supervision of emitters, and the design of additional supporting programs. The pilot program was applied to the entire lake basin (ADB 2018; Tai and Ellis 2008). A trading center was established in Jiangsu Province, led by the Jiangsu Provincial EPB. Whereas trading information is provided by the center to companies, companies can be involved in negotiation with other parties, face to face and determine their trading price (Zhang et al. 2012).

Jiaxing City's Water Pollution Trading Center was the first institution in the country, November 2007, and Shanghai established the Environment Renewable Trading Center in 2008. Zhejiang Province's center was established in March 2008, Changzhou City's in 2009, and Suzhou City's in December 2012. Tai Lake became the starting point and hub of the Water Pollution Trading Scheme in China (Shen et al. 2017).

The scheme has been widespread since the official launch in 2014 by the central government, and there are positive developments in water pollution control and management. Conspicuous levels of water pollution load have decreased in scheme-applied areas, and local environmental agencies have developed their monitoring and implementation capacities for running the scheme. It appears that the overall capacity of environmental management at localities has enhanced. In addition, the scheme has provided companies with additional financing vehicle for preparing water pollution abatement facilities. For instance, in Zhejiang and Hunan Provinces, diversified transactions of Water Pollution Trading have been introduced, i.e., Mortgage Lending based on Water Pollution Trading Right, and Water Pollution Trading Right Lease (Shen et al. 2017).

Compared with the relative success of water rights trading, the Water Pollution Trading Scheme has not developed well. Total traded allowances in water quality markets in 2013 were recorded at 175,600 tons of Chemical Oxygen Demand (COD), 10,000 tons of Total Phosphorus (TP), and 160,000 tons in Ammonia Nitrogen (NH3-N). Considering the scale of pollution loads at the national level, these trading volumes are negligible. No similar institution like the China

Water Exchange related to the water rights trading scheme has been established for helping water pollution trading take off and spread out in larger areas or to facilitate financing for water pollution control technology (Moore and Yu 2020).

The WPT program encompasses several challenges. First, the most salient problem is the allowance allocation. The official title of the WPT program is 'the Compensated Use of Tradable Emission Permit Program (CUTEPP)'. An indicative emission cap was determined by the government for every company, however, the emission allowance was not freely allocated as in other emissions trading programs. According to the CUTEPP, all companies should purchase an allowance from public authorities within their allowance cap. Emitters are allowed to trade purchased permits in the program. But the emission allowance in the CUTEPP does not necessarily represent a legally tradable property right by law but an administrative license because property rights pertinent to emission allowances are not mentioned in law.

Compared with the CUTEPP, there are several caps in China's conventional environmental regulatory systems, which are the EIA system, the Three Synchronizations, and the Water Pollutant Discharge Permit System (WPDP). These systems intend to monitor and regulate the total maximum load of pollutants. An EIA report submitted by a company prior to the onset of the construction of a facility embraces possible emission discharges and potential environmental impacts of the facility. Local EEBs closely monitor if the company would construct the facility according to the report and comply with the maximum load of pollutants written in the report. The Three Synchronizations scheme ensures the compliance of requirements embedded in the EIA report during the trial production period of the company. Together with these instruments, the Chinese authorities added another regulatory instrument, the WPDP in 2008, according to the Water Pollution Prevention and Control Law, which is dedicated to regulating the total maximum load of water pollutants.

There has been little discussion on the coordination of diverse caps between the old and the new environmental regulatory instruments in water pollution control. If a company discharge more pollutants than the cap determined by the EIA or its permits, this action is regarded as an illegal behavior even though the company keeps an emission allowance purchased from the market. Companies would be in a noncommittal position to purchase emission allowances from the market before or after an EIA report, because there is no guarantee that there would be sufficient allowance in the market, or the EIA report would be approved by the government.

It is necessary for Chinese environmental authorities to provide a clear definition of discharge allowance and implement it in a consistent manner, and the only cap (the maximum total amount of pollutant discharge) should be allowed in the WPDP system in which the emission allowance is sold and purchased by companies. At the macro level, it is plausible to establish a national pollution reduction target of the FYEP based on the total caps of companies (Zhang et al. 2012).

Second, the newly introduced the environmental protection tax system in 2018 may conflict with the current pollution discharge fee and permit systems. Companies have to pay an emission discharge fee as well as acquire a pollution discharge permit. It is an important homework for Chinese environmental authorities to put these systems in harmony avoiding the double charging issue.

Third, critical challenges remain related to the matter of monitoring and enforcement. There are different kinds of discharge data of companies in China, which triggers a series of problems in environmental regulatory works. For instance, environmental accounting measures have been developed through the current monitoring system in the CUTEPP, however, the system is not compatible with other pertinent accounting methods and systems. Diverse water pollutant discharge data can be utilized for recording the emissions of a company taking part in the CUTEPP, which would entail confusions and difficulties for companies as well as environmental authorities.

In addition, the WPT program is not firmly embedded in a national-level law, and local regulations to punish polluters often do not conform to the level of punishments imposed by national-level laws, which leads to the inconsistency of regulatory works. To make things worse, if the Pollution Discharge Fee System remains in effect and the WPT system is actively implemented, companies may face the situation to be fined by both systems. Therefore, it is crucial to have a unified database of environmental statistics that can be used for the CUTEPP and become the yardsticks for both fees and fines (Zhang et al. 2012).

Fourth, trading entities are primarily administrative units rather than enterprises. Company-to-company transactions are dubbed as 'public trades' and observed as rare cases. Fifth, the level of prices is still very low, and sixth, Water Pollution Trading is mostly undertaken within the same province-level units rather than cross-provincial or other administrative boundaries (Moore and Yu 2020; World Bank 2019).

Case Study: Tai Lake Algae Incident in 2007

Tai Lake is the third largest freshwater lake in China, situated at the southern Yangtze Delta Plain with the length of 68 km, the width of 35.7 km, and the water depth of two meters on average. The size of the surface water of the lake is as large as 2,425 km^2 whose catchment areas occupy 36,900 km^2. The lake basin is shared by four administrative units, Jiangsu Province (52.6%), Zhejiang Province (32.8%), Shanghai Municipality (14%), and Anhui Province (0.6%). Some of the most developed and wealthy cities in China are part of the basin, such as Shanghai, Hangzhou, Wuxi, and Suzhou. The lake basin provides water to 30 million people, which is equivalent to approximately 4.4% of China's total population and contributed 9.8% of China's GDP in 2019 (Dai 2019; Ortigana et al. 2019; Wang et al. 2009; Zhang et al. 2012) (see Fig. 6.4).

A sudden bloom of blue-green algae hit Tai Lake on 30 May 2007, and more than two million people of Wuxi City were affected by the incident, having no access to tap water for a few days due to colored and foul-smelled water. Since the outbreak of the incident, Wuxi people, who were reliant on drinking water from Tai Lake, became panicked and rushed for grabbing more bottled water at shops. The price of an 18-L bottle skyrocketed from RMB 8 (US$1.2) to RMB 50 (US$7.7) in a single day (Jing 2007).

Various explanations were given on why such an unusually large-scale blue-green algae occurred in the summer of 2007. The excessively hot and dry weather around the lake from April 2007 triggered

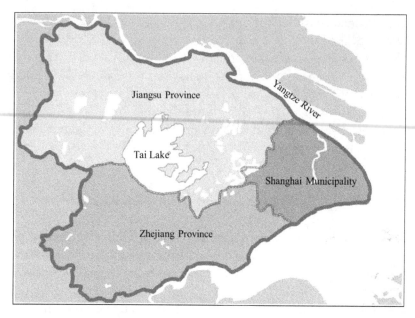

Fig. 6.4 Tai Lake Basin (*Source* Author)

the build-up of tremendous amounts of organic matter and sulfur-containing compounds in the water and sediments of the lake prior to the crisis. More specifically, industrial and domestic wastes with high concentrations of organic matter seem to have entailed a rapid deterioration of water quality, consuming the dissolved oxygen, and nourishing heterotrophic micro-organisms. Wuxi City has grown rapidly based on the development of chemical industries, and innumerable small chemical companies were active along the city's over 3,000 water courses and rivers and had dumped waste directly into adjacent rivers that are connected to the lake (Zhang et al. 2010).

The blue-green algae bloom in 2007 was not the first or one time occurrence in the lake. Whilst 86.5% of the surface water of the lake was classified as Grade II or III in 1990, more than 87% of the surface water was classified as Grade IV in 2000, and there was a clear indication of acute eutrophication. The intense concentration of Total Phosphorus (TP), Permanganate COD (CODMn), and Total Nitrogen (TN) had continued to soar up from 1981 to 2000. The concentration of TP increased from 0.02 to 0.12 mg/L (5 times), that of CODMn grew from 2.9 to 5.1 mg/L (75% increase), and that of TN rose from 0.9 to 3.0 mg/L (234.7% increase) in the same period. The concentration of these posed a grave threat to both population of living creatures and biodiversity and helped some species extinct in the lake (Wang et al. 2009).

The annual duration of cyanobacterial blooms lasted longer between 1987 and 2007 alongside more frequent outbreaks of cyanobacterial blooms in spring and summer months. In 1987, organic pollution was impacting upon around 1% of the water surface but the figure soared to 29.2% in 1994. The phenomenon of eutrophication in the lake entailed killing fish and other living organisms in the lake (Liang and He 2012). The trend of population growth, urbanization, and industrialization has engendered tremendous pressures on the environment in the lake, and the lake has become heavily damaged by agricultural pollution, flooding, eutrophication, aquaculture, and overfishing. As a result, the water quality of the lake and its nearby rivers has continued to deteriorate, partly associated with an inadequate level of regulation and law enforcement against polluters in the lake basin. This regulatory setback reflects a loophole caused by the failure of transboundary water pollution control between Jiangsu Province, Zhejiang Province, and the Shanghai Municipality (Wang et al. 2009; Zhang et al. 2010).

Aware of the deteriorating water quality of Tai Lake, the central government embarked on its first project to clean up the lake in 1991 and poured more than RMB 10 billion (US$1.6 billion) for about 10–15 years to make the lake cleaner and rehabilitating ecosystems of the lake. The Tai Lake Environmental Management Plan was introduced in 1998, and joint efforts of three local governments, Jiangsu Province, Zhejiang Province, and the Shanghai Municipality, led 1,035 major polluting units to comply with emission standards in the late 1998. Despite these efforts, the overall ecosystems and environmental conditions of the lake were not necessarily improved around the turn of the century. Then, the 2007 incident occurred in the lake and alarmed Chinese society about the grave condition of water quality in Tai Lake and its ensuing impacts on public health and environmental sustainability (Liang and He 2012).

Particular attention and immediate responses were necessary right after the incident occurred in May 2007, and the State Environment Protection Administration (SEPA), the precursor of the Ministry of Ecology and Environment, was determined to shut down all factories that discharged pollutants into the lake as an immediate step for controlling the situation. The SEPA confirmed that the amounts of nitrogen in the lake in 2006 tripled that of 1996 and the amounts of phosphate pollutants had soared by 150%. All non-compliant companies which discharged nitrogen and phosphate pollutants into the lake were investigated, and any factory which dumped beyond the threshold of emission standards should be closed with the application of production restrictions (Jing 2007).

More than 6,000 tons of algae were removed from the lake, and drinking water pipelines were protected with polyvinyl chloride (PVC) barriers. Additional water pollution abating works were introduced, including augmenting water supply through cloud seeding for diluting the level of pollution in the lake, and the diversion of water from the Yangtze River to the lake by the Jiangsu Provincial Government. Local governments in the lake basin were prepared to bring in more bottled water from neighboring areas (Jing 2007; Tai and Ellis 2008).

Zhang Lijun, then Vice Minister of the SEPA in June 2007, elucidated the major causes of the incident of Tai Lake, pointing out the discharge of excessive amounts of industrial and household sewage in conjunction with the overexploitation of fertilizers in the lake basin. Stricter pollutant discharge standards for industries should be necessary together with the betterment of sewage treatment facilities and the reduction of chemical fertilizer use. The SEPA officially banned the construction of new factories discharging nitrogen and phosphate pollutants.

It took four days for local authorities to confirm that tap water in Wuxi City would be safe to drink after testing the quality of tap water in the city. However, most residents of the city had to rely on bottled water for a while even after the announcement, since polluted water would be left in water pipelines. The massive scale of algae bloom in 2007 entirely faded way around the winter. Five local officials in charge of water pollution control faced punishment from the Wuxi City Government (Jing 2007; Tai and Ellis 2008).

Special measures to remove blue-green algae blooms in the lake were implemented by the Wuxi City Government, including strict standards for waste discharges, the close monitoring of major industrial polluters, and the restoration of wetlands along the shore of the lake. In addition, Wuxi EPB closed sewage drain outlets along the river channels into the lake, which forced companies in the city to emit their wastewater into the sewage collection system and increased the rate of sewage treatment. Around 600 small chemical factories out of 20,000 were shut down in 2007 for reducing the influx of wastewater into the lake. Together with chemical factories, textile and food processing industries contributed to raising the level of COD in the lake (Tai and Ellis 2008; Zhang et al. 2010).

In addition to Wuxi's special measures, the Jiangsu Provincial Government launched a five-year rehabilitation plan with the investment of US$14.5 million and revised the 1996 Regulation of Preventing Water Pollution in Tai Lake in the province. The regulation embraces strict measures to tackle pollution problems with the four key points as follows. First, the government should set up the Regional Approval Restriction system, and the Environmental Impact Assessment (EIA) approval of major construction projects should be suspended for regions that discharge effluents beyond the total pollutant cap. Second, the water quality monitoring system should be enhanced through total discharge caps, and national- and local-level monitoring stations, higher penalties for non-compliance, and inter-regional law enforcement. Third, the Drinking Water Protection Program should be established, which encompasses a public notification mechanism and daily patrolling of drinking water sources. Fourth, the government will clarify the liability of water pollution.

The Jiangsu Provincial Government created the Working Plan of Water Pollution Prevention in Tai Lake in 2007 for controlling the eutrophication in the lake within five years and remove pollution in the lake in next eight to ten years. Specific policy measures were introduced and pushed local governments to raise the bar of emission standards and suspend approving projects whose emission discharges would be more than the total emission discharge control threshold in the area. These policy measures serve as significant foundations to take off the water pollution trading program (Tai and Ellis 2008).

Vast amounts of nitrogen discharged in Tai Lake contributed to about 20–30% of eutrophication, and urban households were responsible for around 50% of the nitrogen in the lake. In 2008, 127 sewage treatment plants in Jiangsu Province could treat three million tons of sewage per day, however, the actual sewage treatment rate was estimated at 1.98 million tons. This signifies 66% of the operation rate and implies that 30% of wastewater would be dumped into the lake without proper treatment. Agriculture contributed about 60% of organic pollution to the lake, and the remaining 40% stemmed from aquaculture. Farmers in irrigated areas used heavy amounts of chemical fertilizers and pesticides, and in 2002, farmers in the lake basin exploited 553.5 kg of fertilizer and 34.5 kg of pesticide per hectare, which were more than the average national use of 411.0 kg fertilizer and 11.3 kg pesticide per hectare. This trend had continued before the incident took place in 2007 (Tai and Ellis 2008; Wang et al. 2009).

In March 2008, the hierarchical status of the SEPA was elevated, becoming the Ministry of Environmental Protection. Although it is unclear whether such a change stemmed from the Tai Lake incident, the creation of the MEP highlights the renewed commitment of the central government to an emphasis on environmental protection. The revised Water Pollution Control Law in 2008 leveled up the punishment for a water pollution accident from no more than RMB 200,000 (US$30,769) to as large as 30% of direct losses due to an accident (Zhang et al. 2010).

For more than three decades' efforts, the water quality of Tai Lake shows a decent progress as the case of Lake Wuli's water quality improvement, which is part of Tai Lake, in terms of the level of the concentration of Total Nitrogen (TN), Total Phosphorus (TP), and Permanganate Index (CODMn) between 2000 and 2005 (Wu et al. 2019). However, the phenomenon of eutrophication is still observed together with poor water quality in the lake. Rampant discharges of household and industrial wastewater into the lake are to be blamed, which demonstrates more room for revamping the capacity of regulatory settings, relevant organizations, specific planning and implementation, and substantial investment. A key issue to be resolved is fragmented regulatory frameworks in the lake basin. Since the lake basin stretches over Jiangsu Province, Zhejiang Province, and the Shanghai Municipality, it was necessary to establish an overarching authority tackling numerous issues within the basin and coordinating conflicts of interests between the local governments (Wang et al. 2009).

As an inter-provincial authority, the Tai Basin Authority (TBA) was established initially in 1964 with the name of the Tai Lake Water Resources Bureau, and in 1984, the title of the TBA was officially given (Tai Basin Authority 2019). However, no regulatory or law enforcement mandate for water pollution control was given to the authority in the lake basin, and the works of monitoring, regulating, and penalizing polluting companies are in the hands of the local governments. What the TBA can do is to establish the basin-wide planning accommodating opinions and views from the local governments in the basin. Each local government seeks to undertake its own socioeconomic development and land use plans which do not correspond to or conflict with regional planning for the lake basin (Liang and He 2012; Wang et al. 2009). This can bring about the ineffectiveness of planning, monitoring, and regulatory works in water quality enhancement in the lake basin. It will be imperative to empower the TBA for facilitating inter-provincial cooperation to enhance the water quality of the lake in the future.

Conclusion

The chapter has appraised the development of water quality management in China with a primary focus on the two major policy initiatives, the River (Lake) Chief System and the Water Pollution Trading Scheme together with the review of the general framework for water policy control. The case study of the blue-green algae incident of Tai Lake in 2007 has unveiled the environmental damage derived from development-centered policies around one of the most industrialized areas in the country and regulatory efforts after the incident to enhance water pollution control at the local and national levels.

It is true that water quality control policies and measures in China have played a significant role in decreasing loads of water pollution into water bodies, struggling with pro-growth policies and programs that have been dominant in the reform era since 1978. A series of major regulatory programs in water quality management, i.e., the EIA, the Three Synchronizations, the Pollution Discharge Fee, and the Pollution Discharge Permit System, have served as the essential tools for environmental agencies and have been effective in 'diluting' the level of water pollution in rivers, lakes, and groundwater resources at the national level. However, the sustainability of numerous water bodies has not necessarily been improved although the social demand for good water quality is increasing.

The study has highlighted the River (Lake) Chief System as a newly committed response of Chinese environmental authorities to address severe water pollution in rivers and lakes. The system has turned out to be successful in many localities since the onset of the system in 2007 in Wuxi City of the Tai Lake Basin. The influence of the CCP has been critical in the effectiveness of the system in terms of safeguarding the soundness of local water bodies because river (lake) chiefs are usually appointed among either CCP cadres or high-ranking officials at local governments. Such a high-profile status of chiefs has helped business and industrial units well comply with water pollution regulations. This typically top-down nature of the system, however, seems to require revision in due course, such as the introduction of consensus building process for allowing public participation and the systemization of regulatory measures within the system, not relying on personal charisma or influence.

Discussions on Water Pollution Trading have disclosed that the scheme needs improvement in the coming decades. The scheme is similar to the water rights trading and the carbon emission trading as an economic instrument for reducing loads of water pollution in local water bodies with an emphasis on voluntary participation of business and industrial players. Although the scheme was first devised back in the middle of the 1980s in Shanghai, a series of huddles have been identified for further development. These include the confusion between the old and the new environmental regulatory instruments in water pollution control, including the potential conflict of environmental protection tax, and pollution discharge fee and permit systems. The discrepancy may involve confusion for environmental regulatory authorities in terms of water pollution discharge data and information collection. More carefully designed principles and guidelines have to be prepared for taking off the water polluting trading in the future.

The case study of the Tai Lake blue-green algae outbreak in 2007 has demonstrated the evolving trend of China's water pollution control policy. Whereas long-term and macro-scale planning and strategies are suggested at the central level, the Tai Lake case explicitly illustrates the extent to which an emblematic event has become a wake-up call for water pollution control authorities, which have endeavored to rectify wrongdoings and to introduce stringent law enforcement measures and preventive policies for not repeating similar mistakes and improve the water quality of rivers and lakes. It still requires more time, investment, and efforts to bring the clean Tai Lake back together with the implementation of the integrated lake basin management through the Tai Basin Authority.

More attention is now paid to the enhancement of water quality management amongst diverse dimensions of water resources management in China in the era of ecological civilization. Environmental regulatory programs in China, however, appear to be too many and complex so that the effectiveness of implementing those programs needs attention of concerned authorities. Chinese environmental authorities are required to streamline numerous water pollution control-related programs and prioritize a few of them for better regulatory results. A good coordination of old and new water pollution control programs will determine the success of water quality management for achieving ecological civilization.

References

Asian Development Bank (ADB). 2018. *Managing Water Resources for Sustainable Socioeconomic Development—A Country Assessment for the People's Republic of China*. Manila: Asian Development Bank.

CEIC Data. 2020. China Disposable Income Per Capita: Jiangsu, Wuxi. Available online: http://www.ceicdata.com. Accessed 27 July 2020.

Chan, Woody. 2018. Key Water Policies 2017–2018. *China Water Risk*, 16 March 2018.
China Council for International Cooperation on Environment and Development (CCICED). 2019. The Shift to High-Quality, Green Development. CCICED Issus Paper 2019.
China Water Risk. 2016. China's 13th Five Year Plan for Ecological & Environmental Protection (2016–2020). *China Water Risk*, 9 December 2016.
Dai, Liping. 2015. A New Perspective on Water Governance in China: Captain of the River. *Water International* 40 (1): 87–99.
Dai, Liping. 2019. *Politics and Governance in Water Pollution in China*. Cham, Switzerland: Palgrave Macmillan.
Fang, Guohua, Zhaojie Li, and Zhexin Lin. 2018. *River/Lake Chief System Assessment* (河 (湖) 长制考核). Beijing: China Hydraulics and Hydropower Press.
Han, Dongmei, Matthew Currell, and Guoliang Coa. 2016. Deep Challenges for China's War on Water Pollution. *Environmental Pollution* 218: 1222–1233.
Hao, Yaguang. 2017. Establishment of Local Water Management Responsibilities Under the River Chief System ("河长制"设立背景下地方主官水治理的责任定位). *Journal of Henan Normal University (Philosophy and Social Sciences)* (河南师范大学学报 – 哲学社会学版) 44 (5): 13–18.
He, Guizhen, Yonglong Lu, Arthur Mol, and Theo Beckers. 2012. Change and Challenges: China's Environmental Management in Transition. *Environmental Development* 3: 25–38.
Jing, Xiaolei. 2007. Green Disaster. *Beijing Review*, June 21. Available online: http://www.bjreview.com. Accessed 27 July 2020.
Lau, Lynia. 2014. China's Newly Revised Environmental Protection Law. *Clyde & Co*. 16 May 2014. Available online: http://www.clydeco.com. Accessed 21 January 2020.
Lee, Seungho. 2006. *Water and Development in China: The Political Economy of Shanghai Water Policy*. New Jersey: World Scientific.
Liang, Guorui and He, Hanfu. 2012. Long Struggle for a Cleaner Lake Tai. *China Dialogue*, February 14.
Liu, Liuyuan. 2019. River and Lake Chief System. Presented at the 2019 International Specialty Conference, jointly organized by the American Water Resources Association and the Center for Water Resources Research, Chinese Academy of Sciences. Beijing, September 17, 2019.

Ma, Xiaoying and Leonard Ortolano (2000) *Environmental Regulation in China*. Boulder, New York and Oxford: Rowman & Littlefield.

Minzner, Carl. 2011. China's Turn Against Law. *The American Journal of Comparative Law* 59: 935–984.

Moore, Scott, and Winston Yu. 2020. Environmental Politics and Policy Adaptation in China: The Case of Water Sector Reform. *Water Policy* 22: 850–866.

OECD. 2006. *Environmental Compliance and Enforcement in China: An Assessment of Current Practices and Ways Forward*. Paris: OECD.

Ortigana, Angela, Chaochao Chen, Yifeng Liu, and Aihui Yang. 2019. Taihu (Lake Tai) Basin China. WWF China.

Shen, Manhong, Huiming Xie, and Yuwen Li. 2017. *Chinese Water Policy Research* (中国水制度研究), vols. 1 and 2. Beijing: Renmin Press.

Smith, Sam. 2018. China's River Chiefs—An 'Ingenious' Approach to Water Governance in China? *International Rivers*, August 30.

State Council. 2016. 13th Five Year Environmental Plan (生态环境保护规划). State Council, 24 November 2016.

Tai Basin Authority. 2019. History of the Tai Basin Authority (太湖流域管理局 历史沿革). 18 December 2019. Available online: http://www.tba.gov.cn. Accessed 29 July 2020.

Tai, Pei-yu, and Linden Ellis. 2008. Taihu: Green Wash or Green Clean? A China Environmental Health Project Research Brief, October 2008.

Tan, Debra. 2014. The War on Water Pollution. *China Water Risk*, 12 March.

Tilt, Bryan. 2015. *Dams and Development in China: The Moral Economy of Water and Power*. New York: Columbia University Press.

Wang, Liang, Yongli Cai, and Liyan Fang. 2009. Pollution in Taihu Lake, China: Causal Chain and Policy Options Analyses. *Frontiers of Earth Science in China* 3 (4): 437–444.

World Bank. 2019. *Watershed: A New Ear of Water Governance in China—Thematic Report*. Washington, DC: World Bank.

Wu, Fengshi, and Richard Edmonds. 2017. Chapter 7: Environmental Degradation. In *Critical Issues in Contemporary China*, 2nd ed., ed. Czeslaw Tubilewicz, 105–119. London: Routledge.

Wu, Junli, Zishi Fu, Hongxia Qiao, and Fuxing Liu. 2019. Assessment of Eutrophication and Water Quality in the Estuarine Area of Lake Wuli, Lake Taihu, China. *Science of the Total Environment* 650: 1392–1402.

Xinhua. 2020. China Won't Relax Ecological, Environmental Protection: Minister. *Xinhua Net*, 25 May. Available online: http://xinhuanet.com. Accessed 10 March 2021.

Xu, Yuanchao. 2017. China's River Chiefs: Who Are They? *China Water Risk*, 17 October.

Zhan, Yunyan. 2019. River Chief System: Advantage and Disadvantage, Argument and Improvement (河长制的得失, 争议与完善). *China Environmental Management (中国环境管理)* 4: 93–98.

Zhang, Jiaqi. 2018. 2 New Environmental Laws to Go into Effect in 2018. *China.org.cn*, January 1. Available online: https://china.org.cn. Accessed 29 February 2020.

Zhang, Xiao-jian, Chao Chen, Jian-qing Ding, Aixin Hou, Yong Li, Zhangbin Niu, Xiao-yan Su, Yan-juan Xu, and Edward Laws. 2010. The 2007 Water Crisis in Wuxi, China: Analysis of the Origin. *Journal of Hazardous Materials* 182: 130–135.

Zhang, Youngliang, Bing Zhang, and Jun Bi. 2012. Policy Conflict and the Feasibility of Water Pollution Trading Programs in the Tai Lake Basin, China. *Environment and Planning c: Government and Policy* 30: 416–428.

Zhang, Yongxi. 2019. Summary of Experiences in Compiling the Implementation Scheme of River Chief System in Haxi Prefecture, Qinghai. Presented at the 2019 International Specialty Conference, jointly organized by the American Water Resources Association and the Center for Water Resources Research, Chinese Academy of Sciences. Beijing, September 17, 2019.

7

Water Resources Development

Introduction

The chapter sheds light on China's policies on water resources development, not only for augmenting substantial amounts of water for agricultural, household, and industrial uses but also energy generation, flood prevention and protection, and inland navigation. Large dam development has been pivotal in bolstering socioeconomic development through water supply for food production, hydropower generation, flood prevention and control, and ecosystem rehabilitation in the reform era. Major attention was placed on dams for irrigation and flood control prior to the open-door policy, but more investment has been made on large hydropower dams from the 1980s to the present, powering the rapid growth of economic centers in the eastern and southern coastal areas. The Three Gorges Dam is the epitome of the hydraulic mission for China's modernization to prevent floods intensified by climate change, generate green energy that would replace thermal power plants, and facilitate inland navigation that expand inter-provincial trade for local economies.

While debates on the sustainability of inter-basin water transfer projects are ongoing, there have been various scales of inter-basin water transfer projects galore in China over the past decades, and many cases are found in water shortage areas which are more developed and politically powerful for channeling more water from less developed and relatively water abundant areas. The South North Water Transfer Project has been delivering vast amounts of water from the Yangtze River to water-scarce areas in North China, particularly Beijing and Tianjin, since 2014, through the eastern and the middle routes. Although the construction of the western route has not been begun yet, the project appears to have developed smoothly for fulfilling its own mission, quenching the thirst of North China. It is important to notice that progress of the project has been made because of the heroic sacrifice of people in South China who might need more water under uncertain hydro-meteorological regimes impacted by climate change.

In-depth discussions will be made on the two cases of water resources development, the Three Gorges Dam, and the South North Transfer Project. The exploration of these projects unpacks pros and cons of the 'nationally significant development projects,' often emphasized by the central government of China. Whereas those projects are touted as the CCP's monumental achievements for legitimizing their rule over the masses, little voice has been heard on critical challenges in socioeconomic, political, and environmental senses. Dissident views on the projects are unwelcome although a series of unsolved problems and newly emerging challenges, such as climate change driven uncertainties, need to draw more attention of decision makers as well as hydraulic engineers. Water resources development policies and projects in China should closely be monitored and assessed based on their socioeconomic and environmental performance in accordance with more mature levels of contemporary Chinese society.

The first discussion in the chapter goes to the development of dams in China. Sketching the patterns of dam development at the national level, the study pays special attention to the case of the Three Gorges Dam for its socioeconomic, political, and environmental implications. Water transfer projects will be focused on in the next section, evaluating inter-basin water transfer projects in different parts of the country, and special attention is placed on the South North Water Transfer Project with the appraisal of achievements and challenges.

Dam Development

Development of Large Dams

China's dam development has more than 3,000 years of history, and large-scale irrigation projects and agricultural production have been possible by damming large rivers such as the Yellow River and its tributaries in the North China Plain. Even though many of the early great empires with massive-scale irrigation projects and hydraulic works ebbed away, China has been one of the exceptions to continue to develop megascale hydraulic projects strongly encouraged by the state, including dams, aqueducts, water diversion projects, and waterways.

Prior to 1950, there were only 23 large dams in China whereas 5,196 existed at the global level (Jiang et al. 2020; World Commission on Dams 2000). Large dams are defined as the dams with more than 15 m in height or with a storage capacity greater than three million m^3 (ICOLD 2011). Since the establishment of the PRC in 1949, the communist China inherited the long history of hydraulic bureaucracies from the dynasties and began to envisage major flood control and hydropower projects with the help of Soviet engineers. Among many dam projects, the Sanmenxia Dam unfolds a complexity of problematic issues related to dam construction, such as the relocation of more than 400,000 people, siltation, salinization, and financial losses (Molle et al. 2009; Shapiro 2001).

Strong commitments from the central government to dam building have been made with a special emphasis on hydropower dams to meet the growing demands of electricity since the onset of the open-door policy. The strategic shift in dam development of China from irrigation and flood control to hydropower generation is the outcome of the unprecedented scale of socioeconomic development. As Tilt (2015) maintains, water is not simply as a resource for various uses but as an origin of kinetic energy which can be distributed and utilized for scaling up socioeconomic development in electricity-hungry areas thousands of kilometers away. The wealthier and industrialized China has had more financial resources available for large-scale dams. The Three Gorges Dam and the Xiaolangdi Dam Projects, some of the largest dams in the world, were commenced and completed in this phase.

China's water supply infrastructure development had demonstrated a remarkable increase of 600 dams on average every year from 1949 to 1979. Vast amounts of governmental investment in water infrastructure were mobilized, RMB 92 million (263 million in 2018 prices) in 1950 and RMB 3.7 billion (US$ 10.6 billion in 2018 prices) in 1979, which occupied 7.1% of all national construction investment (Jiang et al. 2020). Small- and medium-sized dams were primarily built, and irrigated areas were expanded in the former three decades, and these types of dams were recorded at 83,387 by 1990 (Shui 1998).

Large-scale dams with the larger storage and hydropower generation capacity were advocated in the two decades after 1978. Chinese investment in water infrastructure soared in the 1990s, amounting to RMB 4.9 billion (US$ 10.65 billion in 2018 prices), however, the portion of Chinese investment in water infrastructure reached only 3.3% of national spending in the 1990s. The Three Gorges Dam was one of large-scale water infrastructure projects whose construction lasted from 1993 to 2009. The central government has never been tired of building gigantic hydraulic structures in the new millennium, pumping in over RMB 2 trillion (US$ 307 billion) investment in water infrastructure in the 12th FYP (2011–2015) (Jiang et al. 2020; Shui 1998).

A long-term development plan for hydraulic projects was established in China in the 1980s with a special emphasis on hydropower dams. China's hydropower potential is estimated at 694 GW, which is equivalent to a 15% of the global share, ranked the first in the world. In terms of the total hydropower installed capacity of each country, the top ten countries are shown in 2020 in Fig. 7.1.

The exploitable installed capacity of China's hydropower can reach 542 GW, a 17% global share, with an annual electricity generation of 2,474 TWh. Major features in China's hydropower are threefold: (1) uneven spatial distribution; (2) uneven temporal distribution; and (3) high concentration. Southwest China boasts the largest hydropower development potential, which accounts for two-thirds of total technically exploitable hydropower capacity (Sun et al. 2019). In 2020, the total installed capacity of China's hydropower generation reached 356.4 GW,

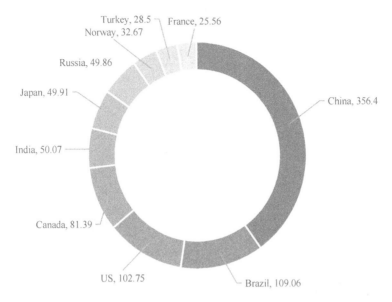

Fig. 7.1 Hydropower installed capacity (GW) of the top 10 countries in 2020 (*Source* IHA [2020])

the world's number one hydropower capacity country, which accounted for more than 27% of the total installed hydropower capacity in the world, 1,308 GW (IHA 2020).

The hydropower development plan encompasses the geographical distribution of hydropower bases in different river basins across China, identified by the Ministry of Water Resources (MWR). The MWR has actively promoted hydropower dams, because hydropower is one of the most significant green energy options that replaces numerous coal-based power plants in the country, which can generate electricity, reduce CO_2 emission, and improve air quality. Continuous socioeconomic development, urbanization, and the change of consumption patterns in households have driven the central government to push forward the construction of more large-scale hydropower dams all over the China (Sun et al. 2019; Wang et al. 2014). Table 7.1 summarizes the detailed data on the hydropower bases.

China's ambition for hydropower generation continues. According to the National Energy Administration, the conventional installed capacity

Table 7.1 Summary of 13 hydropower bases in China

No	Base	Location (Province)	Total installed capacity (GW)	Annual electricity generation (TWh)	Development status (%)[a] based on the installed capacity		
					Constructed	Under construction	Planning
1	Jinsha River	Sichuan, Xizang, Yunnan	58.58	282.6	0	35	65
2	Upstream of the Yangtze River	Sichuan, Chongqing, Hubei	33.197	143.8	73	12	15
3	Yalong River	Sichuan	25.7	125	13	33	54
4	Mainstream of the Lancang River	Yunnan	25.11	120.3	19	40	41
5	Dadu River	Sichuan	24.92	113.6	13	18	69
6	Nu River	Yunnan	21.99	103.7	0	0	100
7	Upstream of the Yellow River	Qinghai, Gansu, Ningxia	20.93	75	65	6	29
8	Nanpan River-Hongshui River	Yunnan, Guizhou, Guangxi	14.3	63.5	47	49	4
9	Northeast China	Heilongjiang, Jilin, Liaoning	13.26	35.5	43	4	53
10	Fujian-Zhejiang-Jiangxi Area	Fujian, Zhejiang, Jiangxi	12.2	31.5	68	3	29

No	Base	Location (Province)	Total installed capacity (GW)	Annual electricity generation (TWh)	Development status (%)[a] based on the installed capacity		
					Constructed	Under construction	Planning
11	Wu River	Guizhou, Chongqing	11.22	39.6	82	18	0
12	Western Hunan	Hunan	10.8155	37.8	68	9	23
13	North mainstream of the Yellow River	Shanxi	6.408	17.8	20	6	74
Total			278.6305	1189.7	511	233	566

Remark [a]Based on the data of 2010
Source Modified based on Sun et al. (2019)

Table 7.2 China's hydropower development plan from 2020 to 2050

Year		2020	2030	2050
Installed Capacity (GW)	Large- & medium-sized hydropower	270	340	410
	Small hydropower	80	90	100
Total		350[a]	430	510

Remark [a]the goal of 2020 was already achieved since the total installed capacity of China's hydropower generation amounted to 356.4 GW in 2020
Source Modified based on Sun et al. (2019)

of hydropower will grow to 430 GW by 2030. The Chinese government endeavors to achieve the conventional installed capacity of hydropower, 510 GW by 2050. Table 7.2 provides the details of China's hydropower development plans from 2020 to 2050 (Sun et al. 2019).

Economic liberalization has prevailed over hydropower development in the country, particularly since 1996 when the Electric Power Law was passed. The law stipulates the separation of electricity producers and distributors for establishing a nationwide market for electricity. Another step was taken for dissolving the State Electric Power Corporation in 2002 which was responsible for both power generation and distribution. The division of labor was given to the two main electricity SOEs, the one responsible for electricity generation, and the other focusing on electricity transmission and distribution. The group of SOEs involved in electricity generation comprise five SOEs, often dubbed as the 'Five Energy Giants (China State Grid Corporation, China Huadian Corporation, China Huaneng Corporation, China Three Gorges Corporation, and China Datang Corporation),' which show a variety of energy portfolios, i.e., hydropower, coal, wind, and solar.

The largest SOE in the fields of electricity transmission and distribution is the China State Grid Corporation that owns several regional subsidiaries. China Huadian Group and China Huaneng Group have their state-granted monopolistic right to develop hydropower dam projects in major rivers of China as well as diversified portfolios oversees. For instance, China Huadian Group has its exclusive right to construct hydropower dams on the Nu-Salween River through its subsidiary Yunnan Huadian Nu River Hydropower Development Company. Regarding the Lancang-Mekong River, China Huanneng

Group is given the exclusive right for hydropower development through its subsidiary company, Yunnan Huaneng Lancang River Hydropower Company. The China Three Gorges Corporation was first founded in September 1993 for undertaking the construction and management of the Three Gorges Dam. Since then, the company has not only specialized in hydropower development and generation but also involved the provision of other clean energies, i.e., solar photovoltaic and wind power. Four mega-scale hydropower dams have been built and in operation by the company in the upstream of the Yangtze River, the Jinsha River, including the Xiluodu, Xiangjiaba, Wudongde, and Baihetan Dams (CTG 2020a; Habich 2016; Tilt 2015).

It is the National Development and Reform Commission (NDRC) that leads hydropower planning and policy by examining national economic development and energy situations and makes decisions on the feasibility of dam construction. Alongside the NDRC, relevant ministries and bureaus are involved, such as the Ministry of Water Resources, the Ministry of National Resources, Ministry of Agriculture and Rural Affairs, and the Ministry of Ecology and Environment, which approve pre-feasibility and feasibility studies, design plans, and a project application report for each dam project.

The Seven River Basin Commissions under the auspices of the MWR oversee the management of river projects in China if these are associated with the transboundary rivers between provincial boundaries. Basin plans are established by the commissions that embrace the assessment of hydropower potential. The five energy giants have been given the exclusive right to build and operate hydropower dams along the country's major rivers since 2002. The companies' plans for hydropower development should be compatible with the basin plans of the commissions, being subject to the commissions' approval. Eventually, the commissions should prepare and submit a report to the NDRC for further approval (Habich 2016; Mertha 2008).

The next two steps for the hydropower companies after a project is approved are to prepare specific design plans for dam construction and to undertake feasibility studies for the dam. The hydropower companies hire one of the design institutes that is part of the China Hydropower Engineering Consulting Group (Hydrochina) for developing a detailed plan for the dam, i.e., the dam site and size, and the details plans for resettlement. In addition, the companies have a contract with agencies to undertake feasibility studies that provide the information on the technological, economic, environmental, and social aspects of the dam project. On the approval of the studies, the project application reports are ready to be submitted coupled with the EIA. At the local level, local bureaus involved in hydropower development are working together with the five electricity giants, such as water resources bureau, development and reform commission, and ecology and environment bureau, for instance (Habich 2016; Mertha 2008).

A decision on dam construction does not necessarily go smooth. Whereas the NDRC advocates the expansion of hydropower plants for bolstering socioeconomic development, non-economic or pro-environment ministries and bureaus demand more cautious approaches to such large-scale and potentially damaging projects, including the Ministry of Ecology and Environment, Cultural Relics Bureau, and Seismology Bureau. It is true that such a conflict of interests between ministries and bureaus depending upon their given mandates occurs in other countries. However, the prestigious status of the NDRC under China's bureaucratic hierarchy and pro-growth units prevail over in such debates.

In addition, the State-Owned Assets Supervision and Administration Commission is in charge of the power companies, which is under full control of the State Council, and therefore, it is difficult to envisage that other ministries might argue against the will of the State Council. Another interesting dimension is that top management positions are often filled with influential personnel which are well connected with powerful figures in the central government. For instance, Li Xiaopeng, the former General Manager of China Huaneng Group was the son of China's ex-premier, Li Peng (Habich 2016; Mertha 2008). Figure 7.2 describes the current governance structure of hydropower generation in China focusing on the five large SOEs.

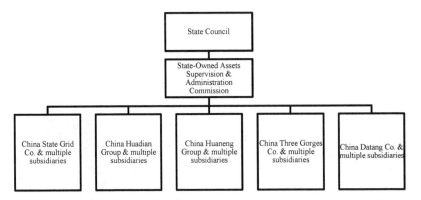

Fig. 7.2 Current governance structure of hydropower generation in China (*Source* Modified based on Habich [2016], Mertha [2008], and Tilt [2015])

Three Gorges Dam

More than 220,000 casualties were recorded in the unprecedented scales of flood events of the Yangtze River Basin in 1931, 1954, and 1998. Taming the river has been a major assignment for top leaders of contemporary China. First envisaged by Sun Yat-Sen, the ambition of damming the Three Gorges in the middle reaches of the Yangtze River was eventually achieved when Premier Li Peng-led CCP endorsed the controversial Three Gorges Dam (TGD) project in 1994 (Chang et al. 2018; Pyo 2020; Wang et al. 2014).

In spite of such blessings and attention from top leaders of China, loads of debates and critique abounded prior to the actual launch of the project. Concerns were centered on not only financial and technical difficulties but also long-term ecological and human impacts, particularly in relation to relocating more than one million people around the newly created reservoir. Such a mood against the dam project engendered unusual opposition votes within the CCP. There was a voting about the launch of the project at the National People's Congress (NPS) in 1992, and although the project managed to receive a green light in the 1992 NPS, the voting result unveiled that the total votes of opposition and abstention amounted to 31.5%. Considering the role of the NPS and

Fig. 7.3 a. Façade of the Three Gorges Dam. b. The Ship Lock of the Three Gorges Dam. c. Transmission Lines from the Three Gorges Dam (*Source* Author [photo taken in October 2019])

c

Fig. 7.3 (continued)

other political events as the rubber stamp, the result sums up the sensitivity of the dam project that can cause many ensuing problems (Pyo 2020; Wang et al. 2014).

Relevant issues of the TGD appeared to be too politically sensitive even until the end of the construction, and there was no eye-catching event to celebrate the completion of the dam construction on 20 May 2006 in addition to the absolute absence of any Chinese top leaders, including President Hu Jintao and Premier Wen Jiabao. It is plausible to maintain that the leaders would like to keep away from any negative assessment of the dam project recognizing the complexity of socioeconomic, technical, and environmental challenges entrenched in the project (Pyo 2020).

President Xi Jinping is the first leader of China to be outspoken about the achievements of the TGD Project. In April 2018, President Xi visited the dam site, praising the achievements of the dam in the context of realizing the China's Dream. His visit to the dam was interpreted in various ways but should be associated with his intention to emphasize technological advancement for national development and overseas expansion in the framework of the Belt and Road Initiative (Pyo 2020).

The primary purposes of the TGD are threefold. First, flood control is the major mission of the dam, which has enabled Chinese water managers to improve the capacity of flood control over the Yangtze River, particularly in the middle and lower reaches. Ever since the first consideration of the TGD, large-scale flood events had been a lingering and formidable challenge to Chinese authorities. Frequent floods took place in the river almost every summer that had often triggered many human losses and the destruction of major infrastructures and buildings.

Chinese hydraulic engineers have been committed to preventing human casualties as well as economic losses by floods through the dam. The height of the dam reaches 185 m above the sea level, the length of this massive infrastructure is 2.3 km, and the total volume of the dam reservoir is 39.3 billion m^3. More importantly, the dam can store 22.15 billion m^3 of water in the flood season with the maximum water level of 145 m above the sea level. The dam can protect downstream areas from 100-year return-period floods and effectively tackle 1,000-year floods in alliance with other flood protection projects.

Second, the installed power capacity of the dam is enormous, 22.5 GW, and this scale of generation capacity has been able to replace the burning of 30 million tons of coal per annum (Stephens 2016). The left and right bank power stations include 26 sets of 700,000 kW hydroelectric generating units in addition to two sets of 50,000 kW hydroelectric generating units in the power supply station and six sets of 700,000 kW hydroelectric generating units installed in the right bank underground power station. On the basis of these generating units, the generating capacity of the dam is as large as 22.5 million kW with an annual average generation capacity, 88.2 billion kWh, which confirms the world's largest hydropower station. Hydropower generated from the dam can be transmitted to the radius of 1,000 km with the power primarily sent to ten provinces in the East, central China, Guangdong, Shanghai, and other areas (Three Gorges Hydropower Station 2019; Wang et al. 2014).

Massive amounts of hydropower generation by the dam can lead to reducing annual emissions by 100 million tons of CO_2, two million tons of SO_2, and 0.37 million tons of NO_2 compared with equivalent production from coal-based power plants and the emissions to produce and transport the coal from North to South China (Andreas and Yang 2011).

The dam plays a pivotal role in improving inland navigation with the reservoir, stretching from Yichang City to the Chongqing Municipality, directly creating 660 km of easily navigable waterway. Poor navigation conditions between the two cities were attributed to sharp bends, narrow channels, shoals and rapids prior to the project. The TGD has paved the way for the Yangtze River to become the real 'Golden Waterway,' bridging downstream and upstream (Andreas and Yang 2011; Chang et al. 2018; Three Gorges Hydropower Station 2019). Boats and ships travel through the dam's unique navigational systems: (1) vertical ship lifter; and (2) ship locks. The vertical ship lifter allows passenger boats with less than 3,000 tons to travel from the downstream to the upstream of the Yangtze River or vice versa. Large vessels between 3,000 and 10,000 tonnage use the ship locks that allow them to sail upstream or downstream in the river, particularly to and from major cities in Sichuan Province, including the Chongqing Municipality (Three Gorges Hydropower Station 2019; Wang et al. 2014). Table 7.3 summarizes the technical specifications of the dam, and Fig. 7.3a, b, and c show the façade of the dam, the ship lock, and the transmission lines, respectively.

The official budget for the project was as large as RMB 200 billion (US\$ 30.7 billion), however, the costs of the project have increased due to additional costs of providing new homes for relocated people with RMB 69.3 billion (US\$ 10.6 billion), which brings the total costs of the project to RMB 269.3 billion (US\$ 41.3 billion). Many experts suspect that the final cost estimation of the project would be even larger than this (Andreas and Yang 2011; Hvistendahl 2008).

Despite the large project costs, it appears that the Chinese government has not been so much worried about it because of a unique structure of financing for the project. The Three Gorges Dam Construction Fund has been created based on a levy on each household and company in China since 1992 when the Prime Minister's Office of the State Council decided to levy 0.003 cents on every kWh of electricity consumed at the national level for additional financing of the project. The level of levy has increased twice since then, up to 0.004 cents in 1994 and 0.007 cents in 1997 for 16 provinces and major cities. The total amounts of the fund

Table 7.3 Technical specifications of the Three Gorges Dam

Item		Content
Dam type		Concrete gravity dam
Total length		2,309.6 m
Storage capacity		39.3 billion m^3
Flood control capacity		22.15 billion m^3
Normal headwater level		175 m
Flood control level		145 m
Dry season control headwater level		155 m
Flood control (100-year frequency flood)	Max headwater level	166.9 m
	Max discharge	56,700 m^3
Flood control (1,000-year frequency flood)	Max headwater level	175 m
	Max discharge	69,800 m^3
Three Gorges Hydropower Plant	Total installed capacity	22,500 MW
	Number of units	32 (2 sets)
	Average annual power generation	> 88.2 billion kWh
Ship Lock	Navigation capacity	10,000 tonnage fleet
	Annual (one-way) navigation capacity	50 million tons
Vertical Ship-lifter	Max capacity	3,000 tonnage passenger ship or barge
	Annual (one-way) navigation capacity	3.5 million tons

Source Three Gorges Hydropower Station (2019)

reached RMB 103.4 billion (US$ 15.9 billion) from 1993 and 2009. But there was no transparency about the scale of the fund as well as the way it had been used (Chang et al. 2018).

The first criterion of assessment of the TGD is the effect of flood control. The function of flood control of the dam has appeared to work well. There has been no major flood disaster downstream since the dam was completed (Chang et al. 2018). A group of experts belonging to the Chinese Academy of Engineering undertook an in-depth stage-based evaluation on the TGD Project and produced the final result in 2010. For instance, the flood control capacity of the Jing River, one of the branch rivers in the Yangtze River, had improved thanks to the TGD which can control about 95% of an annual average amount of flood water in the Jing River and approximately two-thirds of flood water volumes beyond the upstream of Wuhan City. Prior to the project, flood control facilities in the Jing River were able to protect people and economic assets against 10-year frequency floods but the dam allowed the population, cities, and industrial sites along the river to be safe against 100-year frequency floods. The flood water diversion sub-project within the TGD Project has relieved the pressure to control flood water for the Jing River as well as Dongting Lake.

On the basis of the hydrological data between 1860 and 2007, the team of experts concluded that the TGD would be able to reduce an annual average of 28,000 people affected by flood and to produce economic benefits equivalent to RMB 7.61 billion (US$ 1.17 billion) according to the average prices of 2007 in China. The effectiveness of flood control of the dam would be much larger if this were applied to the 1998 Flood in China, which would be able to decrease the number of people affected by the flood, 600,000, and to downscale economic losses equivalent to RMB 78.84 billion (US$ 12.13 billion) (Chinese Academy of Engineering 2010).

Hydropower generation by the TGD has brought about enormous socioeconomic benefits to China. The designed hydropower capacity of the dam is 82 billion kWh per annum, which is equivalent to RMB 22.3 billion (US$ 3.43 billion) in revenue every year (25 cents per unit price for 1 kWh). The total installed hydropower capacity reaches 22.5 million kW, which accounted for 3% of the national electricity generation as well as 14% of the total amount of hydroelectricity in China in 2007. The multi-year amount of hydroelectricity generation is 88.2 billion kWh, which reached 2.6% of the national electricity generation and 18% of the total amount of hydroelectricity in China in 2007. Until the end of 2013, the dam had produced 700 billion kWh, which is worth approximately RMB 182 billion (US$ 28 billion) (Chang et al. 2018; Chinese Academy of Engineering 2010).

Hydropower has been transmitted to the Shanghai Municipality, and Hubei, Hunan, Henan, Jiangxi, Jiangsu, Zhejiang, Anhui, Guangdong, Chongqing, and Sichuan Provinces, and has accelerated socioeconomic development in those areas. The dam's hydropower began to be generated in 2003, and by the end of 2008, the accumulated amounts of hydroelectricity generation were 288 billion kWh, which contributed to China's GDP by RMB 1,683.2 billion (US$ 258.9 billion).

The dam has made substantial contribution to generating massive amounts of hydropower itself and has paved the way for central, eastern, and southern parts of the country to enhance transmission networks. Plans of a series of dam projects in the upstream of the Yangtze River began to be considered thanks to the flow control capacity of the dam, and the quality of the national electricity grid has been improved. In addition, the dam has served as a major alternative to replace the important role of coal-based thermal power plants in China. Based on the data from 2007, the annual average volume of hydropower generation of TGD, 84.7 billion kWh, would be able to reduce the emission of 626 billion tons of CO_2, 7 billion tons of SO_2, and 3.7 billion tons of NOx (Chinese Academy of Engineers, 2010). The dam decreased coal consumption by 31 million tons per annum, resulting in avoiding 100 million tons of Green House Gas (GHG) emissions and one million tons of SO_2, 370,00 tons of NOx, 10,000 tons of CO, and large amounts of mercury (Chang et al. 2018).

Inland navigation in the Yangtze River has improved thanks to the dam. The dam reservoir which stretches over 660 km from Yichang City to the Chongqing Municipality provides favorable conditions for large vessels sailing and has expanded the upstream-to-downstream transport annual carriage capacity from 10 to 50 million tons in the river. Such an improvement of inland navigation has entailed socioeconomic development in the southwestern parts of China as well as the boom of inland navigation industries in the Yangtze River. By the end of 2008, the cargo volumes through the dam amounted to 685.7 billion tons, which were more than three times the volumes in 2003. Sailing conditions became even better after the impoundment of the reservoir so that there had been a gradual decrease of boat or ship accidents on the river (Chinese Academy of Engineers 2010).

In addition to these three major benefits, the dam has brought about a myriad of other benefits. For example, downstream areas receive more volumes of water even in drought periods, and cities and counties along the river are provided more water for economic development, social well-being as well as agricultural production. Water diversion for the middle route of the South North Water Transfer Project is stemming from the Danjiangkou Reservoir, which is adjacent to the dam, and therefore, the dam helps quenching the thirst in North China, including Beijing and Tianjin (Chinese Academy of Engineers 2010).

A variety of potential problems related to the project were discussed even at the early phase of the project, and major concerns have been associated with human and environmental impacts of the dam. The most controversial challenge for the project has been the relocation of more than 1.3 million people near the dam reservoir. At the early stages of the dam project, a series of new policies were introduced to implement resettlement-related policies and programs, including economic development of resettled areas and individual and household-level compensation. These policies were labeled as 'Resettlement with Development.' Primary purposes of the policies were to support relocated people with higher income levels and a better quality of life coupled with bolstering the surrounded areas' economy (Chang et al. 2018).

A total of 19 counties in the Chongqing Municipality and one county in Hubei Province were directly affected by the dam. In addition, numerous residents from other provinces and counties around the reservoir were involved in the resettlement project. For instance, 25,000 people in affected counties in Hubei Province were relocated to other counties in the same province, and 20,000 residents in affected counties in Chongqing were evacuated and resettled in other counties in Chongqing. 70,000 people were moved to other provinces, i.e., Sichuan, 9,000, Jiangsu, Zhejiang, Shandong, Hubei, and Guangdong, 7,000 each, Shanghai and Fujian, 55,000 each, and Anhui, Jiangxi, and Hunan, 5,000 each. There was often a discrepancy between the planned and the actual numbers of relocated people, and for example, 8,094 residents from Chongqing were eventually relocated to Anhui Province although the original plan was to relocate 5,000 people (CTG 2020a; Wang et al. 2014).

Relocated people often faced a variety of difficulties once settling in new places. Tough living conditions were attributed to the lack of farmland and other resources which were somewhat different from the promises given by local governments, and newly settled people had extreme difficulty mingling with local communities. It is difficult to estimate social costs of 1.3 million relocated people in addition to the total costs of the dam construction (Chang et al. 2018; Wang et al. 2014).

The project wreaked havoc on the remaining residents near the reservoir. Concerns were related to landslides, triggered by increased pressure on the surrounding land, a growth of waterborne diseases, and deteriorating biodiversity. Landslides frequently took place in the course of filling up the reservoir, and in November 2008, 3,050 m^3 of earth and rock tumbled onto a highway in Badong County near a tributary of the Yangtze River to the reservoir, burying a bus and killing at least 30 people. More serious issue related to the impoundment of the dam reservoir is a possibility of earthquakes. This phenomenon is called, 'reservoir-induced seismicity.' There were 822 tremors in seven months after the September 2006 reservoir-level increase according to the China Three Gorges Power Company (Mvistendahl 2008). Chinese Academy of Engineers (2010) also admitted that it should be critical to continue to pay close attention to the possibility of earthquakes around the dam site because of unstable geological structures.

The 2020 floods in China became the primary testing ground for the capacity of the TGD in terms of flood control. The volume of water flow into the dam's reservoir amounted to 75 million liters per second in August 2020, which was nearly approaching the maximum capacity of water storage in the dam. Despite the rumor that the dam might collapse due to substantial amounts of water flowed into the reservoir, the Chinese authorities reassured that the dam would perfectly work well against the once-in-a-century flooding. Chinese media companies claimed that the dam had prevented even worse flooding in large cities downstream, including Wuhan (Myers 2020; Pyo 2020).

Regarding inland navigation, the TGD has enhanced inland navigation, particularly allowing 3,000 to 10,000 tonnage vessels to travel between the upstream and the downstream of the river. However, large boats and ships are able to be in operation only for up to six months due to less water volumes in dry periods (Pyo 2020).

Attention is paid to the altered flow regime of the Yangtze River due to the dam, which has resulted in worsening water quality. As for water quality, the concentration of nitrate increased twice downstream, and many heavy metal ion such as Pb (Plumbum, lead), Cu (Copper), Cd (Cadmium), and Cr (Chromium) have turned out to be concentrated much more in the river. Massive amounts of pollutants discharged from nearby cities and counties have jeopardized the water quality of the dam reservoir, which would be estimated at around ten million tons of plastic bags, bottles, animal corpses, trees, and other detritus. These pollutants would have otherwise been flowed downstream and eventually to East Sea but have been trapped and accumulated behind the gates of the dam. More than 1,600 deserted factories, mines, dumps, and potential toxic waste sites were flooded, and untreated wastewater and solid wastes have continued to pollute the Yangtze River together with NPS pollutants from agriculture as well as various pollutants from industries.

The study on the water quality of the dam reservoir from 2006 to 2011 discloses that although the overall water quality of the reservoir remained sound, water quality at the outlet in Yichang City remarkably declined over time with higher pH and Permanganate COD (CODMn) (chemical oxygen demand through potassium permanganate index) values and lower concentrations of dissolved oxygen (DO) and ammonia nitrogen (NH3-N) than the inlet area in Zhutuo. This confirms the dam's impacts on the deterioration of water quality by blocking water flow downstream and continuously receiving large volumes of pollutants from adjacent areas. In addition, it is important to note that the decomposition of vegetation and organic materials on the bottom of the reservoir becomes a major source of GHGs (Wang et al. 2014; Zhao et al. 2013).

One of the common challenges embedded in dam development is siltation that takes place behind the water gates of dams. The TGD is no exception but more concern about this issue is attributed to the scale of siltation. In 2007, an estimated 530 million tons of silt accumulated in the reservoir due to the reduced flow velocity behind the dam. Soaring levels of sediment would possibly hamper water flow through the sluice gates, which could engender more flooding to occur upstream in the event of heavy rainfall. In addition, less water flow downstream in the river would lead to less water as well as less sediment, which could damage floodplain downstream, such as the Jiangshan (Yangtze River and Han River) Plain and Dongting Lake Plain (Wang et al. 2014).

It is cautious to make the conclusion that the TGD would trap vast amounts of siltation behind its gates, thereby causing various environmental damages in the long term. Even though there was little change in the volumes of water in the upstream of the Yangtze River, a variety of anthropogenic activities such as hydraulic structure construction, water conservation works including afforestation projects, and sand mining entailed a reduction of siltation downstream after the new millennium. In the 1990s, the annual average volume of water in the upstream of the river reached 391.3 billion m^3, which showed a slight reduction compared with previous years by 1.8%. The amounts of siltation, 377 million tons, in the same period were much less than previous years by 23%. This trend intensified at the advent of the new millennium, and from June 2003 to December 2007, after the impoundment of the dam reservoir, the total volumes of stored water in the reservoir were as large as 1,790 billion m^3 with the total amounts of siltation, 951 million tons. The annual average volume of water reached 358 billion m^3 with the annual average amount of siltation, 190 million tons, and these figures ranged within the designed capacity of the dam for water storage and siltation, 85% and 37%, respectively (Chinese Academy of Engineers, 2010). Multi-year data are available in Table 7.4.

China boasts rich biodiversity and is home for 10% of the world's vascular plants (those with stems, roots and leaves). The reservoir area occupies 20% of Chinese seed plants with more than 6,000 species. The TGD Project has flooded some habitats and reduced water flow

Table 7.4 Siltation in the Yangtze River in the different time periods

Year	Annual average water flow volume (billion m^3)	Annual average siltation volume (million ton)
1950–1986	398.6	493
1961–1970	419.6	509
1991–2000	391.3	377
2003–2007	358	190

Remark This was measured in Chuntan and Wulong in the Chongqing Municipality
Source Chinese Academy of Engineers (2010)

to other places, especially downstream areas and resulted in changing weather patterns. Less sediment downstream has triggered the morphological evolution of the river channel and the hydrological regime of lake wetlands, including Poyang Lake. Ecosystems in Poyang Lake have badly been damaged, which poses threats to the survival of migrant birds and aquatic fish and gives negative impacts on human health in the lake basin. Changing inundation patterns due to the dam project have induced the transmission of schistosomiasis, exacerbating contamination, and engendered the eutrophication of Poyang Lake.

Even prior to the dam, specific fish populations were damaged due to other dams, such as the Gezhouba Dam in 1981, 40 km downstream from the TGD. It is believed that the Gezhouba Dam caused a sharp decline in the populations of the Chinese sturgeon, the Yangtze sturgeon, and Chinese paddlefish (Wang et al. 2014; Zhang et al. 2016). Other fish species in the Yangtze River have been in jeopardy because of the dam, because less flooding takes place downstream in the river, disconnect the network of lakes around the middle reaches and decrease water volumes downstream. Consequently, this alteration of ecosystems has made fish difficult to survive, including the Baoji dolphin (white dolphin). In the period between 2003 and 2005, the timing of impoundment, the commercial harvest of four carp species, and the number of drift-sampled carp eggs and larvae reduced remarkably (Wang et al. 2014).

It is worthwhile to pay attention to possible droughts caused by the dam. Whereas the dam has long been planned in order to control floods, the dam would entail droughts in central and eastern China. In January 2008, the water level of the Yangtze River reached the lowest in 142 years, resulting in stranding ships along the waterway in Hubei and Jiangxi Provinces. The phenomenon was partly attributed to the dam that triggered the micro-level climate change, reducing the flow volume by 50% (Mvistendahl 2008). Another severe drought occurred in the Yangtze River Basin in 2011, and after the drought, torrential rains caused floods and landslides, giving tremendous economic losses, human casualties, and environmental destruction (Wang et al. 2014). Long-term impacts of the dam on the hydrodynamics of the river can be combined with those of the South North Water Transfer Project, particularly associated with the middle route that would take 13 billion m³ per annum for the parched North China since 2014.

Questions on the TGD Project have partly been answered, perhaps related to technical issues, but social and environmental questions remained to be answered in the long term although it is still doubtful whether those who are able to answer the questions are willing to do so.

Water Transfer Projects

Water Transfer Projects in China

There is no official report or dataset to provide the information on all water transfer projects in China. Different sources partly reflect details of a variety of water transfer projects, including planning documents for nine major water basins, the annual report of the water authority in each city, environmental impact assessment reports for transfer projects, the official website of each water transfer project and academic studies. Yu et al. (2018) collected the data on inter-basin water transfer projects that embrace mega projects with total pipeline (or man-made canal) length of more than 50 km or annual water transfer volume of larger than 100 million m³, and as a result, the total number of such projects is 59 in China.

China boasts 18 large-scale water transfer projects whose transferred water volumes were as large as 51 billion m³ in 2015. Half of the projects will be completed in the future. Among them, the most renowned project is the South North Water Transfer Project, and the eastern and middle routes of the project were completed in 2014, which accounted for approximately 70% of the total large-scale diverted water volume in the country.

Large-scale water transfer projects are aimed at addressing one of the imperative challenges in China's water resources management, the imbalance of water endowment. At the embryonic stage of water transfer projects in China, the Hai and the Huai River Basins were the major beneficiaries of those projects, importing 2.7 billion m³ and 0.6 billion m³ of water per annum from the Yellow River, respectively. It is important to note that the Yellow River increasingly faces water insecurity in terms of water flow. The average water resources per capita in the Hai, the Huai, and the Yellow River Basins were all lower than 1,000 m³ in 2015. The Yangtze River, which abounds in vast amounts of water, has become a major conduit which the thirsty three river basins rely on. The Hai, the Huai, and the Yellow River Basins imported 10 billion m³, 12.3 billion m³, and 19.5 billion m³ of water per annum from the Yangtze River, respectively. More water had been transferred from the Northwest River Basin to the Yellow River which was equivalent to 0.4 billion m³. A noteworthy fact is that the Yellow River also exports water to the other river basins despite its water shortage problem (Yu et al. 2018). Shandong Province is reluctant to tap in diverted water from the Yangtze River through the eastern route of the SNWT project due to the high price of the water and would like to keep sourcing water from the Yellow River (Chen et al. 2020).

It is maintained that most of the large-scale water diversion projects in China benefit the three river basins thanks to their significance in socioeconomic and political senses, particularly food security. While the three river basins are major beneficiaries, the Yangtze River Basin seems to have been sacrificing itself for the other river basins by exporting its water, contributing to more than 88% of water diverted through large-scale projects (Yu et al. 2018).

Attention is paid to multi-dimension impacts of water diversion projects in the country. First, a general challenge of economic aspects for China's water diversion projects is the cost overrun. For instance, the estimated construction cost of the eastern and the middle routes of the SNWT project was US$ 18 billion in 2002, however, the final bill for the two routes amounted to US$ 31 billion in 2014, almost twice the initial budget estimate (Yu et al. 2018).

Another issue is the balance between water donor and receiving regions. The Yangtze River Basin has made tremendous contributions to achieving water security in the Hai, the Huai, and the Yellow River Basins. By doing so, however, an opportunity of US$ 830 billion increase in GDP per annum would be lost to the Yangtze River Basin whereas the three river basins which import water from the Yangtze River can increase annual GDP by US$ 1,190 billion all together (Yu et al. 2018).

It can be accepted that water diversion projects in China can contribute to China's continuous socioeconomic development, primarily by resolving severe water shortages in North China. Special attention should be placed on environmental consequences of these projects in the long term. Long-distance water transfer projects generally require large amounts of electricity for pumping, which are estimated at more than 50 billion kWh, and the mean electricity use per m^3 of transferred water was estimated at 0.6 kWh in 2015, more than twice that of local water production that is shown in the case of Qingdao with reference to transferred water from the Yangtze River through the eastern line of the SNWT project (Wen et al. 2014). The calculated average electricity use for the SNWT project was 1.4 kWh per m^3 of water, which amount to approximately four times that for local water production. In addition, inter-basin water transfer projects lead to a growth of GHG emissions by 48 million tons per annum (Yu et al. 2018).

Ecological degradation has occurred due to inter-basin water transfer projects, particularly a decrease of water volumes. The total amounts of 3.3 billion m^3 of water were diverted from the Yellow River, which was equivalent to 11% of its annual flow, and the Han River serves as a main source for the middle route of the SNWT project and divert water for additional two water diversion projects, the Han River-to-Wei River Water Transfer Project in Shaanxi Province and the North Water Diversion in Hubei Province. With no surprise, these three diversion projects require the same water source, the Han River, so that there would be about one billion m^3 of water in deficit for local people. In order to fill the gap of water deficit, the construction of the Yangtze River to the Han River diversion project is envisaged, which sounds environmentally unfriendly (Guo et al. 2016; Yu et al. 2018).

A growing number of water diversion project have been introduced for enhancing ecological conditions, including the Peacock River Diversion Project that is aimed at remedying degraded wetland ecosystems in Tarim River (Zhuang 2016). Nevertheless, it is important to note that water donor regions often face environmental degradation because of the change in water depth, river flow and substrates, and water transfer projects transport not only water resources but also alien species of flora and fauna, which can engender the disruption of ecosystems (Yu et al. 2018; Zhuang 2016).

Social impacts of inter-basin water transfer projects are measured based on the fairness of social welfare between water donor and receiving regions. There is a clear demarcation between the Yangtze River and the Huai, the Hai, and the Yellow River Basins in terms of social welfare fairness. The Huai, the Hai, and the Yellow River Basins would be able to meet the water demand of 43 million, 32 million, and 37 million people by diverting water from the Yangtze River, respectively, whereas the Yangtze River Basin would not be able to meet the water demand of 70 million people per annum because of water diversion.

At the national level, water transfer projects drove more than 500,000 people to leave their homes in 2015, which demonstrates disadvantageous situations that residents have to face in water donor areas. On the contrary, water receiving areas took advantage of transferred water, irrigating more than 17,000 km^2 of arable land. Another significant issue is the cost of construction and ensuing high water prices. In Shanxi Province, the designed capacity of the Wangjiazhai Project is to divert 1.2 billion m^3 of water, however, the annual transferred volume of water reaches only 4.2 million m^3. The major reason behind this situation is associated with the high water price, and the cost of transferred water is as high as RMB 8 (US$ 1.2)/m^3 whereas the cost of local surface or groundwater is only RMB 3 (US$ 0.5)/m^3. More water through transfer projects requires more taxpayers' money (Yu et al. 2018).

The South North Water Transfer Project

There is a basic assumption behind the early envisagement of the South North Water Transfer Project, 'too less water in North, too much water in South.' For instance, the conflict between water resources capacity and socioeconomic development is seriously acute in the Yellow/Huang-Huai-Hai River Basins in which the total amount of water resources per capita is estimated as low as 21% of the national average. The river basins have made substantial contribution to the Chinese economy, occupying 35% of its population and 35% of its GDP. But there are only 7.7% of China's total water resources available in the river basins (Zhao et al. 2017). Water shortage in North China was already serious in the early 1950s due to population growth, industrialization, and urbanization, particularly near Beijing and Tianjin. Growing pressures of these socioeconomic phenomena pushed forward water engineers to seek for dramatic solutions. One of ambitious but risky options was to divert water from the Yangtze River to the Yellow River, suggested by the Director of the Yellow River Conservancy Commission, Wang Huayun. Chairman Mao Zedong agreed with his idea, and the Chinese government began to take into serious consideration the launch of the project (Construction and Administration Bureau of the SNWT Middle Route Project 2019; Shui 2008).

It took almost half a century to put this project idea into practice and hold a ground-breaking ceremony, and the construction of the eastern and middle routes managed to embark on in 2003. The reason why the project was postponed was primarily associated with continuous sociopolitical turmoil between the 1950s and the 1970s, such as the Great Leap Forward Movement (1958–1962) and the Cultural Revolution (1966–1976). In addition, the open-door policy kicked off in 1978, and there was little room for the central government to launch such a large-scale inter-basin water transfer project which would require intense planning, substantial amounts of public funds, political bargaining between the center and local governments, and institutional reforms, e.g., as water rights and water pricing. The project managed to kick off in 2003 when the government was confident of making planned achievements and coping with expected challenges based on more than two decades' reform experiences.

Vast volumes of water will be diverted from the Yangtze River to North China with the distance of approximately 1,500 km, mainly to relieve water shortages of politically and economically important areas in North China, including Beijing and Tianjin. This massive-scale project consists of three man-made canals from South to North, namely the eastern route, the middle route, and the western route. These artificial canals will be linked to the four major rivers in China, the Yangtze River, the Huai River, the Yellow River, and the Hai River, which makes Chinese hydraulic engineers' ambition come true, the water grid, featuring 'four latitudes and three longitudinal water courses.' Such a gigantic scale of water grid can regulate and allocate water from south to north and from east to west and gives an impact on almost one-third of China's landmass (Chen et al. 2002; Construction and Administration Bureau of SNWT Middle Route Project 2019; Liu 2000; Liu and Zheng 2002; Zhang 2009).

The project plans to transfer 44.8 billion m^3 of water per annum, including 14.8 billion m^3 via the eastern route, 13 billion m^3 via the middle route, and 17 billion m^3 through the western route. Invaluable transferred water through the project will be allocated for household, agricultural, and industrial uses in addition to the enhancement of ecological rehabilitation. The first phase of the eastern route was completed and began supplying water on 15 November 2013, and the first phase of the middle route started to deliver water on 12 December 2014 (Construction and Administration Bureau of SNWT Middle Route Project 2019). Construction works of the 2nd Phases in both routes are underway. As for the western route, the central government pledged to begin construction during the 12th FYP (2011–2015) and the 13th FYP (2016–2020), however, the plan does not seem to materialize soon. The completion of all the construction works will be expected in 2050 (see Table 7.5).

Primary attention of top leaders has been paid to the project because the project manifests itself in revealing the political legitimacy and the triumphant achievement of the CCP through the hydraulic mission. The Office of the Construction Council of the South North Water Transfer Project was established within the State Committee which has been led by Premier Li Keqiang since 2008. Such a structure of project construction and management at the central level confirms the national magnitude of the project. The project encapsulates the CCP-led central government's intention to claw back its power over water resources allocation and exploitation, which has been devolved into the hands of local governments since the 1980s.

Government offices of the project have been created at all levels of government in water donor and recipient areas, and the offices are

Table 7.5 Construction plan of each phase of the three routes of the South North Water Transfer Project

Phase	Route	Construction period	Additional volume in completion (billion m^3)	Accumulated volume in completion (billion m^3)
1st Phase	Eastern 1st	2003–2013	18.5	18.5
	Middle 1st	2003–2014		
2nd Phase	Eastern 2nd & 3rd	2015–2030	15	33.5
	Western 1st & 2nd	Postponed		
3rd Phase	Middle 2nd	2031–2050	12	45
	Western 3rd			

Source Modified based on Li and Xu (eds.) (2004), and Construction Administration Office of the SNWT Middle Route Project (2019)

supervised by one upper-level project official and one local government official. The network coordinates the project's operation with other related government bureaus and agencies. In addition to government offices of the project, there are three state-owned enterprises (SOEs) involved in the project, the SNWT Project Eastern Route Corporation in Beijing, the Construction and Administration Bureau of the SNWT Project Middle Route in Beijing, which are managed by the SNWT Project central government office, and Water Source Corporation Ltd, which is administered by the Yangtze Water Resource Commission in Wuhan. These SOEs are in charge of routine water supply matters, which represents the phenomenon of 'de-bureaucratization of water' in China. This unique system encompasses extensive outsourcing, the creation of companies (often SOEs), and cost recovery from end-users (Nickum 2010; Rogers et al. 2020).

The 2018 administrative reform bolstered the influence of the Ministry of Water Resources related to the SNWT project, because SNWT offices are now under the auspices of the MWR except for pollution responsibilities that are taken care of by the Ministry of Ecology and Environment. In late October 2020, the creation of the South-to-North Water Diversion Group was announced by Premier Li Keqiang in order to strengthen the management of the project and improve its engineering system as well as the allocation of water resources. The management structure of the SNWT Project seems to be constantly evolving for optimizing planning, construction, and management of the project (Rogers et al. 2020; Xu 2020).

The total cost of the SNWT project may amount to about US$ 40 billion. Whereas the construction of the eastern and middle routes may require US$ 10 billion each, that of the western route may need approximately US$ 20 billion. These figures reflect the calculation result of the late 1990s, and in 2011, the state media suggested that the total cost of the project would soar up to US$ 76 billion (*China Daily* 2011).

The responsibility of financing for the total cost of the project was shared between the central government and beneficiary provinces coupled with debt financing from banks in China. Initially, the central government would like to contribute to 20% of the total cost, local governments would channel 35%, and bank loans would amount to 45%. However, complaints and discontent of local governments which argued large financial burdens had forced the central government to renegotiate with local governments about the distribution of financing responsibilities (Liu and Zheng 2002).

Attention is placed on the financing structure of the first phase of the eastern and middle route. According to the 2008 estimate based on the prices of 2004, out of the total investment, RMB 254.6 billion (US$ 39.1 billion), the central government would contribute to RMB 41.4 billion (US$ 6.3 billion) (16.26%), the South North Water Transfer Fund, RMB 29 billion (US$ 4.4 billion) (11.39%), bank loans, RMB 55.8 billion (US$ 8.6 billion) (21.92%), self-financing of local governments and companies, RMB 4.3 billion (US$ 0.6 billion) (1.69%), and the Important Hydraulic Project Fund, RMB 124.1 billion (US$ 19 billion) (48.74%). Although the direct capital input from the central government seems to account for only 16.26% of the total investment, the real commitment of the central government should include contributions from the Important Hydraulic Project Fund, 48.74%, which is the leftover from the Three Gorges Dam Fund since 2009. Consequently, the central government is now responsible for financing 65% of the total investment. Local governments which receive water through the two routes are supposed to keep the South North Water Transfer Fund ongoing by collecting urban water bills from different water users, which will amount to 11.39% of the total investment. Bank loans seems to play a less role in financing the project, 21.92% (China South-to-North Water Diversion Project Editorial Board 2018a).

It is argued that the South North Water Transfer Project is a centrally planned, financed, engineered, and controlled hydraulic infrastructure. Whereas official documents have unveiled the increased sum of investment for the project in a relatively transparent fashion (at least the eastern and middle route so far), there is no clear indication of payback schedule in detail. This demonstrates the conventional approach of project financing to this hugely expensive and techno-centric hydraulic project in a top-down policymaking manner. Table 7.6 summarizes the specifications of the three routes, and Fig. 7.4 shows the location of the three routes.

Prior to the final touches of the eastern and the middle routes, the Global Financial Crisis occurred in 2008, which made China's annual GDP growth rate plummeted from 14.2% in 2007 to 9.7% in 2008, even further down to 9.4% in 2009 (World Bank 2020). Although the era of more than the double-digit GDP growth rate was over, the way how China managed to sustain its growth rate at around 9% in the midst of the financial crisis was to quickly provide the stimulus package that was estimated up to RMB 4 trillion (US$ 615 billion). One of the positive impacts of the package was to speed up the construction of water infrastructures in the country, and the government paid special attention to the SNWT project as a significant water infrastructure example, allocating RMB 2 billion (US$ 307 million) as part of the stimulus package (Jiang et al. 2020).

Eastern Route

The eastern route is to divert 14.8 billion m^3 per annum, and the route delivers this volume of water from the Yangtze River to North China with the distance of 1,466 km through a system of pumps, rivers, lakes, reservoirs, and canals. An average volume of water supply after the completion of the route will amount to about 9.06 billion m^3, which

Table 7.6 Specifications of the three routes in the South North Water Transfer Project

	Eastern	Middle	Western
Recipient areas	Jiangsu, Anhui, Shandong, Hebei, Tianjin	Henan, Hebei, Beijing, Tianjin	Gansu, Ningxia, Inner Mongolia, Shaanxi
Construction period	2003–2013 (1st and 2nd Phase)	2003–2014 (1st Phase)	Under consideration
Volume delivered (billion m^3/year)	14.8	13	17
Length (km)	1,466	1,432[a]	320
Dam construction	N/A	Danjiangkou dam elevating from 162 m to 176.6 m in height	New dams of more than 200 m in height
Capital Investment[b] (RMB billion)	1st Phase: 38.3 (US$ 5.9 billion) 2nd Phase: 22.4 (US$ 3.4 billion) 3rd Phase: 11.6 (US$ 1.8 billion) (Total 72.3/US$ 11.1 billion)	1st Phase: 201.3 (US$ 31 billion)	130–240 (US$ 20–37 billion)
Usage	Agricultural, domestic, industrial & navigation	Agricultural, domestic and industrial	Agricultural, domestic, industrial and environmental

Remarks [a]This includes 155 km in the sections of Hebei Province and the Tianjin Municipality, [b]In 2014, the State Council decided to revise the total amount of investment for construction of the first phase of the eastern and middle routes, amounting to RMB 308.2 billion (US$ 47.4 billion) (China South-to-North Water Diversion Project Editorial Board (2018a)

Source Construction and Administration Bureau of the SNWT Middle Route Project (2019), China Eastern Route Corporation of South-to-North Water Diversion (2018), SNWT Project website. Available Online http://nsbd.mwr.gov.cn/, accessed 12 December 2020, and Zhang (2009)

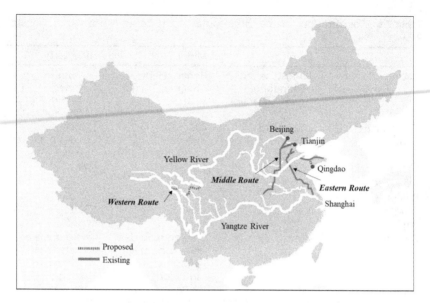

Fig. 7.4 Location of the Three Routes in the South North Water Transfer Project (*Source* Author)

excludes the amount of water losses in the course of delivery. A total of 54 large-scale pumping stations are required to lift water 65 m from the Yangtze River and deliver water to Dongping Lake in Shandong Province where the route is divided into the two sub-routes. The one goes north, crossing the Yellow River via the underground tunnel and flows into Tianjin. The other one is linked to the Shandong Peninsula through the existing Yellow River-to-Qingdao Water Diversion Project (China Eastern Route Corporation of the South-to-North Water Diversion 2018; China South-to-North Water Diversion Project Construction Yearbook Editorial Board 2006; Li and Xu 2004; Zhang 2009) (see Table 7.7 and Fig. 7.5).

As seen from Table 7.7, each phase in the eastern route requires approximately US$ 2 billion to 6 billion, totaling US$ 11.1 billion. The immense scale of investment is subject to change, depending upon socioeconomic circumstances, i.e., inflation, water demands, or industrial structural change (China Eastern Corporation of South-to-North Water Diversion 2018).

Table 7.7 Water transfer volume in the three phases of the eastern route (Unit: billion m^3/year)

Province	1st Phase (2003–2007)	2nd Phase (2011–2014)	3rd Phase (2031–2036)
Jiangsu	1.93	2.21	2.82
Anhui	0.32	0.34	0.52
Shandong	1.35	1.69	3.72
Hebei	N/A	0.7	1
Tianjin	N/A	0.5	1
Total	3.6	5.44	9.06

Source China Eastern Route Corporation of the South-to-North Water Diversion [2018], China South-to-North Water Diversion Project Construction Yearbook Editorial Board [2006], Li and Xu [2004], Zhang [2009])

The beneficiary provinces are Jiangsu, Anhui, Shandong, Hebei Provinces, and the Tianjin Municipality. The construction of the route is to be completed in three phases, and the first and the second phases of construction were completed, and the route began to divert water in November 2013 to the Provinces of Jiangsu, Anhui, and Shandong. The third phase has not been in operation and will convey water to cities in Hebei Province and the Tianjin Municipality (Chen et al. 2020).

Diverted water will alleviate severe water shortages in North China for household, agricultural, industrial uses as well as ecological rehabilitation. For instance, 260,000 residents of Qixian county in Hebei Province have had access to the water diverted through the eastern route in recent few years, and prior to the project, many households managed to have a bath once a year because of water shortages.

More water available in North China has helped rehabilitate ecosystems, because additional water resources can guarantee additional water for ecological water flow. Better access to surface water through the eastern route leads households and farmers to resort to less groundwater, which results in addressing the over-exploitation of groundwater resources. Drought-hit regions in the northern parts of Jiangsu Province in April 2020 were provided emergency water for rice production in 1.1 million ha through transfer projects connected to the eastern route (Hou 2020).

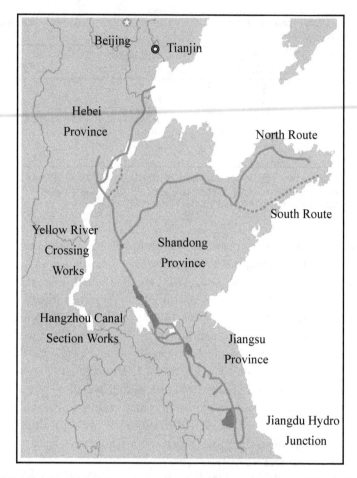

Fig. 7.5 Roadmap of the eastern route of the South North Water Transfer Project (*Source* Author)

Numerous existent canals have been connected to the eastern route, including the Grand Canal, which is part of the route and was first constructed back to the Sui Dynasty in around sixth century, more than 1,400 years ago, and Jiangsu Province's existing south-to-north project that has been expanded and extended northward (Cho 2011). Thanks to these existing infrastructures, engineers in charge of the route have not necessarily invested their time, fund, and energy to cope with a complexity of socioeconomic and cultural challenges related to migration. Major works have been undertaken to restore the function of inland navigation in the Grand Canal so that various construction works were associated with revamping or renovating dilapidated waterway facilities along the route (Freeman 2010). The total number of forced migrated people was estimated around 10,000. From Chinese authorities' points of view, there was no need to be worried about high transaction costs of resettlement issues in the eastern route (Chen et al. 2020; Freeman 2010; Liu 2000; Liu and Zheng 2002; Webber et al. 2017).

There are several challenging issues embedded in the eastern route. Ecologists have expressed their concerns about long-term disruption of ecosystems due to the reduction of water volumes in the downstream of the Yangtze River, which amounts to 14.8 billion m^3 per annum, equivalent to 2–3% of the annual flow of the river. Less water means less sediment for lower reaches of the river, and the sediment is essential for maintaining riparian and coastal wetlands. Less water may also cause the acceleration of estuary salinization, which would badly affect water supply in one of the major urban centers, Shanghai. Another concern is the mixture of chemicals and biota from different ecosystems watered into the lakes along the route, namely Hongze, Luoma, Nansi, and Dongping Lakes, which can entail disruption in wildlife. Consequently, the breakdown of biogeographic demarcations may occur, inducing the northward migration of alien species and the proliferation of parasitic diseases such as schistosomiasis (Crow-Miller 2014; Crow-Miller and Webber 2017; Liu 2000; Liu and Zheng 2002; Zhang 2009).

Water pollution control has been a daunting task for central planners in charge of the route. The middle route is relatively less prone to water pollution thanks to fences off from public access, however, many sections of the eastern route are open, consisting of canals and regulating lakes (Rogers et al. 2020). The eastern route goes along heavily industrialized cities and areas whose canals and waterways are connected to each other, serving not only as inland navigation routes but also dumping ground for polluted water. In addition, NPS pollution in suburban and rural areas has become acute along the route. For example, more than 20% of water pollution monitoring stations along the route reported severe water pollution even though primary water pollution abatement works had been completed in March 2011 (*China Daily* 2011; Zhang 2009).

Major cities such as the Tianjin Municipality have demanded the central government to ensure the enhancement of the quality of water transferred through the eastern route. The Tianjin Municipality and Hebei Province explicitly conveyed their concerns about the quality of transferred water through the route from 2011, and Tianjin has decided to rely more on desalinated water rather than diverted water through the route, expanding the capacity of desalination plants by converting 4 billion m^3 of seawater to 150 million m^3 of freshwater.

Such a policy shift securing additional water through desalination rather than transferred water is attributed to the issue of water quality of the eastern route as well as of price differences between the two options. The production cost of desalinated water plummeted to RMB 5 (US$ 0.77)/m^3 whereas that of transferred water through the eastern route was estimated at almost RMB 20 (US$ 3)/m^3. Qingdao and Yantai Cities, which are situated in the eastern coast and are supposed to receive water through the eastern route, took into serious consideration the introduction of desalination plants (Lei 2009; Wong 2011).

Confronted with strong requests from local governments about water pollution abatement, 424 brand-new wastewater treatment plants have been constructed for abating water pollution along the route. The central government introduced strong measures to require local authorities to enforce water pollution regulations more strictly, which has been implemented in accordance with the Three Red Lines in 2011 (China Eastern Route Corporation of the South-to-North Water Diversion 2018).

Eye-catching improvements of water quality along the route are often reported. In the 11th Five Year Plan (2006–2010) and the 12th Five Year Plan (2011–2015), Jiangsu and Shandong Provinces made many efforts to improve water quality by strengthening water pollution management, blocking pollutants, and enhancing water pollution monitoring. The completion of 426 water pollution control projects of the two provinces culminated in reducing the total amounts of pollutants into the route by more than 85%. Consequently, such efforts kept the water quality of the route Grade III and above (China Economic Net 2017). The Chinese National Environment Monitoring Center in November 2020 informed that the portions of Grade II and Grade III water quality in the eastern route reached 17.6 and 76.5%, respectively, which demonstrates good achievements in water quality enhancement of the route (China National Environment Monitoring Center 2020).

Middle Route

The middle route has diverted 9.5 billion m^3 of water per annum since 2014 from the middle reaches of the Yangtze River to North China, particularly to Beijing and Tianjin through the 1,432 km canal. It takes 15 days to transfer water from Danjiangkou Reservoir to water treatment plants in Beijing. Tap water in Beijing consists of almost 70% of transferred water and 30% of water allocated from local sources. The transferred water is also used for irrigating 5 million ha of croplands, divided evenly between the southern and northern regions of the Yellow River (China South-to-North Water Diversion Project Construction Yearbook Editorial Board 2006; Economist 2018) (see Table 7.8 and Fig. 7.6).

Construction of the route comprises two major components: (1) the elevation of the existing Danjiangkou Dam on the Han River; and (2) the construction of the water conveyance system. The height of the dam has been elevated from 162 to 176.6 m, which has expanded the dam's storage capacity from 11.6 billion to 29.05 billion m^3, inundating an additional land area of 370 km^2. Danjiangkou Reservoir has been

Table 7.8 Water transfer volume in the two phases of the middle route (Unit: billion m^3/year)

Location/Phase	1st (2014)	2nd (2030)
Beijing	1.2	1.7
Tianjin	1	1
Hebei Province	3.5	4.8
Henan Province	3.8	5.5
Total	9.5	13

Remark The total volumes of transferred water in the 2nd and the 3rd Phases indicate the accumulated amounts of transferred water.
Source China South-to-North Water Diversion Construction Project Yearbook Editorial Board (2006) and Li and Xu (2004)

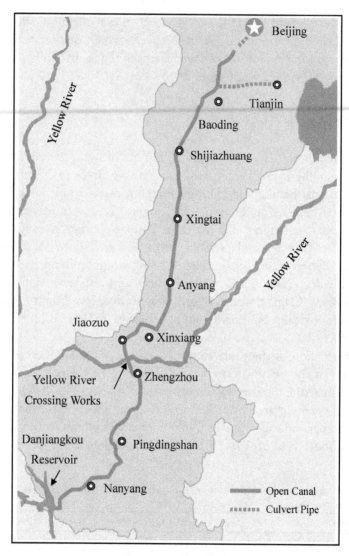

Fig. 7.6 Roadmap of the middle route of the South North Water Transfer Project (*Source* Author)

connected with the water conveyance system toward Beijing with the length of 1,277 km (155 km inside Hebei Province and the Tianjin Municipality). Water is diverted from high terrain in the reservoir and flows by gravity (Zhang 2009).

As a consequence, those who lived near the dam had to be relocated, and the number of relocated people has been counted 345,000 (181,000 from Hubei Province and 164,000 from Henan Province). The resettlement process was primarily implemented in 2010 and 2011, and the final process was completed in 2013. The official policy of migration related to the SNWT Project is to provide upfront compensation and continuous support. Continuous technical and financial support from local governments will be given according to the resettlement plan. This project is regarded the first water resources project in the country which has improved the compensation standard of acquired cultivated land, from 10 to 16 times of its annual output. In addition, compensation for each household is based on a standard of at least 24 m^2 of floor area per person for the house reconstruction (China South-to-North Water Diversion Project Editorial Board 2018b; Construction and Administration Bureau of South-to-North Water Diversion Middle Route Project 2019; Zhao et al. 2017) (see Table 7.8 and Fig. 7.6).

Similar with the eastern route, the middle route has contributed to alleviating severe water shortages in North China. Water deficit per annum in the North China Plain can be as large as 6 billion m^3 in order to support Beijing and Tianjin, the political, socioeconomic and cultural centers, and Henan and Hebei Provinces, the significant grain producers and industrial bases.

Until now, more than 120 billion m^3 of groundwater resources have been exploited in the North China Plain, and the deep aquifer is overpumped so that diverted water through the middle route and the eastern route will ameliorate such dire situations of water shortages in North China (Construction and Administration Bureau of South-to-North Water Diversion Middle Route Project 2019). For instance, the average groundwater level in the capital rose to 22.49 m by late September 2020, which was 3.68 m higher than in 2015. Numerous dried-up springs in the city revived thanks to the middle route (Hou 2020).

Between December 2014 and June 2018, the middle route had diverted 15 billion m^3 of water, benefiting 19 large- and medium-sized cities, including Beijing and Tianjin. More than 53 million people have been drinking water transferred through the middle route. Among various beneficiaries in the middle route, Beijing is the largest beneficiary. Beijing's per capita water resource has grown from 100 to 150 m^3, and less water has been allocated from the Miyun Reservoir, which has been one of the major water supply sources for the capital, eventually saving 2.3 billion m^3. This enhances water security for the capital in terms of water supply for various purposes as well as water supply in flood and dry seasons (Construction and Administration Bureau of South-to-North Water Diversion Middle Route Project 2019).

In 2019, more than 73% of Beijing's tap water stemmed from Danjiangkou Reservoir, providing tap water service to more than 12 million residents, and 14 out of 16 districts Tianjin benefited from the water through the middle route. From 2014 to 2018, Beijing received more than 5.2 billion m^3 of water through the middle route, and approximately 70% of the transferred water were allocated to waterworks for being stored in reservoirs and for replenishing urban rivers and lakes. By late October 2019, the average depth of groundwater resources in Beijing's plain areas was recorded at 22.78 m, which was 2.88 m higher than the average depth of groundwater resources prior to the transferred water through the middle route in late 2014. Eventually, groundwater reserves in the city increased by 1.62 billion m^3 (Wang and Zhao 2019; Xinhua 2019).

More than four million people in Hebei Province no longer suffer from drinking highly concentrated fluorine and salty-tasted water because of the middle route. More than 10 cities in Henan Province, including Nanyang, Luohe, Pingdingshan, and Xuchang Cities, have had access to water diverted through the route.

Economic benefits through the middle route have grown at RMB 50 billion (US$ 7.7 billion) per annum in terms of the industrial and agricultural output value, and the government argues that approximately 500,000 to 600,000 new jobs per annum have been created (Construction and Administration Bureau of South-to-North Water Diversion Middle Route Project 2019).

Ensuring a good level of water quality for the route has been one of the major works with the establishment of 12 automatic monitoring stations. The water quality along the route has met or surpassed the Grade II-level national standard of surface water thanks to the sophisticated water quality management (Construction and Administration Bureau of South-to-North Water Diversion Middle Route Project 2019). In addition, compared with the eastern route, the middles route embraces less developed areas with less industrial and agricultural activities, which naturally guarantees the quality of water transferred along the route.

There are several challenges in the middle route. Large-scale upgrading works for the Danjiangkou Dam resulted in inundating an area of 370 km^2 around the reservoir, and 345,000 residents had to be relocated out of their homes because of the rising waters. In addition to the resettlement process, the project planners were concerned about pollution on water source areas in Danjiangkou so that numerous industrial units along canals and reservoirs were shut down, including major industries near Danjiangkou, fish farming and turmeric processing. The administrative measures have engendered the rise of resettlement costs and the reduction of tax base for local governments.

Relocated people do not appear to be entirely happy about circumstances they have to face. Local officials paid a local village person RMB 450 (US$ 69)/m^2 for his or her old house but charged RMB 1,000 (US$ 154)/m^2 for his or her new house. The government charged 40% of what was paid for his or her land, arguing that the portion should belong to the government. Relocated people are given a new house but with no savings, no job and RMB 600 per annum of income support, which is insufficient to live on (Crow-Miller 2014; Crow-Miller and Webber 2017; Economist 2018).

Furthermore, a large gap related to the amount of compensation between geographical regions is a significant issue in the resettlement process, and the complexity of different levels regarding urban and regional areas has triggered confusion and misunderstanding for local residents related to the standards and compensation arrangements. Another thorny issue is that compared with highway and energy projects, the compensation standard for water resources project is lower. There are gaps between different types of projects in terms of compensations for two parcels of lands near each other so that relocated people often find the system not equitable (Zhao et al. 2017). In addition, adjacent farm areas around the reservoir were inundated, and the size of farmland per capita is reduced by 753 m^2 (Barnett et al. 2015; Yu et al. 2018).

Another resettlement project has been undertaken in order to protect water source areas near Danjiangkou Reservoir for people in southern Shaanxi's Ankang, Hanzong, and Shanluo prefectures. 880,000 people had to be relocated by 2014 for the purposes of poverty relief, disaster relief and pollution control upstream of the reservoir, and the resettlement process was expected to continue until 2020. It is too early to judge the extent to which this resettlement project is smooth in terms of compensation, job training, overall living conditions, and income levels (Rogers et al. 2020).

The water quality issue in the middle route is closely related to pollution near water source areas that are supposed to be tightly controlled and protected. The upper Han River Basin, the water supplying areas of the middle route, occupies 95,000 km^2 with the population of 13 million, and there are few adequate wastewater treatment facilities. NPS pollution is severe around the river basin because more than 80% of the total population in the upper river basin are farmers who are actively engaged in diverse farming activities, triggering NPS pollution.

Less volumes of water in the Han River mean a possible rise of phosphorus and nitrogen levels due to the annual transfer of 14 billion m^3 of water to north, which is equivalent to 35% of the total discharge at the point of transfer. The change of hydrological regime can induce widespread blooming of green algae. Similar with the eastern route, alien species can migrate along the canal to the north and cause disruption of local ecosystems (Zhang 2009).

In November 2020, the water quality of the middle route was assessed as Grade I (57.1%) or II (28.6%) based on the seven environmental monitoring stations along the route, which indicates good management of water quality (China National Environment Monitoring Center 2020). Danjiangkou Reservoir has three monitoring stations, which inform a high centration of nitrogen, making the water quality of the outlet of the reservoir Grade IV so that many efforts have been made to reduce the level of nitrogen in the reservoir. The other concerns are linked to pollution spills from vehicles near bridges and phytoplankton (Rogers et al. 2020).

Western Route

The construction of the western route has not kicked off yet in July 2021. No specific schedule has been released although the original plan was to begin and complete the construction of the first phase between 2010 and 2020 (Li and Xu 2004; Lu 2006; MWR 2001). The 14th Five Year Plan (2021–2025), which was announced in March 2021, includes the serious discussion of action plans for the route although no specific plan about the beginning of the project has been announced (Xinhua 2021).

The western route plans to transfer 17 billion m^3 of water per annum from three tributaries (Tongtian, Yalong, and Dadu Rivers) of the Yangtze River to the upper reach of the Yellow River including the northwestern provinces and a few autonomous regions, including Qinghai, Shanxi, Gansu Provinces and Ningxia and Inner Mongolia Autonomous Regions. Diverted water through the western route will benefit an area of 684,100 km^2 and more than 53 million people. A noteworthy fact is that 30–40% of planned transfer water per annum will be opted for ecological restoration. Industrial and household uses of the transferred water will account for more than 60%, and the agriculture, 6% until 2050.

The total investment for the construction of the 1st phase in the western route was estimated at RMB 89.5 billion (US$ 13.7 billion) in 2000. A newly calculated figure in 2006 indicates that the total sum of the investment for all phases of the route would increase to RMB 207 billion (US$ 31.8 billion) reflecting inflation and rising costs of construction materials, wages and financing (Lu 2006; MWR 2001). The western route is closely linked to the 'Go West' campaign that will bolster socioeconomic development in the least developed regions of China (Li and Xu 2004; Zhang 2009) (see Table 7.9 and Fig. 7.7).

Whereas the eastern and middle routes terminate adjacent to Beijing and Tianjin in the North China Plain, the western route arrives in

Table 7.9 Water transfer volume of the three phases of the western route (Unit: billion m³/year)

Location/Phase	1st (not decided)	2nd (2030)	3rd (2050)
Lanzhou-Hekou	2	4.5	7
Longmen-Sanmenxia	1	2.5	5
Yellow River Tributaries	1	2	5
Total	4	9	17

Remark The total volumes of transferred water in the 2nd and the 3rd Phases indicate the accumulated amounts of transferred water
Source China South-to-North Water Diversion Construction Project Yearbook Editorial Board (2006), and Li and Xu (2004).

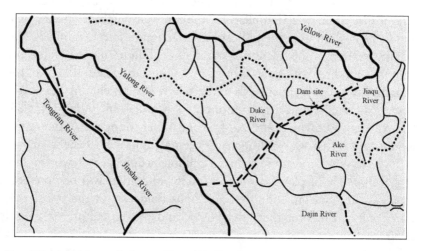

Fig. 7.7 Roadmap of the western route of the South North Water Transfer Project (*Source* Author)

totally different terrains and geophysical areas. The water source areas of the route will be situated on the Qinghai-Tibetan Plateau whose average elevation is as high as about 4,000 m above the sea level, spanning 220,000 km². Weather conditions are harsh, and more than 72% of the region are covered with permafrost whereas general landscapes are fraught with mountains, and frequent seismic activities have been recorded.

There will be sub-routes of tunnels through mountains of the eastern Tibet, western Qinghai, and Sichuan Provinces. Water transfer from the

Tongtian River to the Yalong River will be carried out by gravity with a 302 m high dam on the Tongtian River. The tunnel from the Tongtian River to the Yalong River will be 158 km long. Water will be diverted from the Yalong River to the Dadu River by gravity with a 175 m high dam on the Changxu reach of the Yalong River and a tunnel with a length of 131 km. The next step is to divert water from the Dadu River to the Yellow River through pumping and a 296 m high dam on the Dadu River. Then, water will be pumped again 458 m to the Jiaqu River, a tributary of the Yellow River through a 28.5 km tunnel. Hydraulic engineers have proposed the designs that involve the construction of at least six large dams and seven tunnels with the total length of more than 250 km. Four dams out of the planned six dams are as high as more than 120 m and elevated canals, and those infrastructures are extremely costly and difficult to build (Chen 2006; Ghassemi and White 2007; Ma et al. 2006; Zhang 2009).

Water recipient areas of the western route are spanning 715,000 km^2 with a population of 62 million people. It is anticipated that water demands from agriculture, households, and industries will rise by 40% from 56.2 billion m^3 per annum in 2009 to 78.4 billion m^3 per annum by 2050. Transferred water of 17 billion m^3 per annum will flow more than 3,762 km along the Yellow River and the average annual discharge of the river will soar by more than 80% with 274% increase during the dry season. Additional amounts of water can improve erosion control efforts by irrigating vegetation in the upper reaches of the Yellow River, however, careful attention is placed on the alteration of biota in the riverine ecosystems of the upper river basin (Zhang 2009).

Considering more than 3.3 billion m^3 of water transferred per annum through the route, an annual economic benefit would reach RMB 13.5 billion (US$ 2 billion). This can be calculated from the total economic benefit, RMB 17.2 billion (US$ 2.6 billion) minus economic losses of hydropower generation in the upper reaches of the Yangtze River, RMB 3.7 billion (US$ 570 million) (Lu 2006).

Attention is paid to an array of challenges embedded in the western route. The first challenging issue is linked to geophysical conditions along the route. Harsh terrains and numerous high mountains are a major barrier for construction in general in addition to difficulty in delivering construction materials. A prerequisite prior to major construction works of the route is to develop relevant infrastructures, such as railways and roads, which can badly affect the fragile landscape.

Second, the planned construction of a large reservoir, which can be as large as 289.9 km^2 or 710.4 km^2, depending upon the final design, may trigger earthquakes and landslides and a rise of regional evaporation and precipitation by 0.4–0.7%. Complex and unstable geological structures may hinder construction works for dams, tunnels, and canals. Even though the increase can be marginal, this new phenomenon may aggravate the frequency and intensity of summer storms, thereby giving devastating impacts on mountainous landscapes and altering the biophysical conditions along the valley downstream (Chen 2006; Huang 2006; Ma et al. 2006; Zhang 2009).

Third, it is worth investigating ecological impacts of water transfer through the route. Whereas the eastern route diverts only 2–3% of water source areas (14.8 billion m^3 out of 960 billion m^3) and the middle route, about 33% (13 billion m^3 out of 38 billion m^3), the western route plans to divert more than 66% (17 billion m^3 out of 25.5 billion m^3). This comparison epitomizes possible impacts of the western route on water donor areas in the upper reaches of the Yangtze River, and there must be systematic and careful assessments of the project's long-term impacts on the environment, ecosystems, and geological structures (Ma 2006).

Fourth, compared with the eastern and middle routes, the western route includes a subtle challenge that is linked to international relations with China's neighboring countries, India, Bhutan, and Bangladesh. The route can give impacts on water flows of the Yarlung Tsangpo-Brahmaputra River, which is one of the 15 major transboundary rivers of the country. There are four riparian countries in the river, including China, India, Bhutan, and Bangladesh. Once the route is in full swing, vast amounts of water will be diverted from the Yarlung Tsangpo-Brahmaputra River for the route. In spite of India's repetitive requests about the information of the dam, China went ahead for the construction of the Zhangmu Dam prior to its formal acknowledgment of the existence of the dam and completed the first phase of the Zhangmu Dam in November 2014. Environmental or ecological impacts of the dam would not be far-reaching due to the relatively smaller scale of installed power generating capacity of 510 MW compared with that of the Three Gorges Dam, 18,000 MW (Xinhua 2020; Zhang 2016).

China plans to build more dams on the river, however, India also envisages to build more than 160 dams on the mainstream as well as tributaries of the Yarlung Tsangpo-Brahmaputra River. As Zhang (2016) maintains, more concerns have been reflected about water transfer projects by China rather than hydropower dam building on the river, because less water volumes would entail serious impacts on local livelihoods and ecosystems of downstream communities in India.

Further Implications

The South North Water Transfer Project Construction Bureau (2003) provided the information on expected benefits in three areas: (1) social; (2) economic; and (3) ecological areas. Social benefits are associated with the alleviation of water shortage challenge in the northern parts of China, which can facilitate socioeconomic development and urbanization and resolving the problem of high degrees of fluorine and saline in drinking water for more than seven million people. This diversion project helps establish the system of a balanced water allocation between North and South and of mutual support between East and West.

Regarding economic benefits, a vast amount of investment for the construction of the eastern and middle routes will boost the national GDP by 0.2–0.3%. Water recipient regions in North China after the completion of the two routes are expected to have an increase of agricultural production value by RMB 50 billion (US$ 7.7 billion). In particular, wheat and cotton production industries will be major beneficiaries that increase their productivity thanks to additional water through the project. Furthermore, in the course of the construction of the eastern and the middle routes, large investments for the project entailed a growth of China's GDP by 0.12% per annum with almost 100,000 workers employed every day during the apex of the construction. New jobs will be created for 500,000 to 600,000 people per annum until the completion of all the phases for the eastern and middle routes (SNWT Project Construction Bureau 2003; Xinhua 2019).

Most of socioeconomic benefits in the project are concentrated on urban households and industrial users. The estimated economic benefits of both routes would range between RMB 30 billion (US$ 4.6 billion) and RMB 60 billion (US$ 9.2 billion) per annum. Those of the middle route may reach approximately RMB 30 billion (US$ 4.6 billion) per annum, and out of them, 92%, RMB 27.7 billion (US$ 4.3 billion) can be given to urban households and industrial users. The remaining RMB 2 billion (US$ 300 million) will be given to the irrigation industries and RMB 500 million (US$ 77 million) to the flood disaster sector. This estimation demonstrates the unbalanced allocation of socioeconomic benefits between urban households and industries, and rural areas and agricultural industries (Nickum 2006; SNWT Project Construction Bureau 2003).

Looking at ecological benefits, the transferred water through the eastern and middle routes will contribute to alleviating the problem of over-exploitation of groundwater resources in North China. The amounts of transferred water for ecological and agricultural purposes will reach approximately six billion m^3 per annum, which should be crucial in retarding ecological degradation and rehabilitating ecosystems. As a consequence, ecological rehabilitation through the project can pave the way for the country to be better equipped with its capacity to cope with unprecedented extreme weather events caused by climate change (SNWT Project Construction Bureau 2003).

There are further challenges regarding the SNWT project. The most salient question is associated with the political legitimacy of the project. Since the stage of envisaging the project, the central government has stressed the magnitude of the project for socioeconomic development of the country. The final destinations of the eastern and middle routes are economically and politically powerful urban bases in China, centered on Beijing and Tianjin. This discloses the reason why the successful implementation of the project has been imperative. Crow-Miller (2014) and Crow-Miller and Webber (2017) maintain that the configuration of water donor and recipient regions demonstrates the Chinese government's perspective on where the country's resources should be concentrated for its socioeconomic development. The SNWT project, to some extent, is likely to deepen the gap between rich and developed urban areas (water recipient areas) and marginalized cities and rural towns (water donor areas).

The notion, 'sacrifice' has often been heard from interviews with local officials and media reports in Henan and Hebei Provinces and the entire Han and Middle Yangtze River Basin. Those who live in the water donor areas can feel marginalized and lose opportunities for their future. Such a centrally controlled and planned engineering project can create a new spatial pattern of inequality and ensuing consequences can be far-reaching for Chinese society in the long term (Crow-Miller 2014; Crow-Miller and Webber 2017).

Second, the feasibility of the project can be questioned related to economic soundness. The expected cost of the project is exceeding US$ 62–79 billion (three routes), much more than the cost of the Three Gorges Dam (estimated more than US$ 40 billion). Although Berkoff (2003) maintains that the project is economically feasible and politically significant, many water experts cast a doubt on the economic feasibility of the project, and it is difficult to envisage how the central and local governments under the current underpriced water tariffs in China would pay back the massive project cost.

The economic feasibility of the project has already been under close investigation at the central government level since the revised scale of investment for construction of the first phase of the eastern and middle routes has amounted to over RMB 308.2 billion (US$ 47.4 billion) as of 2014. This figure is far more than the originally planned amount of investment, RMB 124 billion (US$ 19 billion) in 2002 and the revised figure in 2008 was RMB 254.6 billion (US$ 39.1 billion) considering the prices of the third quarter in 2004. Additional costs are associated with project complexity, delays of design drawings, design errors, poor constructability, inadequate budget, escalation of material costs, inflation, project variations, and rising costs related to resettlement, including land acquisition (China South-to-North Water Diversion Project Editorial Board, 2018a; Zhao et al. 2017).

One of the fundamentals to safeguard the economic feasibility of the project is to levy an adequate level of water tariffs to water users along the routes. The beneficiary provinces along the eastern and middle routes have so far been the Beijing and the Tianjin Municipalities, Hebei, Henan, and Shandong Provinces. Whereas Beijing and Tianjin have not introduced separate water pricing systems for the transferred water, Hebei and Henan Provinces have set an independent water pricing system for the transferred water. Shandong Province has decided to launch an integrated water pricing reform for water recipient areas with regard to the transferred water.

Crow-Miller and Webber (2017) argue that the centralization of project operation is explicit in water pricing regarding the SNWT project. Water pricing in urban and rural areas of China is generally in the hands of local governments, however, the price of water supplied by the project is controlled by the central government. The National Development and Reform Commission (NDRC) in 2003 and 2006 introduced a two-part pricing system. In the system, a basic price will be paid by local water authorities which covers the capital and overhead costs of the project, and a measured price is charged to consumers to cover running costs and make some portions for profit based on volumes. Prices for raw water delivered through the middle route were determined by the NDRC in December 2014, slightly lower than the basic price plus calculated prices. For instance, Beijing people are supposed to pay RMB 2.33 (US\$ 0.36)/m^3 and Tianjin RMB 2.16 (US\$ 0.33)/m^3 instead of RMB 2.96 (US\$ 0.45)/m^3 and RMB 3.03 (US\$ 0.46)/m^3. This implies the centralization of water resources management related to the SNWT project and represents economic benefits given to the rich and powerful in water receiving regions in the form of subsidy for the water tariff.

Whereas a strong public resistance is expected against the upsurge of domestic water tariff, it is the local governments that should bear the extra cost purchasing the expensive water and might fall in huge deficit, not able to recoup their investment. Indeed, Shandong Province has experienced public resistance against receiving water through the eastern route, and many cities within the province have refused to source the water through the eastern route due to high prices. Eventually, such a phenomenon has engendered idling of diversion projects and wasting public investments with heavy subsidies (Chen et al. 2020). It is imperative for the local governments to incrementally introduce rational water pricing, which can provide them with financial soundness related to the project as well as enhance dilapidated water supply and sanitation systems in urban areas.

Third, another critical question for the viability of the project is whether it is worthwhile to opt for the project without considering alternatives for augmenting water supply. Recycled water has no longer been an alternative but a main contributor to water-scarce cities, including Beijing. In 2018, diverted water through the middle route contributed to 24% of total water supply of the capital, local surface water 8%, recycled water 27%, and groundwater 41%. This indicates that water supply of the capital would be better off if there were policy efforts to improve water use efficiency through either demand management, i.e., water pricing, or technical advancement, i.e., water saving techniques (Beijing Water Authority 2018; Rogers et al. 2020).

It is increasingly imperative to adopt and implement the water and energy nexus related to the transfer project. The eastern and the middle routes are supposed to divert an amount of 20.9 billion m^3 per annum by 2030, and this figure does not reflect the required amount of power for pumping up the water over 60 m high in the east and the electricity required for running 474 wastewater treatment plants (Parton 2018). Rogers et al. (2020) calculate that energy consumption for delivering water along the eastern route would account for 2.1%, 6%, and 2.6% of total electricity consumption in Jiangsu, Anhui, and Shandong Provinces, respectively, in 2016. In addition, the total amount of carbon emissions was calculated in the eastern route, the highest reaching 0.88 Mt CO2e (million metric tons of carbon dioxide) from 2016 to 2017 and amounting to 0.94 Mt CO2e by 2030. These results lead to a question if China should continue to depend upon such energy and carbon-intensive water resources through the project.

Fourth, the Chinese government should take into serious account possible impacts of climate change on the project (Liu 2008; NDRC, 2007; Piao et al. 2010). The extreme drought around the middle reaches of the Yangtze River between January and May 2011 led to questioning the appropriateness of the project and shaking the hypothesis in the project, 'too less water in North, too much water in South.' The amounts of rainfall in four months managed to reach only 40% of a mean rainfall in Hubei Province, and the drought badly affected Hubei and Jiangsu, Hunan Provinces, and the Shanghai Municipality. This drought also damaged crop production, wild animals and plants, and drinking water for over four million people (Watts 2011). Numerous environmentalists and experts warn that this unusual situation could take place again because of climate change (Wang and Zhang 2011). In this context, the SNWT project might aggravate droughts caused by climate change in South. The uncertain hydrological regime can make the feasibility of the project reconsidered, thereby hampering the political legitimacy of the project. Furthermore, water donor regions in South, including Jiangsu Province and the Shanghai Municipality, which have been uneasy about the project, can express their concerns and give a political pressure to Beijing against the project.

Fifth, it is crucial to highlight the relationships between different levels of governments regarding the SNWT project. Since large-scale water transfer projects often divert water between provinces, prefectures, and cities, close attention should be placed on the overall picture of the extent to which such a project can be incorporated into the total water supply system in local areas. Regarding the SNWT project, it is necessary to shed light on the role of the central government which has envisaged and implemented the project, and the river commissions which take responsibility for IWRM, i.e., water allocation to provinces, flood control, drought relief, and hydraulic projects. Provincial governments are responsible for piping transferred water through the SNWT project into their own water supply systems.

Sixth, socioeconomic and ecological benefits at the national level through the project accompany little evidence. Although the central government has advertised the magnitude of the project as the nationally imperative development project, the most beneficial area is the capital, Beijing. Considering the area around the reservoir in Danjiangkou appears comparatively poor and backward, the project, particularly the middle route, appears to take water from the poor to the rich.

Beijing seems to be better off benefiting from transferred water from the Yangtze River, providing water for various water users. The current level of water consumption of the capital reaches 3.6 billion m^3 of water per annum, which can be sourced from local reservoirs and rivers (2.1 billion m^3) and the middle route (1.1 billion m^3). This means that the middle route project cannot meet water demands of the capital, and its water consumption level would grow to more than 4 billion m^3 per annum by 2020. Looking at a larger picture, North China may require approximately 200 billion m^3 per annum by 2050, and the eastern and middle routes would be able to satisfy one-eighth of that.

More problematic issue is that the official figure of annual water transfer volume in the middle route should reach 9.5 billion m^3, however, water managers in Danjiangkou Reservoir admit that less than half of the planned extraction was flowed out of the reservoir in 2017 thanks to lower demands than expected, which has occurred because of the high price of reservoir water (*Economist* 2018).

Conclusion

The chapter has examined water resources development in China, paying special attention to large dam development and inter-basin water transfer projects. Large dams of the country have made tremendous contribution to food security, flood prevention and control, and hydropower generation together with their role to facilitate inland navigation and to rehabilitate ecosystem services. The magnitude of dam development in contemporary China since 1949 has continued to be emphasized in terms of its path to achieving modernization. The hydraulic mission has been further emphasized by the CCP for hydropower generation, which is critical for fueling the continuous socioeconomic development that can legitimize the rule of the CCP over the country. Numerous large dams in major rivers, i.e., the Yangtze River and the Yellow River, over the past four decades have primarily been commissioned for hydropower generation, including the Three Gorges Dam.

It is undeniable that hydropower generation through the dam has produced vast scales of socioeconomic benefits to Chinese society in general. Flood prevention and control, facilitation of inland navigation as well as more water to revive damaged ecosystems are additional benefits by such a nationally significant dam. But it is important to notice that these benefits have not evenly been shared among Chinese people, particularly for those have been relocated from their homes because of the dam construction and ecosystems that have badly been affected by the alteration of hydrological regime. In addition, socioeconomic gains through the dam seem to have been concentrated in highly industrialized eastern and southern coastal areas.

In-depth discussions on the South North Water Transfer Project have informed that the project encompasses more multifaceted challenges than it looks. The central government has appeared to tackle the many of problems at the early stage of the eastern and middle route effectively, such as water pollution issues in the eastern route and relocation problems in the middle route. However, a number of problems remain unsolved, such as potential ecological damage due to less water available in southern lakes and the estuary areas of the Yangtze River, possible public resistance to high price tags of transferred water, and economic viability of project costs, to name a few. In addition to these domestic matters, China might be confronted with its influential neighbor, India, related to the construction of the western route, which relates to the transboundary river, the Yarlung Tsangpo-Brahmaputra River.

The trajectory of water resources development in China demonstrates the fundamental role of water in achieving socioeconomic development over the past few decades through additional water supply for North China, flood prevention and protection, hydropower generation, and inland navigation. Although the monumental infrastructures for water resources development should be touted as the symbol for China's modernization, there would be the people and the environment behind the shadow of such achievements. More attention should be placed on the marginalized people and ecosystems that should not be forgotten or ignored if China were willing to achieve a 'quality growth' that has explicitly been declared at the 14th FYP (2021–2025) in March 2021.

References

Andreas, Lars, and Guowei Yang. 2011. *Recent Progress of Enabling Framework for IRBM in China—Some Knowledge from Our Perspective*. EU-China River Basin Management Program Technical Assistance Team. 17 March. Available Online http://www.euchinarivers.org. Accessed 4 January 2021.

Barnett, J., S. Rogers, M. Webber, B. Finlayson, and M. Wang. 2015. Sustainability: Transfer Project Cannot Meet China's Water Needs. *Nature* 527: 295–297.

Beijing Water Authority. 2018. Beijing Water Resources Bulletin (北京市水资源公报) 2018. Beijing Water Authority. Available Online http://swj.beijing.gov.cn/zwgk/szygb/201912/P020191219479807999291.pdf. Access 5 January 2021.

Berkoff, Jeremy. 2003. China: The South-North Water Transfer Project—Is It justified? *Water Policy* 5: 1–28.

Chang, Chun Yin, Zhanyang Gao, Amanda Kminsky, and Tony Reames. 2018. Michigan Sustainability Case: Revisiting the Three Gorges Dam: Should China Continue to Build Dams on the Yangtze River? *Sustainability* 11 (5): 204–215.

Chen, Dan, Zhaohui Luo, Michael Webber, Sarah Rogers, Ian Rutherfurd, Mark Wang, Brian Finlayson, Min Jiang, Chenchen Shi, and Wenjing Zhang. 2020. Between Project and Region: The Challenges of Managing Water in Shandong Province After the South-North Water Transfer Project. *Water Alternatives* 13 (1): 49–69.

Chen, Xiqing, Dezhen Zhang, and Erfeng Zhang. 2002. The South to North Water Diversions in China: Review and Comments. *Journal of Environmental Planning and Management* 45 (6): 927–932.

Chen, Zhiliang. 2006. Activities of the Earth's Crust in the 1st Phase of the Western Route of the South North Water Transfer Project (西线一期工程区地壳的活动性). In *Memorandum of the Western Route Construction Process of the South North Water Transfer Project (南水北调西线工程备忘录)*, ed. Leung Lin and Baojun Liu, 9–23. Beijing: Economy and Science Press.

China Daily. 2011. Pollution Blocking Water Path. *China Daily*, March 9. Available Online http://www.china.org.cn. Accessed 21 December 2020.

China Eastern Route Corporation of South-to-North Water Diversion. 2018a. Introduction to the Eastern Route Project (工程介绍). China Eastern Route Corporation of South-to-North Water Diversion (南水北调东线总

公司). 26 October. Available Online http://www.nsbddx.com. Accessed 18 December 2020.

China Economic Net. 2017. Impact of the South North Water Transfer Eastern Route First Phase Strategy More and More Apparent (南水北调东中线一期工程战略作用日益显现). 6 June. Beijing, *China Economic Net*. Available Online http://district.ce.cn. Accessed 5 January 2021.

China National Environment Monitoring Center. 2020. Monthly National Surface Water Quality Report, December 2020 (中国环境监测总站). Available Online: http://www.cnemc.cn/. Accessed 5 January 2021.

China South-to-North Water Diversion Project Construction Yearbook Editorial Board. 2006. *China South-to-North Water Diversion Project Construction Yearbook (中国南水北调工程建设年鉴) 2005*. Beijing: China Power Press.

China South-to-North Water Diversion Project Editorial Board. 2018a. *China South-to-North Water Diversion Project: Economy and Financing Volume (中国南水北调工程:经济财务卷)*. Beijing: China Water Power Press.

China South-to-North Water Diversion Project Editorial Board. 2018b. *China South-to-North Water Diversion Project: Resettlement Volume (中国南水北调工程:征地移民卷)*. Beijing: China Water Power Press.

China Three Gorges Corporation (CTG). 2020a. Introduction to the Three Gorges Dam Project (三峡工程). China Three Gorges Corporation Website. Available Online https://www.ctg.com.cn. Accessed 15 January 2021.

China Three Gorges Corporation (CTG). 2020b. Overview. China Three Gorges Corporation Website. Available Online https://www.ctg.com.cn. Accessed 22 February 2021.

Chinese Academy of Engineering. 2010. *Three Gorges Project Step-by-Step Evaluation Report (三峡工程阶段性评估报告)*. Beijing: China Water Power Press.

Cho, Younghun. 2011. *The Grand Canal and Chinese Merchants (대운하와 중국상인)*. Seoul: Minum Press.

Construction and Administration Bureau of the South North Water Transfer (SNWT) Middle Route Project. 2019. *South North Water Transfer Project: the Middle Route Project. Introductory Booklet*. Beijing, Construction and Administration Bureau of the SNWT Middle Route Project.

Crow-Miller, Britt. 2014. Diverted Opportunity: Inequality and What the South-North Water Transfer Project Really Means for China. *GWF Discussion Paper* 1409, Global Water Forum, Canberra, Australia. Available

Online https://globalwaterforum.org/2014/03/04/diverted-opportunity-ine quality-and-what-the-south-north-water-transfer-project-really-means-for-china/. Accessed 25 December 2020.

Crow-Miller, Britt, and Michael Webber. 2017. Of Maps and Eating Bitterness: The Politics of Scaling in China's South-North Water Transfer Project. *Political Geography* 61: 19–30.

Economist,. 2018. China Has Built the World's Largest Water-Diversion Project. 5 April.

Freeman, Carla. 2010. Quenching the Dragon's Thirst: The South-North Water Transfer Project—Old Plumbing for New China? *China Environment Forum*. Woodrow Wilson Center, Washington, DC.

Ghassemi, Fereidoun, and Ian White. 2007. *Inter-Basin Water Transfer—Case Studies from Australia, United States, Canada, China, and India.* Cambridge: Cambridge University Press.

Guo, Jing, Jia-Li Guo, Dan-Ying Wang, and Jin-Hua Liu. 2016. Vulnerability of Water Resources System in Hanjiang River Basin to Climate Change. Proceedings of 2016 International Conference on Modern Economic Development and Environmental Protection, 207–214.

Gupta, J., and P. van der Zaag. 2008. Interbasin Water Transfers and IWRM: Where Engineering, Science and Politics Interlock. *Physics and Chemistry of the Earth* 33: 28–40.

Habich, Sabrina. 2016. *Dams, Migration and Authoritarianism in China.* Abingdon: Routledge.

Hou, Liqiang. 2020. 120 Million Benefit from Project That Diverts Water. *China Daily*, 14 December. Available Online http://www.chinadaily.com.cn. Accessed 23 December 2020.

Huang, Yunqiu. 2006. Geological Challenges of Major Works in the Western Route Construction Process of the South North Water Transfer Project (西线工程的重大工程地质问题). In *Memorandum of the Western Route Construction Process of the South North Water Transfer Project (南水北调西线工程备忘录)*, ed. Leung Lin and Baojun Liu, 24–35. Beijing: Economy and Science Press.

Hvistendahl, Mara. 2008. China's Three Gorges Dam: An Environmental Catastrophe? *Scientific American*, March 25. Available Online http://www.scientificamerican.com. Accessed 6 January 2020.

International Commission on Large Dams (ICOLD). 2011. Constitution. Paris, ICOLD. Available Online https://www.icold-cigb.org/userfiles/files/CIGB/INSTITUTIONAL_FILES/Constitution2011.pdf. Accessed 21 February 2021.

International Hydropower Association (IHA). 2020. *2020 Hydropower Status Report*. London: IHA.
Jiang, Min, Michael Webber, Jon Barnett, Sarah Rogers, Ian Rutherfurd, and Mark Wang. 2020. Beyond Contradiction: The State and the Market in Contemporary Chinese Water Governance. *Geoforum* 108: 246–254.
Lei, Tian. 2009. Why Is the South-North Water Project Being Postponed? *Probe International*, October 1.
Li, Shantong, and Xinyi Xu, eds. 2004. *South North Water Transfer Project and Chinese Development (南水北调与中国发展)*. Beijing: Economy and Science Press.
Liu, Changming. 2000. Environmental Issues and the South-North Water Transfer Scheme. In *Managing the Chinese Environment*, ed. Richard L. Edmonds. Oxford: Oxford University Press.
Liu Changming and Zheng Hongxing. 2002. South-to-north Water Transfer Schemes for China. *Water Resources Development* 18 (3): 453–471.
Liu, Chunzhen. 2008. Study on Climate Change and Water in China. Presented at the Climate Change and Global Water Cycle Conference, Beijing, 23–27 November. Available Online http://www.mairs-essp.org. Accessed 13 June 2011.
Lu, Jiaguo. 2006. Input-Output Analysis and National Economy Assessment of the 1st Phase Western Route of the South North Water Transfer Project (西线一期工程投入产出分析及国民经济评价). In *Memorandum of the Western Route Construction Process of the South North Water Transfer Project (南水北调西线工程备忘录)*, ed. Leung Lin and Baojun Liu, 155–160. Beijing: Economy and Science Press.
Ma, Dongtao, Chen, Guojie, Chen, Shutao, Shang, Yuming, and Tu, Jianjun. 2006. Environment and Geological Disaster Challenge of the Western Route of the South North Water Transfer Project (西线工程区环境地质灾害问题). In*Memorandum of the Western Route Construction Process of the South North Water Transfer Project (南水北调西线工程备忘录)*, ed. Leung Lin and Baojun Liu.Beijing: Economy and Science Press.
Ma, Huaixin. 2006. Additional Discussion of the Western Route of the South North Water Transfer Project (也谈南水北调西线工程). In*Memorandum of the Western Route Construction Process of the South North Water Transfer Project (南水北调西线工程备忘录)*, ed. Leung Linand Baojun Liu.Beijing: Economy and Science Press.
Mertha, Andrew. 2008. *China's Water Warriors: Citizen Action and Policy Change*. Ithaca and London: Cornell University Press.

Ministry of Water Resources (MWR). 2001. *Outline of the Western Route Construction Plan in the South North Water Transfer Project (南水北调西线工程规划纲要)*. Beijing: Ministry of Water Resources.

Molle, Francois, Peter Mollinga, and Philippus Wester. 2009. Hydraulic Bureaucracies and the Hydraulic Mission: Flows of Water. *Flows of Power. Water Alternatives* 2 (3): 328–349.

Mvistendahl, Mara. 2008. China's Three Gorges Dam: An Environmental Catastrophe? *Scientific American*, March 25. Available Online https://www.scientificamerican.com/article/chinas-three-gorges-dam-disaster/. Accessed 27 April 2021.

Myers, Steven. 2020. After Covid, China's Leaders Face New Challenges From Flooding. *The New York Times*, 21 August. Available Online: https://www.nytimes.com/2020/08/21/world/asia/china-flooding-sichuan-chongqing.html. Accessed 11 February 2021.

National Development and Reform Commission (NDRC). 2007. *China's National Climate Change Program*. NDRC, June.

Nickum, James. 2006. The Status of the South to North Water Transfer Plans in China. In Report to the United Nations Development Programme, Human Development Report. New York, UN.

Nickum, James. 2010. Water Policy Reform in China's Fragmented Hydraulic State: Focus on Self-Funded/managed Irrigation and Drainage Districts. *Water Alternatives* 3 (3): 537–551.

Parton, Charlie. 2018. China's Looking Water Crisis. *Chinadialogue*. April.

Piao, Shilong, Philippe Ciais, Yao Yuang, Shen Zehao, Shuishi Peng, Shengjun Li, Liping Zhou, Hongyan Liu, Yuecun Ma, Yihui Ding, Pierre Friedlingstein, Chunzhen Liu, Kun Tan, Yongqiang Yu, Tianyi Zhang, and Jingyun Fang. 2010. The Impacts of Climate Change on Water Resources and Agriculture in China. *Nature* 467: 43–51.

Pyo, Nari. 2020. Debates on the Three Gorges Dam and the Response of the Chinese Government (싼샤댐을 둘러싼 논란과 중국 정부의 대응). Institute of Foreign Affairs and National Security (IFANS) Focus 2020–17K. September 14.

Rogers, Sarah, Dan Chen, Hong Jiang, Ian Rutherford, Mark Wang, Michael Webber, Britt Crow-Miller, Jon Barnett, Brian Finlayson, Min Jiang, Chenchen Shi, and Wenjing Zhang. 2020. An integrated assessment of China's South-North Water Transfer Project. *Geographical Research* 58 (1): 49–63.

Shapiro, Judith. 2001. *Mao's War Against Nature: Politics and Environment in Revolutionary China*. Cambridge and New York: Cambridge University Press.

Shui, Fu. 1998. Chapter 2. A Profile of Dams in China. In *The River Dragon Has Come: The Three Gorges Dam and the Fate of China's Yangtze River and Its People*, ed. Qing Dai, 18–24. International Rivers Network, Probe International.

Shui, Qingshan. 2008. *South North Large-Scale Redistribution of Water Resources (水资源的南北大调配)*. Beijing: Five Continent Press.

South North Water Transfer (SNWT) Project Website. Available Online http://nsbd.mwr.gov.cn/. Accessed 18 December 2020.

South North Water Transfer (SNWT) Project Construction Bureau. 2003. Expected Benefits of the South North Water Transfer Project (南水北调工程的预期效益). 26 August 2003. Available Online http://nsbd.mwr.gov.cn/zw/gcgk/gczs/200308/t20030826_1128050.html. Accessed 18 December 2020.

Stephens, Leah. 2016. Three Gorges Dam: Masterpiece or Impending Disaster? *Interesting Engineering*. April 26. Available Online http://interestingengineering.com. Accessed 15 January 2021.

Sun, Xingsong, Xiaogang Wang, Lipeng Liu, and Ruizhi Fu. 2019. Development and Present Situation of Hydropower in China. *Water Policy* 21: 565–581.

Three Gorges Hydropower Station. 2019. Three Gorges Hydropower Station Booklet.

Tilt, Bryan. 2015. *Dams and Development in China: The Moral Economy of Water and Power*. New York: Columbia University Press.

Wang, Keju and Yimeng Zhao. 2019. Diversion Project Slakes Northern Need for Water. *China Daily*, 25 December 2019. Available Online: http://www.chinadaily.com.cn (accessed 21 December 2020).

Wang, Pu., Shikui Dong, and James Lassoie. 2014. *The Large Dam Dilemma: An Exploration of the Impacts of Hydro Projects on People and the Environment in China*. Dordrecht: Springer.

Wang, Shourong, and Zuqiang Zang. 2011. Effects of Climate Change on Water Resources in China. *Climate Research* 47: 77–82.

Watts, Jonathan. 2011. China Crisis Over Yangtze River Drought Forces Drastic Dam Measures. *Guardian*, 25 May.

Webber, M., B. Crow-Miller, and S. Rogers. 2017. The South-North Water Transfer Project: Remaking the geography of China. *Regional Studies* 41: 370–382.

Wen, Hua, Lijin Zhong, Xiaotian Fu, and Simon Spooner. 2014. Water Energy Nexus in Urban Water Source Selection: A Case Study from Qingdao. Beijing, World Resources Institute.
Wong, Edward. 2011. Plan for China's water crisis spurs concern. *The New York Times*, June 1.
World Bank. 2020. National Accounts Data. Available Online https://data.worldbank.org/. Accessed 1 July 2020.
World Commission on Dams. 2000. *Dams and Development: A New Framework for Decision-Making*. London: Earthscan.
Xinhua. 2019. Mega water diversion project boon for China with broader spillovers. *Xinhua*, 13 December. Available Online http://www.chinadaily.com.cn. Accessed 21 December 2020.
Xinhua. 2020. The First Large-Scale Hydropower Station in Tibet Has Produced More Than 10 Billion kW so Far (西藏首座大型水电站累计发电突破100亿度). *Xinhua*, 16 April. Available Online http://www.xinhuanet.com/fortune/2020-04/16/c_1125865368.htm. Accessed 13 April 2021.
Xinhua. 2021. 14th Five Year Plan of the National Economic and Social Development and 2035 Long Term Goal (中华人民共和国国民经济和社会发展第十四五年规划和2035年远景目标纲要). *Xinhua*, 12 March.
Xu, Mei. 2020. Premier Calls for Enhanced Water Security, Ecology Work. *China Daily*, 24 October. Available Online http://www.chinadaily.com.cn. Accessed 21 December 2020.
Yang, Xiaoliu, Donzier, Jean-Francois, and Noel, Coralie. 2009. A Comparison between French and Chinese Legal Systems in terms of Integrated Water Resources Management. Presented at the *Yellow River International Forum*, 20–23 October. Zhengzhou.
Yu, Min, Chaoran Wang, Yi. Liu, Gustaf Olsson, and Chunyan Wang. 2018. Sustainability of Mega Water Diversion Projects: Experience and Lessons from China. *Science of the Total Environment* 619–620: 721–731.
Zhang, Hongzhou. 2016. Sino-Indian Water Disputes: The Coming Water Wars? *Wiley Interdisciplinary Reviews: Water* 3: 155–166.
Zhang, Quanfa. 2009. The South-to-North Water Transfer Project of China: Environmental Implications and Monitoring Strategy. *Journal of the American Water Resources Association* 45 (5): 1238–1247.
Zhang, Xiao, Zengchuan Dong, Hoshin Gupta, Guangdong Wu, and Dayong Li. 2016. Impact of the Three Gorges Dam on the Hydrology and Ecology of the Yangtze River. *Water* 8 (590): 1–18.
Zhao, Hongliang. 2013. *Centennial Hydraulic Project: South-North Water Diversion*. Beijing: China Intercontinental Press.

Zhao, Pei, Xiangyu Tang, Jialiang Tang, and Chao Wang. 2013. Assessing Water Quality of Three Gorges Reservoir, over a Five-Year Period from 2006 to 2011. *Water Resources Management* 27 (13): 4545–4558.

Zhao, Zhen-Yu., Jian Zuo, and George Zillante. 2017. Transformation of Water Resource Management: A Case Study of the South-to-North Water Diversion Project. *Journal of Cleaner Production* 163: 136–145.

Zhuang, Wen. 2016. Eco-environmental Impact of Inter-Basin Water Transfer Projects: A Review. *Environmental Science & Pollution Research International* 23: 12867–12879.

8

Water and Wastewater Service Market

Introduction

The chapter appraises the evolution of Chinese water and wastewater service market. Whilst the central government started to promote private sector participation in the water and wastewater service sectors in the late 1990s, the real push for the creation of the water market in China has been possible thanks to the commitment of local governments which opt for enhancing water and wastewater services by inviting private water companies.

Since 1949, one of the fundamental tasks of the Chinese Communist Party (CCP) has been to ensure the universal access to clean water and adequate sanitation services. This task, unfortunately, was inadequately undertaken with the complete domination of the public sector in water and wastewater services whilst private sector participation was never allowed until the late 1990s. Sociopolitical and economic sensitiveness about water and wastewater services hampered effective policies and programs that may bring about the better quality of water supply and wastewater treatment services.

Problems in urban water and wastewater services were explicitly identified and widely discussed at the center and local levels, i.e., underpriced water and wastewater services, inefficient management and operation of facilities, out-of-date technologies and facilities, and a paucity of investment (Fu et al. 2008; Lee 2006, 2010).

The retarding trend of global Public Private Partnership (PPP) projects as been observed in the water sector since 2007. But considering the number of new PPP contracts in the water sector, East Asia and the Pacific have demonstrated a growing trend since 2000, spearheaded by China. From 1990 to 2019, the global water and wastewater service sector had a total of 1,075 PPP financial closure projects with a worth of US$88.67 billion, and the region with the largest investment share was East Asia and the Pacific, accounting for 46% in the world. The number of water and wastewater PPP projects in the same region reached 635, and out of these, Chinese ones occupied 90%, 574 with a worth of US$19.14 billion. The statistics imply that the water and wastewater PPP market of China has still been growing with good potential (Qian et al. 2020; World Bank and PPI 2020).

The projection of the Global Water Intelligence (GWI) for the period between 2020 and 2025 is that the largest water and wastewater PPP market in the world would be China, with a worth of US$39.6 billion, followed by the United States (US$27 billion) and Japan (US$9.8 billion) (GWI 2020b). This indicates that the Chinese water market is and will be one of the most attractive ones in terms of future growth, which is good news for international and Chinese private players. Such a growth potential can serve as an opportunity or challenge for the central and local governments that will manage water and wastewater PPP projects for resolving water shortage and water quality enhancement under the framework of ecological civilization.

The development of China's water and wastewater PPP projects discloses a myriad of unique features in lieu of government policies, PPP contract types, major players (foreign and Chinese companies), and geographical distribution, to name a few. These characteristics are attributed to the country's unique political economy system, the socialist market economy, which embraces egalitarian approaches that provide water and wastewater services almost free of charge, and a strong bureaucratic nature of regulatory frameworks. In addition, the Chinese water market encompasses the commitment of the central and local governments to enhancing water and wastewater services in response to demands from society.

Urban water governance has evolved in order to accommodate the new mode of urban water resources management, and new laws and regulations have been introduced at the center as well as local areas. Foreign and Chinese players have adapted to the highly competitive and volatile market, which is heavily influenced by government policies.

The first part of the chapter highlights the development of PPP projects in water and wastewater service sectors of China since the late 1990s. Attention will be paid to primary policy drives, legal and regulatory settings, and the reshuffling of organizations at the center and local levels. In the second part, the study discusses the overall situations of water and wastewater service sectors, i.e., water supply, wastewater treatment, and desalination industries, and focuses on various models of water and wastewater PPP projects in China. Discussions are made on major players in the market, including foreign as well as Chinese companies. The last part of the chapter sheds light on a myriad of challenges and opportunities in the market.

Development of Water and Wastewater Service Market

From the late 1970s to the late 1990s, the urban water and wastewater service sectors in China epitomized the existence of regulatory authorities and management institutions under the same roof, all the funding and investment from the government, the state monopoly of operating facilities, and water pricing based on government's development policies. Such a firm grip of urban water and wastewater service sectors by the government had eventually engendered ensuing problems, i.e., deterioration of service qualities due to the rapid urbanization, mediocre levels of efficiency in Operation and Maintenance (O&M), and the lack of funding and future investment (Shen and Wu 2017).

As a last sacred cow in the reform era, the water and wastewater service sectors in China were eventually reformed in the late 1990s for improving efficiency, and service quality, and seeking investments. Foreign and Chinese companies reacted enthusiastically for providing water and wastewater services. In line with the upbeat responses from the private sector, the central government responded with institutional rearrangements favoring foreign investors, such as the fixed rate of return that would ensure a certain level of guaranteed profits for a contract period, 10–30 years (Lee 2006, 2010).

Water and wastewater services in Chinese cities had been provided by Chinese companies and owned by local governments. These companies were often incapable of financing and lacking new technologies with unskillful and inexperienced employees that received little training and incentives. In this context, little hope remained in improving water and wastewater services so that local governments increasingly resorted to private sector players on a contract basis (Spooner 2018).

By 1999, only 26 million people out of one billion in China received water and wastewater services by private operators through joint ventures with local partners, either municipalities or financial partners. The first private company's water contract was the Chengdu No. 6 Build-Operate-Transfer (BOT) plant for 18 years that was awarded Veolia Environment in 1998 (Lee 2006, 2010; Spooner 2018).

With the advent of the new millennium, overseas Chinese companies from Hong Kong, Singapore, and Malaysia became active in the Chinese water market through the advantage of their well-established local connections, little barriers of language and culture, and cutting-edge technologies. In the middle of cut-throat competitions between foreign and Chinese companies, the rules of the game were abruptly altered by the central government in 2002, which nullified the fixed rate of return for foreign companies' water and wastewater service projects. Such a policy change cooled down the boom of the market entry by foreign companies and prompted the exit of them, including Thames Water. There is little evidence that such a sudden policy shift occurred for favoring Chinese water companies against foreign invested projects, however, the number of BOT contracts awarded to Chinese companies soared up around this period (Lee 2006, 2010; Spooner 2018).

Veolia Environment made a genuine breakthrough in the market by acquiring the 50-year concession contract of Shanghai Pudong in 2003 through the purchase of equities and closed the similar deal in Shenzhen, 2004 (Gourmelon et al. 2015; Lee 2010; Spooner 2018). Since then, more foreign companies scrambled to the Chinese water market whilst Chinese players were committed to competing with foreign counterparts, taking advantage of understanding better legal and regulatory settings as well as socioeconomic policies. Until 2005, the Chinese water market was divided almost half and half between foreign and Chinese players, and the number of people served by private operators increased to 100 million with an annual growth at 26% (Gourmelon et al. 2015).

Chinese authorities loosened their tight grip on water and wastewater PPP projects by allowing different types of PPP, such as concessions and management contracts for water treatment and supply networks, and sewerage and treatment networks for joint venture companies in some cases. The period between 2005 and 2008 was a boom of investments into the Chinese water market by both overseas and Chinese investors. However, the boom quickly ebbed due to the 2008 financial crisis, and overseas players faced tougher situations because of their difficulty in channeling more fund for new joint venture projects. It is plausible to maintain that the 2008 financial crisis provided a turning point in the Chinese water market in which Chinese companies started to be more dominant in the market than international players in terms of utility business as well as financing capacities. Representative companies are Beijing Capital, Beijing Enterprises Water Group, and Sound Group (Spooner 2018).

Another interesting characteristic since 2005 is the diversification of companies involved in the Chinese water market. Newly participating companies come from property development, and construction and energy industries, i.e., Power China, China Communications Construction Corporation (CCCC), and the China State Construction Corporation (CSEC).

The global financial crisis in 2008 badly impacted upon the Chinese water market, and a number of foreign investors decided to leave the market. To fill the vacuum, the Chinese government was committed to attracting Chinese investors into the market and introduced stricter measures for environmental standards, which helped prompting the invigoration of wastewater treatment PPP projects. These were the new Standards for Drinking Water Quality in 2007 and the revised Water Pollution and Prevention Law in 2008. The governmental policy drives culminated in providing a third wave of the boom of water and wastewater PPP projects in China (Qian et al. 2020).

Dark clouds continued to loom over foreign companies opting for water and wastewater PPP projects in China. In 2017, the Catalogue for the Guidance of Foreign Investment Industries stipulated the obligatory measures that the projects involved with foreign capital should be built through a Sino-foreign joint venture and a majority ownership should be given the Chinese partner. The catalogue provides a specific example that the construction and operation of pipeline networks for gas, heat, water supply, and sewage in cities with a population of more than 500,000 should require Chinese parties to hold a majority of shares in the joint venture company (NDRC and MOC 2017; Spooner 2018).

In retrospect, from 2005 to 2015, there were three distinctive trends worth paying attention to. First, PPP projects had continued to proceed at a slow pace. Second, Chinese players had won over foreign counterparts, and target urban areas had been transferred from Beijing, Shanghai, Guangzhou, and Chongqing with more than 15 million populations to major economic centers in east, central, and west China with 3–15 million people. Third, as the water supply service sector had been saturated, private players focused more on wastewater service contracts through various types of PPP projects, including Transfer-Operate-Transfer (TOT) projects (Gourmelon et al. 2015; Lee and Choi 2015).

It is common to acknowledge that the Chinese water market is a tough one in many senses, including extremely low water tariffs in cities and the rise of costs. Another risky factor is too much competition for contracts with municipalities, which means a disadvantaged position of private players to pay high premiums to municipal governments for winning contracts. In the meantime, the downturn of Chinese economy since the 2008 financial crisis had not seemed to rebound, which puts private players in a difficulty position for accessing capital and investment (Gourmelon et al. 2015).

Nevertheless, the recent observation of the Chinese water market unveils that the market does not necessarily follow the downward trends of the global water industry. The central government has been committed to improving water supply and wastewater services in the coming years. In the 12th Five Year Plan (FYP) (2011–2015), the government pledged to make an investment of RMB 700 billion (US$103 billion) for wastewater treatment services, which is estimated at twice the amount of the 11th FYP (2006–2010). Another promising element is the continuous surge of urbanization rate. In 2019, the degree of urbanization in China amounted to 60.6% (Xinhua 2020). More urban population leads to more demands for water, and the volume of water supply for urban areas should increase by 3% per annum. Local authorities have become increasingly aware of demands for the better quality of water and wastewater services from local residents with higher incomes and consumption pattern changes. This background supports the rationale for water tariff rises in municipalities, which eventually gives good news for private players.

Faced with good opportunities, private companies can have two options: (1) by opting for new contracts; or (2) by revising their existing contracts for expansion. Municipal governments, which were previously unexperienced and unaware of water and wastewater PPP-related issues and risks at the early stage, have been well informed and familiar with a variety of relevant issues and have a good understanding of technical aspects of proposals submitted by private players. Decision makers at local governments are ready to invite private players displaying the potential of urban growth and strive to collect the payment of investment premiums from private players. As for private players, the Chinese water market appear to be still promising regardless of formidable challenges embedded in the unique socialist-market economy system. The players are willing to pay extra premiums for winning contracts expecting the steep rise of revenues generated through an increase of volumes as well as price hikes (Gourmelon et al. 2015).

Qian et al. (2020) discussed the four development stages of water and wastewater PPP projects in China as seen in Table 8.1. More investments have been made in the wastewater treatment sector recently, demonstrating the policy priority of pollution prevention and environmental

Table 8.1 Four development stages of water and wastewater public-private partnership projects in China

Sectors	1994–2000 Stage 1	2001–2007 Stage 2	2008–2013 Stage 3	2014–2018 Stage 4
Potable water & wastewater treatment plants	2	7	1	3
Potable water treatment plants	27	50	5	2
Wastewater collection	0	1	0	1
Wastewater collection & treatment	0	2	1	0
Wastewater treatment plants	1	143	142	97
Water utilities with sewerage	0	7	1	5
Water utilities without sewerage	2	17	6	0
Other	0	0	0	1
Total Number of Projects	32	228	157	110
Projects with foreign investment	*32 (100%)*	*134 (59%)*	*29 (18%)*	*5 (5%)*

Source Qian et al. (2020)

protection. A distinctive feature is the rapid emergence of Chinese companies in recent decades whereas the share of foreign players drops sharply. This phenomenon occurs because of local players' competitiveness in hedging political risks through a good understanding of local cultures and norms and agreeable relationships with local governments.

Whilst an adequate level of water tariffs is essential for foreign and Chinese companies, a salient feature to focus on is securing future investments for dilapidated or ill-maintained water and wastewater service facilities. Water and wastewater treatment plants and pipelines in China are often out-of-date and poorly maintained so that contract awardees can be in jeopardy to take over such facilities whose problems are sometimes hidden during the time of contract. In particular, considering an average leakage rate of urban water pipelines in the country is as high as 20%, private players should be cautious when assessing existing facilities prior to signing a contract with local governments (Lee and Choi 2015).

In 2011, the State Council announced the newly established target for water resources management, the Three Red Lines in 2011, as discussed in Chapter 4. The first red line is associated with the cap on national annual consumption at 700 billion m^3 by 2030, the second red line is related to water use efficiency, and the third one is linked to the enhancement of water quality in major water bodies. The first and the second red lines are directly relevant to water pricing, which have been emphasized as an effective tool to implement associated targets.

Much room is available for water tariff hikes in China because Chinese water bills are some of the lowest ones in the world. Chinese water and wastewater service bills account for 0.5% of net disposable income per capita compared with 0.7% in OECD countries and 3–4% recommended by the World Bank. Around 10% of the poorest households in China spend 1.6% of their incomes for water use compared with 2.6% in OECD countries (Gourmelon et al. 2015).

Two primary factors may hamper the increase of water tariffs in China. First, the Chinese egalitarian approaches under the socialist market economy prevent rational water pricing from being implemented at the local level. In spite of serious water shortage and pollution-caused water scarcity, strong public resistance would be expected about water tariff hikes. Second, tariff rises are politically sensitive and tricky and can be affected by complicated and long decision-making processes. The concerned local government should review the plan and organize a public hearing. In addition, local associations, experts, consumers, and researchers are expected to put their thoughts, opinions, and other comments. Although the process sounds logical and reflects good governance, the challenge is that little information is available about the cost of water distribution and services so that final decision making may be undertaken based on false or fabricated information.

A good case is available related to water tariff increases, such as Liuzhou City, which has introduced an increasing block tariff system. In the system, the first cubic meter of water use is charged at a low tariff, and the unit tariff surges substantially as use increases. Whereas this type of water tariff system is a common norm in many countries, it can take some time to observe the application of the system into many cities in China. This system can promote conservation goals, manage water demands more efficiently by letting consumers be aware of the value of water, and encourage water-guzzling industries to save water for ensuring their own profit margins (Gourmelon et al. 2015).

Governance Structure and Regulatory Frameworks

Primary Policy Drivers

The most recent policy commitment of China to tackling water and environmental challenges is 'ecological civilization'. This policy drive reflects the urgent needs of China to be transformed into an environmentally friendly society with efficient water resources management and clean technologies and to be geared toward a circular economy.

In line with the introduction of the approach, a series of newly created policy documents have been made related to the water sector, including the 2011 No. 1 Policy Document on Water Reform, the revised Water Pollution Prevention and Control Law in 2018, and the revised Environmental Protection Law in 2015. Swathes of legal and policy documents have led to establishing quotas and targets for water resources use, pollution load controls, water efficiency, permission, and law enforcement with penalties alongside relevant regulations and guidelines. Detailed guidelines and plans were set in the 13th Five Year Plan (FYP) (2016–2020) and sector plans, including the Water Pollution Action Plan or known as the Water Ten Plan in 2015. It is imperative to note that the Chinese Communist Party (CCP) has enhanced the practicability of policy implementation through the top-down decision making and implementation system from the center to localities and the assessment indicators for senior officials at central and local levels that relate to their performance in implementing those plans.

Another significant event for China's water market is the 2018 organizational reform at the central level. Special attention is paid to the empowerment of the environmental watchdog with broader responsibilities and law enforcement power, the establishment of the Ministry of Ecology and Environment (MEE). In addition, the Ministry of

Natural Resources (MNR) has begun to oversee the management of water resources and abstraction permission. The transfer of many other duties from the Ministry of Water Resources and other ministries to the two ministries implies the reshuffling of central ministries for a new form of administrative system tuned to the implementation of ecological civilization-related policies and programs.

Legal and Regulatory Settings

From the late 1990s, innumerable foreign companies scrambled to enter the Chinese water market, such as Veolia, Suez, Saur, and Thames Water. Risk assessment was essentially a key to business success prior to entering into any overseas market, particularly the Chinese water and wastewater services market which was just opened without adequate sets of laws and regulations in the late 1990s. The paucity of relevant legal settings in the market implies that as for foreign companies, there was no institutional safety net to guarantee their business activities, and for the government, no effective tool to regulate companies is available, which would result in allowing those companies to behave badly without any regulatory boundaries. Consequently, the lack of legal and regulatory frameworks or settings in any water market helps creating high risks to both the public and private sectors. A good set of laws and regulations at the central and local levels are prerequisites for the central government and local governments in the course of water and wastewater PPP projects in China since the 1990s.

One of the earliest government initiatives for inviting more foreign investment in water and wastewater PPP projects was the 21st Century Urban Water Management Pilot Scheme in 1997. Water tariffs were determined based on negotiations with foreign partners, which paved the way for foreign investors to secure favorable rates of return for water and wastewater service projects. In 1998, the Urban Water Price Regulation was promulgated, which guaranteed foreign investors to claim a net return rate of 8–10%. Local governments were given the mandate by the regulation to decide water tariffs based on the information of water and wastewater service costs. The critical principle of water pricing was also included in the regulation, which embraces the costs of construction, operation, and maintenance with good returns to investors (Lee 2007, 2010).

There are several important laws and regulations in the Chinese water market. The first legal document was the 1995 BOT Circular, the earliest version of legal dossier on water and wastewater BOT projects. This regulation, however, includes only general terms and lacks specific guidelines for foreign investors, and therefore, water and wastewater BOT projects were not attractive to foreign companies due to the lack of legal support for business. Second, it is important to shed light on the Measures on the Guarantee of Fixed Profit Margins for Foreign Investment Projects in 2002. The fixed rate of return was the solid insurance policy for foreign companies that would take high risks to enter such a volatile Chinese water market. However, the 2002 measures made the previous policy illegal by indicating that preferable terms only for foreign business units are unfair for Chinese water companies.

Third, the Catalogue of Industrial Guidance for Foreign Investment in 2002 allowed foreign companies to construct and operate water supply and drainage networks in large and medium-sized cities. Thanks to the catalogue 2002, foreign investors were able to produce revenues from the management of water supply and drainage networks and to lower revenue risks. Fourth, attention is paid to the Opinions Concerning the Acceleration of the Marketization of Urban Utilities Industries in 2002 and the Measures for the Administration of Concessionary Operation of Urban Utilities Industries in 2004. These policy documents oblige foreign and Chinese investors to undergo a public bidding process for any urban water and wastewater service contracts. Later, the General Procurement Law and the Tender and Bidding Law followed this policy direction.

The fifth regulation is the Circular on Accelerating the Reform of Water Price, Promoting Water Saving and Protecting Water Resource in 2004. The circular 2004 specifically stipulates the four primary components comprising water tariffs: (1) water resources fee; (2) water supply engineering fee; (3) water supply fee; and (4) wastewater treatment fee. An emphasis is placed on the need to adjust the tariff of water supply to a rational level in the 2004 circular (Lee 2010; Zhong and Mol 2010).

The circular provided a watershed for China's urban water and wastewater service tariff system, since many municipal governments began to actively charge wastewater service fees after the issuance of the 2004 circular, although the basic principles in water pricing were mentioned in the Water Law 2002 (Lee 2010; Qian et al. 2020).

Apart from these regulations, national-level water- and environment-related laws have helped the Chinese water market become mature, such as Water Law (1988 and 2002), Water Pollution Prevention and Control Law (1984, 1996, 2008 and 2018), Environmental Protection Law (1989 and 2015), and Water and Soil Conservation Law (1991 and 2010). The Water Pollution Prevention and Control Law was revised in 2008 for strengthening the capacity of law enforcement of environmental watchdogs at the central and local levels by implementing the discharge permit of wastewater and total pollution control and revised again in 2018. The major reason behind the revision in 2018 was geared toward supporting the implementation of the Three Red Lines in 2011, particularly pertaining to tackling agricultural and water pollution control. Also, the 2018 version officially encompasses the River (Lake) Chief System in order to bolster law enforcement against water polluting units and stipulates special articles on the enhancement of drinking water quality and levying more fines on violators up to RMB one million (US$150,000).

In the same vein, the Environmental Protection Law was revised in 2015 for giving more regulatory power to the central and local environmental authorities by levying fines on a daily basis and for providing clear provisions on public interest litigation against polluters. These measures have paved the way for local governments to focus more on urban wastewater services by putting more investments, which has also led to creating more business opportunities for wastewater treatment companies. A series of laws and regulations in urban water and wastewater PPP projects are described in Table 8.2.

Reshuffling of Organizations

Prior to the 2018 administrative reform in China, the Ministry of Water Resources (MWR) was the ministry primarily in charge of water

Table 8.2 Laws and regulations on public private partnership water and wastewater projects in China

Year	Title
1988	Water Law
1988	Provisional Regulations of Private Enterprises
1989	Environmental Protection Law
1989	Rules for Implementation of the Prevention and Control of Water Pollution
1991	Water and Soil Conservation Law
1995	Interim Provisions on Guiding Foreign Investment Direction
1995	Certain Matters Relating to Project Financing by Domestic Institutions Notice
1995	Several Issues Concerning the Examination, Approval and Administration of Experimental Foreign Invested Concession Projects Circular (the BOT Circular)
1995	Circular on Major Issues of Approval Administration of the Franchise Pilot Projects with Foreign Investment
1997	Partnership Business Law
1997	Catalogue for Guiding Foreign Investment in Industry
1997	Administration of Project Financing Conducted Outside China's Tentative Procedures (The Interim Procedures)
1998	Urban Water Price Regulation
1999	Contract Law
2000	Circular on Strengthening Urban Water Supply, Water Saving and Water Pollution Prevention and Control
2001	Several Opinions of the State Development and Reform Commission concerning the Promotion and Guidance of Private Investment
2002	Water Law (revised)
2002	Circular on the Relevant Provisions Concerning the Handling of Quality and Safety Problems of Foreign Contracted Projects
2002	Foreign Investment Industrial Guidance Catalogue
2002	Opinions concerning the Acceleration of the Marketization of Urban Utilities Industries
2003	Decision on Several Issues of the Central Committee of the Communist Party of China to Improve Socialist Market Economy System
2004	Measures for the Administration of Concessionary Operation of Urban Utilities Industries
2004	Circular on Accelerating the Reform of Water Price Promoting Water Saving and Protecting Water Resource

(continued)

Table 8.2 (continued)

Year	Title
2004	Sample Document for the Franchised Operation of Urban Water Supply, Gas Supply and Waste Disposal
2005	Suggestions on Encouragement of Supporting and Directing non-public Ownership Economy
2006	Partnership Enterprise Law
2006	Sample Document for the Franchised Operation of Urban Heat Supply and Wastewater Disposal
2007	Partnership Enterprise Law (revised)
2008	Water Pollution Prevention and Control Law (revised)
2008	Research Reports of PPP Legislation in Infrastructure Development
2009	Administrative Measures for the Establishment of Partnership Enterprises within China by Foreign Enterprises or Individuals
2010	Circular on Strengthening Urban Water Supply, Water Saving and Water Pollution Prevention and Control
2010	Guidelines on Encouraging and Guiding the Sound Development of Non-governmental Investment
2011	Foreign Investment Industrial Guidance Catalogue
2013	Regulation on Urban Drainage and Sewage Treatment
2015	Notice of the Relevant Work concerning the Promotion of the Development Financial Support for Public-Private-Partnership
2015	Environment Protection Law (revised)
2018	Water Pollution Prevention and Control Law (revised)

Source Compiled based on Lee (2010) and Oh (2021)

resources management in China. Since 2018, many responsibilities, particularly related to river basin management, have been transferred from the MWR to the Ministry of Ecology and Environment (MEE). There are gray areas to be clarified in the course of the transition period between water-related ministries.

As for water and wastewater services in urban areas, it is the Ministry of Housing and Urban and Rural Development (MOHURD) that has the overall responsibility, including public private partnership in urban water and wastewater projects. The main mission of the MOHURD is to manage the construction of water supply and wastewater treatment facilities and networks in urban areas. In addition, the MOHURD evaluates and double-check project proposals associated with foreign investment before the proposals being submitted to the National Development and Reform Commission (NDRC).

The MEE and local Ecology and Environment Bureaus (EEBs) are responsible for water quality control in addition to water conservation as well as water environment protection. Supervision and management of the drinking water quality and its suppliers are in the hands of the Ministry of Health (MOH). Apart from these, there are several ministries involved in water and wastewater services PPP projects in China shown in Fig. 8.1.

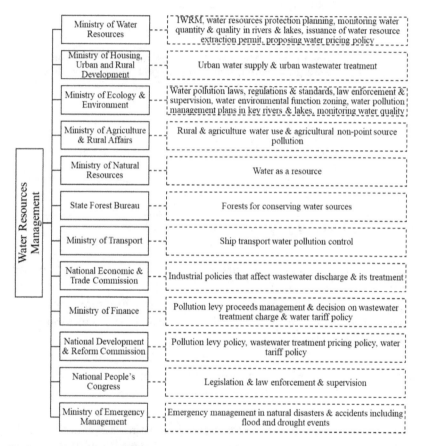

Fig. 8.1 Ministries and bureaus at the national level involved in water resources management in China since 2018 (*Source* Updated based on Lee [2010] and World Bank [2006])

In accordance with a set of laws and regulations by the central government, local governments have continued to establish their laws and regulations regarding urban water and wastewater services favoring private sector involvement. Thanks to devolution in the reform era, laws, regulations, and policy measures at the local level have been heterogeneous depending upon socioeconomic and environmental circumstances in localities. Local governments have been given the mandate to create their own institutions reflecting local conditions. In addition, higher incomes and consumption pattern changes of urban households have prompted local authorities to pay more attention to water and wastewater services by ratcheting up investment, revising institutions and organizations, and strengthening law enforcement.

Urban water and wastewater services are usually in the hands of construction bureaus at the provincial, municipal, prefectural and county levels, similar to the role of the MOHURD at the central government. The structure of Chinese local governments varies, and there are three different types of urban water and wastewater service regulatory bureaus in China. First, one of the most common types is the construction commission or construction bureau, for instance, in Shenyang City and Ma'anshan City. Second, the municipal public utility bureau is found in many local governments, which represents separate bureaus on the function of construction and management. The third type is the system of water authority. Several water-related bureaus were integrated into one for integrated water resources management, which was first introduced in Shenzhen, 1993 and became popular in a number of municipal governments, such as Shanghai in 2000 and Beijing in 2004. Although this type has gained popularity right after the new millennium, it is unclear if such an administrative reform of water authority has given a positive impact on managing and regulating water and wastewater services, especially regarding PPP projects (Lee 2006, 2010). Figure 8.2 describes an example of municipal government's structure related to urban water and wastewater services.

The role of the MWR has been downscaled through the 2018 administrative reform, although the ministry takes its overall responsibility for water resources management policy, protection, investment planning, flood control, drought relief, and soil protection. It seems that

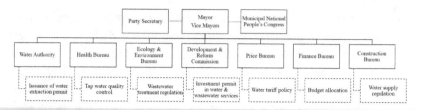

Fig. 8.2 Bureaus involved at the local level in Water Resources Management in China (*Source* Updated based on Lee [2006, 2010])

this reshuffling of water-related ministries at the center may result in altering responsibilities of subsidiary bureaus and agencies at the provincial, municipal, prefectural and county levels but little is known so far (Spooner 2018).

Market Analyses

Market Situations

Urban water and wastewater services in China have gradually been provided by private sector players since the late 1990s. From 1989 to 2008, the ratio of Chinese population served by private companies soared from 8 to 38%, which indicates the eye-catching growth of water and wastewater PPP projects in the country (GWI 2008). In the period between 2008 and 2018, 267 PPP projects in the Chinese water market were recorded, amounting to almost two-thirds of the total water PPP projects in the world. Looking at the statistics of 2019, 51 water and wastewater PPP projects were implemented with a worth of US$4 billion at the global level, and most of the investment commitments in the sectors stemmed from China. China was responsible for 85% of the total investment in the sectors in 2019 with US$3.3 billion for 44 projects (Qian et al. 2020; World Bank 2020).

In 2005, 97% of the urban population had access to centralized water supply systems, however, at best half of this supply satisfied the full set of water quality standards on the basis of 35 measurements that were included in 1995 (GB 5749-1995), which entailed public distrust in the quality of tap water. The new standards (GB 5749-2006) were introduced and in effect in 2007 for gaining public confidence with 106 measurements. The national consensus data in 2011, published by the MOHURD, demonstrates that water quality from waterworks showed a standard compliance rate of 83% whilst the tap water standard compliance rate reached 79.6%. The Ministry of Environmental Protection encouraged local governments to make the investments of more than RMB 430 billion (US$66 billion) for 4,800 projects that would improve the quality of drinking water sources in August 2016. In a nutshell, there is significance room for improvement related to the tap water quality, particularly associated with integrated catchment solutions for water resources quality (Spooner 2018).

As part of the Three Red Lines in 2011, Chinese authorities have pledged to decrease the level of water consumption at 23% by 2020 compared with that in 2015 and develop and upgrade urban wastewater treatment facilities during the 13th Five Year Plan (2016–2020). Regarding water quality, specific targets have also been unveiled, such as the increase of wastewater treatment rate to 95% in urban areas and 85% in non-urban areas. More weight is put on the improvement of water quality in urban areas, and an emphasis is also placed on the extension of basic treatment to more rural areas. There are targets for enhancing water source and drinking water quality. Following the governing philosophy of ecological civilization, the central and local governments have been determined to undertake sponge city projects in large cities which embrace elements of water treatment and storage and urban infrastructure design. This policy initiative can engender higher demands for small-scale package treatment solutions (Spooner 2018).

In 2017, the annual quantity of wastewater discharge in China's urban areas amounted to 49 billion m^3 per annum, and a total of 2,209 in urban wastewater treatment plants treated wastewater with the capacity of 157 million m^3/day with 94.5% of the wastewater treatment rate. There were 2,065 plants, 93%, with the secondary and tertiary treatment facilities with the capacity of 148 million m^3/day (MOHURD 2018). Situations of rural areas looked somewhat different in terms of wastewater treatment. Approximately 40% of rural residents had no access to improved sanitation services in their homes in 2014 (Spooner 2018).

Municipal governments have recently recognized the need to utilize recycled water due to soaring demands for water by different sectors as well as engineered by the sponge city initiative in 2014 that actively promotes water and wastewater reuse and recycle, and rainwater harvesting. In 2017, the total volumes of recycled water production capacity in China's urban areas reached 35.88 million m^3/day, and the actual amounts of wastewater recycled and reused per annum amounted to 7.13 billion m^3 (MOHURD 2018).

Private investors which consider entering the wastewater treatment market of China will be able to find business opportunities in smaller cities and suburban areas regarding the extension of secondary treatment. The other area will be associated with the upgrading of main treatment works to higher levels, particularly enlarging the capacity of wastewater reuse and recycle. The investment of the central and local governments for wastewater treatment plants and networks will continue via the mode of PPP, and rural and decentralized systems are the new target fields from public funding, which will be incorporated into green infrastructure and ecology (Spooner 2018).

As for the desalination sector, the Chinese government was committed to adding 2.2 million m^3/day of new desalination capacity from 2011 to 2015 in the 12th FYP (2011–2015). This option has become popular in the eastern coast areas, including the Tianjin Municipality, and the most common type is Reverse Osmosis (RO) technology-based one. A major barrier of the development of desalination is linked to the heavily controlled urban water supply tariffs because RO-produced water is more expensive than the possible sales price. In addition, demineralized water causes corrosion, and there is a low acceptance of desalinated water in the public. As a consequence, these challenges have made the desalination option under-utilized and often for the emergency use only.

In the 13th FYP, by 2020, the total seawater desalination volume will amount to 2.2 million m^3/day or more, the new seawater desalination volume in coastal areas will be 1.05 million m^3/day or more. The new seawater desalination scale in island areas will be as large as 140,000 m^3/day. Direct seawater utilization volumes will amount to 140 billion m^3/year, and the volume of seawater circulating cooling will be as large as two million m³/hour. The government plans to achieve the scale of direct seawater desalination scale at one million m³/day (NDRC 2016).

Favorable policies have been in pipeline for promoting the desalination industry, including tax reduction for desalination factories, desalinated water that can be accessible through water supply systems, and the encouragement of third parties' investments in desalination facilities. Considering continuously high demands for water in East Coast, the desalination option would be financially attractive, since the production cost of desalinated water is lower than the cost of water from the South North Water Transfer Project (NDRC 2016; Wang 2016).

According to the 2018 National Seawater Utilization Report, by late 2018, there were 142 seawater desalination plants in China with the total capacity of 1,201,741 m³/day. The newly added capacity reached 12,536 m³/day in five new desalination plants. The annual seawater circulation cooling amounted to 139 billion m³/day. At the national level, desalination plants with the capacity of more than 10,000 m³/day were 36, which could produce 1,059,600 m³/day, those with the capacity of more than 1,000 m³/day, 41 which could produce 129,500 m³/day, and those with the capacity of less than 1,000 m³/day, 54 which could produce 12,641 m³/day.

The Tianjin Municipality boasts the largest capacity of desalination production in China with 317, 200 m³/day, followed by Shandong Province with 282,600 m³/day, Zhejiang Province with 232,300 m³/day, Guangdong Province with 89,300 m³/day, and Liaoning Province with 87,700 m³/day. In terms of the type of technology for desalination plants, the most popular technology is RO, which accounts for 68.7% (121 plants), followed by Multiple Effect Distillation (MED), 30.7% (16 plants), Electro Dialysis (ED) (3 plants), Multi-State Flash (MSF) distillation (1 plant), and Forward Osmosis (FO) (1 plant) (MNR 2020).

Modes of Water and Wastewater PPP Projects in China

There are a variety of PPP project types in China's water and wastewater sectors, depending upon different needs from local governments and private sector players. The most popular option has been the Cooperative Joint Venture (CJV) in the Chinese water market. In the embryonic period of the market in the late 1990s, foreign investors had no option but to directly negotiate with local governments for terms and conditions in contracts. In order to remove various market risks, foreign investors chose to work together with Chinese partners as local guide by establishing a joint venture company, such as Sino-French Company that was established and jointly operated between Suez Group and New World Development Company from Hong Kong (Lee 2006, 2010).

The peak years of the mode of CJV did not last long because the guaranteed rates of return for foreign investors became illegal in 2002 by the State Council. Thames Water, back then running the Da Chang BOT water project in Shanghai, decided to exit the Chinese market in 2004, and other overseas companies had to reconsider their strategies of how to do business. As a safer option, the Equity Joint Venture (EJV) emerged, because under the mode of the EJV, profits and losses can be distributed according to each partner's equity shares. Many CJV-based contracts have been switched into EJV-based ones since 2002 (Fu et al. 2008; GWI 2004; Lee 2006, 2007).

Although the mode of BOT was actively promoted by the central government of China in the late 1990s, foreign companies found the option risky due to the paucity of relevant laws and regulations on BOT contracts. There were several unsuccessful BOT cases in the early period of water PPP projects in China. The Da Change Water Project in Shanghai by Thames Water was praised as a successful case with favorable terms and conditions for foreign investors but eventually was sold back to the Shanghai Municipal Government in 2004 due to the issuance of the 2002 rule for the ban of the fixed rates of return (Fu et al. 2008; Lee 2007).

Veolia and Marubeni in 1996 won the contract for the Chengdu No. 10 BOT Water Project but economic losses occurred to the Chengdu Municipal Government because of the wrong prediction of water demand (Chen 2009; Fu et al. 2008; Lee 2007; Wang and Zhao 2004). In 2002, Anglian Water and Mitsubishi acquired the contract for the Beijing No. 10 Water Project, but the project was not closed because of failing to secure debt-financing through local and international financial institutions. The project was revised in five years so that the Golden State Environment Group signed the contract in 2007 (GWI 2004; Lee 2010). Despite such unfavorable track records of water BOT projects in China, the mode of the Wholly Foreign Owned Enterprise (WFOE) has still been feasible. Good news for both foreign investors and local governments is that WFOE projects do not have to be endorsed by the central government any more since 2003 (Lee 2010).

In 2002, Veolia Environment had its successful deal with the Shanghai Municipal Government for purchasing a 50% of equity of the Shanghai Pudong Water Supply Corporation. The contract was the first case of the Joint Stock Company (JSC) in the market, and the company won a similar deal in Shenzhen in 2004 and in Lanzhou in 2007. This model has allowed foreign investors to be involved in not only running water treatment plants but also managing the water distribution network and customer services. The mode of the JSC has enlarged the spectrum of business for foreign investors in water supply services through PPP projects (Fu et al. 2008; Lee 2006, 2010; Owen 2008).

In addition to these options, a unique and peculiar mode of Transfer-Operate-Transfer (TOT) water and wastewater PPP projects is discussed. This business model has become prevalent since the early 2000s, particularly in the Chinese wastewater treatment market. Wastewater TOT projects involve existing wastewater treatment plants at certain local areas. Initially, a local government call for public bidding related to a wastewater treatment TOT project, and if a company is awarded the contract, the government or a public sector agency transfer asset rights, including operation and management rights, to a Chinese or foreign private investor. The private investor has to make a one-off payment to the public sector in order to manage, operate, and maintain the wastewater treatment facilities for 20–30 years. The facilities are returned to the original owner of the public sector at the end of the contract period (Lee and Choi 2015). Table 8.3 and Fig. 8.3 describe a myriad of investment models in the Chinese water market.

In the period between 1994 and 2019, water and wastewater service PPP greenfield projects show the largest number, 362 out of 572 (63%) followed by brownfield projects, 151 (26%), management and

Table 8.3 Investment models in the Chinese Water Market

Options	Description	Examples
Cooperative Joint Ventures (CJV)	Capital contributes by foreign and local partners in cash/assets	Preferred mode of investment for foreign investors by 2002
Equity Joint Ventures (EJV)	Profits and losses distributed depending on each partner's equity shares	Growing numbers since 2002
Wholly Foreign Owned Enterprise (WFOE)	Entirely invested and owned by international players	Dachang, Chengdu, Beijing No.10 (Build-Operate-Transfer – BOT)
Joint Stock Companies (JSC)	Established by Chinese for listing on stock markets	Veolia in Shanghai-Pudong in 2002, Shenzhen in 2004, and Lanzhou in 2007
Transfer-Operate-Transfer (TOT)	Transfer of operation and management rights to the private sector	Hefei Wangxiaoying Wastewater Treatment Project in 2004

Source GWI (2004), Lee (2010), and Lee and Choi (2015)

lease, 46 (8%), and divestiture, 13 (2%) (see Table 8.4). Greenfield projects embrace BOT, and Build-Own-Operate (BOO) projects whereas brownfield projects indicate Build-Renovate-Operate-Transfer (BROT) and Renovate-Operate-Transfer (ROT) projects that require no major construction. Such phenomena indicates that opportunities for building new infrastructure in water and wastewater service sectors are still available in China although construction-related business opportunities would be less available, particularly in the water supply sector. Considering massive investments injected by the central government through the Five Year Plans, the wastewater treatment sector is still promising for PPP projects, especially in mid-level municipalities located in central, western, and northeastern parts of China (World Bank and PPI 2020; Qian et al. 2020).

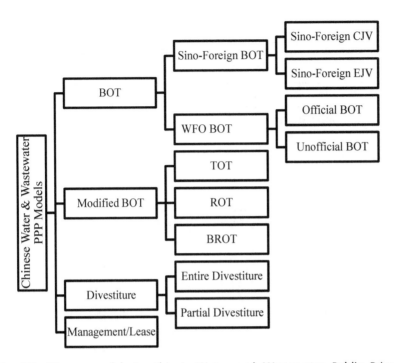

Fig. 8.3 Diverse models in China's Water and Wastewater Public Private Partnership (PPP) Projects. *Remarks* BOT means Build-Operate-Transfer, TOT, Transfer-Operate-Transfer, ROT, Renovate-Operate-Transfer, and BROT, Build-Rehabilitate-Operate-Transfer. WFOE stands for Wholly Foreign Owned Enterprise. CJV indicates Cooperative Joint Ventures, EJV, Equity Joint Ventures (*Source* Modified based on Choi and Lee [2010])

Spooner (2018) introduces a new form of PPP in China with the purpose of local government's debt. This business model is called 'Build-Transfer' arrangements which encompass additional land transfers and residential development approvals for the purpose of payment for major infrastructure constructions. When a private water company builds a wastewater treatment plant, the public sector partner, a local government provides plots of land in which the company can construct apartments and then sell them to pay for the infrastructure construction cost. However, this option does not appear to be feasible for foreign water companies, because they might lack a good understanding of local legal and administrative practices and local contacts.

Table 8.4 Types of water and wastewater public-private partnership projects in China from 1994 to 2019

Type of PPI	Project No	Share (%)
Brownfield (Build-Renovate-Operate-Transfer, BROT & Renovate-Operate-Transfer, ROT)	151	26
Greenfield (Build-Operate-Transfer, BOT & Build-Own-Operate, BOO)	362	63
Management & Lease	46	8
Divestiture (Partial & Full)	13	2
Total	572	100

Source World Bank and PPI (2020). http://ppi.worldbank.org (accessed 7 October 2020)

Major Players

The emergence of the Chinese water market in the late 1990s was attributed to the two major driving forces, namely strong commitments of the central government and local governments of China and proactive responses from major multi-national corporations such as Veolia, Suez, and Thames Water. Following the four different stages of the Chinese water market by Qian et al. (2020), however, the pendulum of the market dominance has been swung to Chinese companies since the beginning of the third stage, 2008 (see Table 8.5).

For instance, in the third stage from 2008 to 2013, the projects awarded to Chinese companies reached 114, accounting for almost 74%, and this percentage would rise to 88%, 137 projects, considering the projects awarded to the combination of Chinese and Overseas Chinese companies without foreign partners. By 2019, the dominance of Chinese players had been more consolidated, and the number of projects awarded to Chinese and Overseas Chinese companies was 150, 97%. Overseas Chinese companies are Cathay International Holdings, NWS Holdings, China Water Affairs Group (Hong Kong), Hyflux, Asia Environment, Asia Water Technology (Singapore), Sime Darby, and Salcon (Malaysia). The share of foreign companies had plummeted to about 3%, five

Table 8.5 Four development stages and classification of private players depending on the nationalities of water companies in China

Four Stages of Development/Nationalities of Water Companies	C	F	OC	Other	C+F	C+OC	F+OC	C+F+OC	Total
1994–2000	0	8	8	3	0	0	13	0	32
2001–2007	95	48	66	0	1	4	14	0	228
2008–2013	114	15	21	1	0	1	3	2	157
2014–2019	127	3	2	21	1	0	1	0	155
Total	336	74	97	25	2	5	31	2	572

Remarks C: Chinese, F: Foreign, OC: Overseas Chinese (Hong Kong, Singapore, Malaysia), Other: no information (perhaps Chinese), C+F: Chinese & Foreign, C+OC: Chinese & Overseas Chinese, F+OC: Foreign & Overseas Chinese, C+F+OC: Chinese, Foreign & Overseas Chinese
Source Updated based on Qian et al. (2020) and World Bank and PPI (2020). http://ppi.worldbank.org (accessed 7 October 2020)

projects between 2014 and 2019, compared with 21 out of 32, 65% from 1994 to 2000 (World Bank and PPI 2020).

The early enthusiasm of foreign companies toward the Chinese water market seems to have faded away due to different kinds of political, economic or financial, legal and regulatory, and socio-cultural risks. Since the middle of the 2000s, it is recognized that the share of foreign water companies in the market has quickly dwindled whereas Chinese companies have taken up more shares of the market. Such an emergence of Chinese companies is partly attributed to massive investments from the central government coupled with its flagship plans and projects, such as the several Five Year Plans, particularly the 12th FYP (2011–2015) and the 13th FYP (2016–2020). In particular, the Water Ten Plan in 2015 epitomized the strong commitment of the central government to implementing policies for 'Beautiful China' by focusing on water quality enhancement, which in turn resulted in the boom of the construction of wastewater treatment plants, utilities, and relevant networks in major cities.

The Chinese water market is primarily led by Chinese companies that have more competitiveness compared with their foreign counterparts. Their competitive areas include favorable policies by the central as well as local governments for Chinese companies, easier access to relevant information, regulations, and policies, better negotiation skills and tactics with local public authorities, better access to debt financing through Chinese banks, and familiarity with practices and norms in legal and regulatory settings, to name a few.

GWI (2019a) unveils the top 50 private water operators in the world based on the number of people served by private water operators. Considering the concrete dominance of Chinese companies in the Chinese water market, it is no surprising that 13 Chinese water companies are included in the top 20 water companies around the globe. A total of 20 Chinese water companies are ranked in this league of table, and there are three overseas Chinese companies, i.e., Hua Yan Water (Hong Kong), Sembcorp Utilities (Singapore), and Hyflux (Singapore). Table 8.6 illustrates the detailed information of Chinese water companies on the amount of revenues, the number of people served and remarks for their distinctive market activities, which are highlighted in gray.

8 Water and Wastewater Service Market 321

Table 8.6 The World's Top 50 Private Water Operators (Chinese water companies highlighted in gray)

Rank	Company	Country	Water Revenues	No. of People served	Notes
1	Suez[a]	France	€9.7bn	135,000,000	Progress in India; ramp-up in US M&A; Spanish risks looming
2	Veolia	France	€10.9bn	129,000,000	Gains in Japan, France & USA; Gabon contract expropriated
3	Beijing Enterprises Water	China	HK$24.6bn	78,193,095	Shifting away from new PPPs to Yangtze River program
4	Shanghai Industrial Holdings	China	HK$6.3bn	62,796,935	Includes SIIC Environment (46.67%) & General Water (45%)
5	Beijing Capital	China	RMB8.42bn	50,000,00	Includes 100% of Hebei Huaguan EP Science & Technology
6	VA Tech Wabag	India	INR27.8bn	48,430,172	Benefiting from One City, One Operator model in India
7	Acciona Agua	Spain	€639m	40,555,549	Loss of ATLL contract cancels out effect of new contract wins
8	Shenzhen Water	China	RMB7.42bn	30,000,00	Provides water & wastewater services in 7 Chinese provinces
9	Sabesp	Brazil	BRL16.1bn	27,900,00	Steady growth, but private sector competition on the way
10	Aqualia	Spain	€1.115bn	25,137,500	Significant expansion in Spain, France, Egypt, Tunisia UAE
11	Beijing Origin Water	China	RMB9.28bn	24,198,427	Around one third of wastewater portfolio is built & operational
12	Shanghai Chengtou Holding	China	RMB10.4bn	23,239,622	Includes Shanghai Chengtou Water & Shanghai Env. Group
13	China Water Affairs[b]	China	HK$7.7bn	22,000,00	Population served more than doubled since 2016 thanks to M&A

(continued)

Table 8.6 (continued)

14	Tianjin Capital Environmental	China	RMB2.07 bn	20,125,025	Water & wastewater utility for Tianjin with prospects nationwide
15	Chongqing Water Group	China	RMB5.17 bn	17,274,499	Acquired 100% of water utility in Jiangjin District in 2019
16	Manila Water	Philippines	PHP19.84 bn	16,959,971	Added significant provincial contracts in the Philippines
17	Chengdu Xingrong Env.Co. Ltd.	China	RMB3.6bn	16,652,429	Chengdu's water & wastewater utility, with prospects nationwide
18	Guandong Investment	China	HK$8.23bn	15,636,822	Acquired 79% of Jiangxi Haihui Utility in April 2019
19	China Everbright Water Ltd.	China	HK$4.77bn	15,587,820	Listed on Hong Kong Stock Exchange in May 2019
20	Anhui Guozhen Env. Protection	China	RMB4bn	15,003,025	Most aggressive growth in towns & rural areas
21	BRK Ambiental	Brazil	BRL2.3bn	15,000,000	Brazil's largest majority privately-owned water service provider
22	Thames Water	UK	£2.1bn	15,000,000	The UK's largest regulated water & sewerage company
23	Hua Yan Water	Hong Kong	HK$2.52bn	14,921,960	Bought 26.67% of Foshan Water in 2018–2019
24	Sound Global	China	RMB4.81 bn	14,519,056	Continues to win contracts, despite financial woes
25	Remondis Aqua	Germany	-	14,400,000	Includes Istanbul; excludes industrial contracts
26	Yunnan Water	China	RMB4.47 bn	14,150,030	Massive capacity ramp-up due to BOT/TOT contract successes
27	American Water Works	US	US$3.44bn	14,000,000	Accelerated regulated growth; exited municipal contract operations
28	Jiangxi Hongcheng	China	RMB2.57 bn	13,819,380	Controls 80% of Jiangxi Province's sewage treatment market
29	Acea	Italy	€880m	12,860,476	33% of population served under long-term concessions in South America
30	Eranove	France	–	12,737,610	Set to lose large contract in Senegal in January 2020
31	Tus Environmental	China	RMB624 m	12,663,844	Acquired 100% of operator Thunip Corp. in April 2018

(continued)

Table 8.6 (continued)

32	Kangda Int'l Environmental	China	RMB3bn	12,569,066	Strong organic growth from new municipal BOT/TOT projects
33	Saur	France	€1.3bn	12,400,000	Pushing into South America & the Middle East under new owners
34	SPML Infra	India	–	11,951,000	Excludes those served by the SAUNI Yojana irrigation project
35	Copasa	Brazil	BRL4.17bn	11,330,000	Continues to renew & secure new concession contracts
36	Sanepar	Brazil	BRL4.16bn	11,000,000	Incremental organic growth in core franchise in Parana State
37	Metro Pacific Investments Corp.	Philippines	PHP22.9bn	10,921,443	Expanding internationally through acquisitions in Vietnam
38	Jacobs	US	US$1.5bn	9,300,000	Rapidly gaining market share in the US contract operation sector
39	ACWA Power/Nomac	Saudi Arabia	–	8,500,000	Total aggregated from desalination plant portfolio in the Gulf
40	Aguas Andinas	Chile	CLP530bn	8,500,000	Population also included Suez total due to majority ownership
41	Vishvaraj Infrastructure	India	–	8,004,000	Largest contract in Nagpur shared 50:50 with Veolia
42	Severn Trent	UK	£1.767bn	8,000,000	Figures include the Hafren Dyfrdwy company in Wales
43	Sembcorp Utilities	Singapore	–	7,578,581	Figures include the Salalah & Fujairah desal plants
44	Hyflux	Singapore	S$290m	7,104,478	Excludes Tuaspring; includes Qurayyat desalination plant
45	United Utilities	UK	£1.818bn	7,000,000	Modest organic growth from existing franchise area
46	Aegea Saneamento	Brazil	BRL1.73bn	6,629,691	Acquired Manaus in 2018; Saqua deal cancelled in July 2019
47	Igua Saneamento	Brazil	BRL745m	6,594,670	Minority stakes in local subsidiaries bought out in 2018
48	Penyao Env. Protection Co., Ltd.	China	RMB772m	6,241,178	Acquired 51% of China Railway Urban-Rural Env. In late 2018
49	GS Inima	South Korea	–	6,234,229	Total reflects GS Inima's concessions & O&M projects
50	Anglian Water	UK	£1.28bn	6,000,000	Serves fastest-growing region of the UK in terms of population

Remarks [a]Includes Aguas Andinas, Lydec and the former GE Water, [b]Excludes CWA's 29.5% stake in Kangda International
Source GWI (2019b)

Whereas the list of GWI (2019b) demonstrates the ranks of global water companies based on the number of people served, GWI (2020c) ranks the top 40 global water companies on the basis of the amount of revenues, which embraces four Chinese water companies, which are Beijing Enterprises Water Group with US$3.6 billion (No. 7), China Lesso Group with US$2.6 billion (No. 13), Beijing Origin Water Technology with US$1.8 billion (No. 21), and Beijing Capital with US$1.5 billion (No. 29).

Since 2003, the E20 Environment Platform in China has announced the top 10 most influential water companies in China. The ranking discloses water companies based on a myriad of selection criteria that embraces the total and incremental capacity of projects that the companies fund and operate, market influence, integrated service capabilities, market position, activeness, strategic core competitiveness, involvement in charitable causes, and brand reputation and influence. The selection was undertaken through a review process by a panel of experts, journalists, and industry analysts (E20 Environment Platform 2020; WaterWorld 2020).

In the 2019 list, Beijing Enterprises Water Group is ranked No. 1, followed by Beijing Capital and General Water of China. Suez is the only foreign water company listed in the ranking (No. 4), which has been included in the list for 13 years and provides its water and wastewater services to more than 32 million people in China (WaterWorld 2020). The other Chinese water companies are already mentioned in the previous table of the top 50 global water companies, which implies that the Chinese water market is no longer subject to the influence of foreign companies that previously had more financing capacities, advanced technologies, and management experiences (see Table 8.7). It seems that the Chinese water and wastewater PPP market continues to be dominated by Chinese companies that are accustomed to sociopolitical and economic norms of the socialist market economy and meet the needs of local governments in tune with policies and programs under the Xi Jinping era.

There are three types of Chinese investors in water and wastewater PPP projects of the country. The first type is Chinese investment companies involved in capital investment, i.e., Beijing Capital Group. These

Table 8.7 2019 Top 10 Most Influential Water Companies in China

Rank	Name	
	English	Chinese
1	Beijing Enterprises Water Group	北控水务集团有限公司
2	Beijing Capital	北京首创股份有限公司
3	General Water of China	中环保水务投资有限公司
4	Suez	苏伊士新创建
5	Guangdong GDH Water Co. Ltd	广东奥海水务股份有限公司
6	Beijing Origin Water Technology	北京碧水源科技股份有限公司
7	China Everbright Water	中国光大水务有限公司
8	Tianjin Capital Environment Protection Group	天津创业环保集团股份有限公司
9	Anhui Guozhen Environment Protection Technology	安徽国祯环保节能科技股份有限公司
10	CSD Water Service	中持水务股份有限公司

Source E20 Environment Platform (2020)

companies are state-owned. Second, privately owned water companies, and a good example of Shenzhen Water Group. This type of companies has had various experiences in working together with foreign companies and are capable of investing and operating as seen from the joint venture project with K-water and Kolon Water in Sayang Prefecture, Jiangsu Province, back in the mid-2010s. The third type of companies are Chinese operators such as General Water of China and Tsinghua Tongfang. These companies have entered the water sector thanks to their technical know-hows and innovative solutions (Qin et al. 2020).

Implications

Challenges

Under the socialist market economy system, the boom of water and wastewater PPP projects in China has been possible thanks to the top-down nature of policymaking and implementation. Deference to hierarchy is the fundamental nature of the Chinese governance system under the strong leadership of the CCP. This system pressures local government officials and CCP cadres to achieve tasks imposed by higher authorities, including attracting substantial investments for local governments. As a consequence, if a project gained favor of a local government, related government officials would be willing to speed up approvals, provide matching resources, and secure subsidies.

There are dark sides of this effective top-down model for water and wastewater PPP projects. This effectiveness can work well in the short-term, however, local government officials may leave the position and be relocated to different areas. If so, coordination between bureaus and agencies at the same level or different levels would become problematic. Other ensuing challenges can take place, such as a lack of long-term strategic consideration that may entail over-investment, precarious investment, and corruption.

The capacity of government officials involved in water and wastewater PPP projects is significant. Water and wastewater PPP projects are normally in operation for 20–30 years, and in the meantime, it is imperative for private companies to have positive relationships with local governments. Uncertainties and unexpected policy shifts can come into being during water and wastewater PPP contract periods, often caused by the change of policy directions of the central government, and then it would be local government officials who should closely communicate with their private partners and tune up the policies for local circumstances, particularly considering their private counterparts. The cases of Ma'anshan and Fuzhou water treatment PPP projects show government officials' capability to tackle policy changes and market conditions whereas the case of Xining wastewater treatment PPP project demonstrates the incapability of government officials to cope with unexpected policy changes (Qian et al. 2020).

Water and wastewater PPP projects in China have rapidly developed over the past two decades. The number of water and wastewater PPP projects has increased, the active engagement of foreign and Chinese companies has been observed, and relevant policies and regulatory settings have incrementally been established. Nevertheless, the sound development of water and wastewater PPP projects in China will be guaranteed by overcoming the following challenges. First, comprehensive legal and regulatory guidelines should be improved for clearing out instability in the operation of water and wastewater PPP projects. A policy vacuum and technical capacity weakness remain critical, concerning regulations, particularly associated with economic regulations and tariff determination. The mandate to decide water tariffs has been given to local governments, however, local governments often have little expertise to implement such an authority.

Second, there is a lack of coordination between different levels of government. As observed, various bureaus and agencies are involved in policymaking and implementation in water and wastewater PPP projects. At the center, the MOHURD takes responsibility for policy development in water and wastewater PPP projects, and provincial development and reform commissions are designing and implementing policies for PPP projects at the provincial level. Municipal and prefectural governments are in charge of project design and implementation at the project level. In this top-down policymaking and implementation system, inconsistencies and contradictions between diverse levels of governments are common considering the complicated operation of PPP projects.

Third, significant decisions are made for water and wastewater PPP projects, not based on their long-term socioeconomic and environmental impacts on local areas but based on short-term economic goals because of the evaluation of government officials. Local government officials are evaluated based on their short-term performance, and therefore, they favor projects with substantial investment that would entail impressive and short-term achievements during their tenure. Such phenomena might trigger over-investment in certain BOT or greenfield projects whereas ignoring more efficient and rehabilitation-involved projects (Qian et al. 2020).

Whilst interests of domestic and foreign investors in the Chinese water market appear to continue, it is necessary to shed light on unsuccessful projects in the market for addressing complicated and unique challenges embedded in the market. There are different types of the so-called unsuccessful water PPP projects in China: (1) the project in which there was no closing of the contract; (2) the project with the contract but has been terminated in the middle of the contract period; and (3) the project with the contract and in operation but has been transferred to the third party.

In any type, a variety of risks are involved for both the public and the private sectors, and it is significant to address different types of risks associated with unsuccessful projects and take policy implications from the project.

According to the Private Participation in Infrastructure (PPI) database of the World Bank in 2021, 2% or 16 out of 586 water and wastewater PPP projects from 1990 to 2020 had failed to be completed until the end of the contract period or have been terminated due to both or one party's responsibility. These phenomena implies that the Chinese water market does not seem to be regarded as a volatile market in the senses of legal settings, law enforcement, pricing mechanisms, and government policies.

As pointed out earlier, it is the private sector that bears a long list of risks regarding water or wastewater PPP projects. In particular, foreign companies often become exposed to sociopolitical, economic, financial, and regulatory risks, which give impacts on the daily operation of concerned plants. Chief engineers or project managers in charge of a water or wastewater PPP project should demonstrate their ability on how to hedge such a complicated set of risks for realizing expected levels of revenues in the contract period, which lingers for more than 20 to 30 years.

Another risk is the trustworthiness of terms and conditions in a contract signed with a local government from a foreign investor's point of view. Despite various national-level laws and regulations, things are somewhat different at the local level, particularly pertinent to a water or wastewater PPP project involved with a Chinese company. In less developed areas, such phenomena may become more serious so that the service fees which are due to the service companies are not paid on time or in full. This occurs because the government may not be able to reach out customers or the level of fees are determined too low to recoup the costs. In some cases, a local government misuses the service fees collected from customers for other purposes, not paying them to service companies (Lee 2010; Spooner 2018).

Opportunities

Newly emerging opportunities in the Chinese water market will be found not in urban but in rural areas soon. In July 2019, the Chinese government announced its plan to invest RMB 3 billion (US$350 million) by the end of the year for enhancing rural wastewater handling services and environmental quality. The plan focuses on 141 counties in central and western regions. The primary purpose of the plan is to narrow the serious regional gap within the country, and in particular, poorly managed and operated wastewater treatment services in rural communities. In 2017, the average rural wastewater treatment rate in the country was estimated approximately 20%, which demonstrates a huge gap with the 2020 goal of 60%. Since 2015, massive amounts of investment have been channeled to rural areas coupled with the Water Ten Plan in 2015 as well as the 13th Five Year Plan (2016–2020), amounting to RMB 46.2 billion (US$6.6 billion) by 2020 for improving wastewater treatment services for 130,000 villages.

In spite of such policy drivers and investment plans, there is no boom or enthusiasm felt in the rural water market of China. This is attributed to a gradual downturn of water and wastewater PPP projects in recent years as well as the paucity of local wastewater discharge standards which are different from urban areas. Instead, technology and system providers can play a significant role in leading the market providing prefabricated, containerized, or modular solutions for decentralized wastewater treatment. The companies involved in this market are Anhui Guozhen, Beijing Capital, Heilongjiang Inter China Water & Shuangliang Eco-Energy Systems (GWI 2019a).

A pursuit of wastewater PPP projects associated with the sponge city initiative in 2014 is a new business area that can generate environmental and economic benefits. Conventionally, projects for providing environmental services have difficulty creating a sound revenue stream, i.e., river remediation, landscaping, and drainage services. The initiative requires a consortium or group of companies to provide green infrastructure solutions at a design or technology supply level to the municipal government. Since the sponge city-related solutions involve design, construction, and water infrastructure elements, a single company is not able to tackle relevant issues. This is the reason why large-scale water companies in China with extensive water PPP project experiences are actively involved in this area, i.e., Tsinghua Holdings Ltd., China Everbright, and Beijing Capital, and these companies are competitive through their good connections with local governments. A recent example is the Sanya Sponge City PPP project in Hainan Island worth US$575 million in December 2016. The awardee was a consortium consisting of Jiangsu Zhongnan Construction, Beijing Urban Construction, and China Wuzhou Engineering (GWI 2016).

As the corona virus pandemic since late 2019 has posed a serious threat to the viability of water and wastewater PPP market in China, such an unprecedented challenge can become an unexpected business opportunity for private players in the market, digitization. Smart water management or technologies have drawn much attention of the water business community for the last few years, and China is one of the leading countries to make fast moves for smart water management solutions. The solutions encompass digital meter installation and water and wastewater treatment plant automation upgrades. Smart city development has been initiated by the government, which includes the promotion of water-related data collection and utilization. Self-isolation at home and social distancing due to the pandemic have prompted flexible working at utilities through remote-control systems and automatic meter reading technologies.

In June 2019, Tianjin Water Group awarded Sanchuan Wisdom Technology the largest digital metering project so far that embraces the replacement of 3.6 million mechanical meters with digital ones through a Narrowband Internet of Things (NB-IoT) network. The percentage of digital meters between the total annual meter installation figure would soar up from 20 to 70% in the next five years, although such a growth is highly subject to two primary factors, namely the commitment of the central and local governments to implementing smart city programs, and the introduction of a rule on who should bear the costs for meter replacements.

Remote control or automation has been put in place since the 1990s in water and wastewater treatment systems of China, and approximately 70% of work in Chinese water treatment plants could be undertaken automatically. Such a rate is likely to increase in the near future, and Beijing Enterprises Water Group renovated 50 wastewater treatment plants with the remote controlling system in 2019, planning to do more in 2020.

The corona virus pandemic has also accelerated the utilization of remote monitoring because more use of chlorine-based disinfection has resulted in a surge and concentration of residual chlorine in both collected and treated municipal wastewater. Potential toxic impacts of residual chlorine would be made on aquatic life and water environments, which leads to more regulatory programs and standards required.

Another interesting change in the composition of municipal wastewater influents is the growth of phosphorus and refractory organic pollutants thanks to the larger use of laundry detergent and hand sanitizer. More time at home means more consumption of food, which entails a rise of domestic wastewater and alters the carbon–nitrogen ratio. All these changes related to influent quality variations should closely be monitored through remote monitoring primarily focusing on river and pipeline monitoring, and such data should be reflected in establishing relevant policies, programs and standards (GWI 2020a).

Conclusion

The chapter has appraised the transformation of water and wastewater service sectors in China from the administrative system dominated by the state to a quasi-market with the mixture of socialist values and capitalist approaches since the late 1990s. The socioeconomic sensitivity of the sectors led the Chinese authorities to ignore the introduction of reform measures for the sectors for almost two decades, but market-based reform policies were gradually introduced by the central government in the late 1990s with the recognition of deteriorating water and wastewater services in urban areas.

A call for private companies into the Chinese water and wastewater service sectors entailed an active participation of foreign companies at the early stage of the market, which was expected to bring in substantial funding, advanced technologies, and management experiences. Favorable policies for foreign investors were introduced simultaneously in the early transition period including a fixed rate of return for foreign companies. But in 2002, the central government decided to lift the safety net for foreign investors, the fixed rate of return, which indirectly signified the beginning of open competitions between foreign and Chinese companies.

Institutional rearrangements have continued for accelerating the marketization of water and wastewater services in the country, including the revision of national-level laws and the establishment of national strategies to enhance water resources management, which have resulted in creating more business opportunities for both foreign and Chinese companies. Proactive engagement with the growth of the market has been easily observed not only in eastern and southern coastal areas, the most industrialized and developed areas of China, but also middle and western parts, which ambitiously opt for PPP projects for improving out-of-date facilities, management skills, and financing capacity for water and wastewater services. As the number of people served by private players in the market increases, there are the emergence and dominance of Chinese companies and the dwindling influence of foreign players. A myriad of advantages of Chinese companies in the market have overwhelmed early movers, foreign investors.

Several challenges need to draw attention, such as the need to enhance legal and regulatory guidelines for eliminating uncertainties for market players, particularly pertinent to financing as well as water pricing. A paucity of coordination between different government ministries and bureaus still gives confusion, inconsistencies, and contradiction in the market. Local governments often tend to make decisions associated with water and wastewater PPP projects for gaining short-term goals, focusing on economic achievements rather than long-term goals, i.e., sustainability or urban development, due to the cadre performance system. The discrepancy between the center and local levels in terms of policy implementation and law enforcement is still pervasive, which is another risky factor for both foreign and Chinese companies.

The message about the Chinese water and wastewater service market is explicitly clear, promising. Whereas most of the urban centers in

China have good access to clean drinking water, the Chinese authorities will allocate more investments and other necessary resources to improve wastewater services in marginalized urban areas as well as rural communities that are far lagging compared with their urban counterparts. The Water Ten Plan in 2015 will be expected to mobilize vast amounts of investment for improving wastewater services at the national level so that there will be large room for private players to contribute to relevant opportunities.

Creative approaches will be required for private players to take advantage of the implementation of the sponge city initiative in 2014, which is aimed at experimenting innovative measures to restore the urban hydrological cycle through green solutions for flood prevention, water quality enhancement, and water supply. The global pandemic of COVID-19 since late 2019 has put the Chinese water and wastewater PPP market in jeopardy, however, the future of the market will provide both risks and opportunities. An unexpectedly rapid pace of digitization in water and wastewater services of China can occur under the pandemic, which can accelerate the introduction of smart water management systems and utilities based on the automation of relevant systems as well as remote control. It appears that the potential of the Chinese water and wastewater PPP market is still large, a lot to be explored in the future.

References

Chen, Chuan. 2009. Can the Pilot BOT Project Provide a Template for Future Projects? A Case Study of the Chengdu No. 6 Water Plant B Project. *International Journal of Project Management* 27: 573–583.

Choi, Jae-ho, and Seungho Lee. 2010. Key Risks and Success Factors on the China's Public-Private Partnerships Water Project (중국 수처리 민관협력사업 사례분석을 통한 시사점 도출: 위험 및 성공요인 도출). *Korean Journal of Construction Engineering and Management* 11 (3): 134–143.

E20 Environment Platform. 2020. 2019 Top Ten Most Influential Water Companies in China (2019 年度水业十大影响力企业). E20 Environment Platform, 8 May 2020.

Fu, Tao, Miao Chang, and Lijin Zhong. 2008. *Reform of China's Urban Water Sector*. London: IWA Publishing.

Global Water Intelligence (GWI). 2004. *Water Market China*. Oxford: Global Water Intelligence.
Global Water Intelligence (GWI). 2008. *Water Market China 2009*. Oxford: Global Water Intelligence.
Global Water Intelligence (GWI). 2016. Global Water Intelligence Magazine—Project Trackers: Sanya Sponge City PPP Project in Hainan Province.
Global Water Intelligence (GWI). 2019a. Chinese Investors Look at Tech Options for Rural Water Focus. *Global Water Intelligence Magazine*, August 28–29.
Global Water Intelligence (GWI). 2019b. The World's Top 50 Private Water Operators. *Global Water Intelligence Magazine*, August 10–11.
Global Water Intelligence (GWI). 2020a. Automation Demand Sparks Digital Acceleration in China. *Global Water Intelligence Magazine*, April, 38–39.
Global Water Intelligence (GWI). 2020b. GWI Water Data: China (Mainland) Key Numbers.
Global Water Intelligence (GWI). 2020c. The Biggest Water Businesses in the World. *Global Water Intelligence Magazine*, August 6.
Gourmelon, Yannig, Qi Wu, and Watson Liu. 2015. *Pouring Profits: Chinese Water Industry*. Beijing: Roland Berger.
Lee, Seungho. 2006. *Water and Development in China*. Hackensack, NJ: World Scientific.
Lee, Seungho. 2007. Private Sector Participation in the Shanghai Water Sector. *Water Policy* 9 (4): 405–423.
Lee, Seungho. 2010. Development of Public Private Partnership Projects in the Chinese Water Sector. *Water Resources Management* 24 (9): 1925–1945.
Lee, Seungho, and Jaeho Choi. 2015. Wastewater Treatment Transfer-Operate-Transfer (TOT) Projects in China: The Case of Hefei Wangxiaoying Wastewater Treatment Project. *KSCE Journal of Civil Engineering* 19 (4): 831–840.
Ministry of Housing, Urban and Rural Development (MOHURD). 2018. *Urban Construction Statistical Yearbook (城市建设统计年鉴) 2017*. Beijing: MOHURD.
Ministry of Natural Resources (MNR). 2020. 2018 National Seawater Utilization Report (全国海水利用报告). Department of Marine Strategy, Plan and Economy, Ministry of Natural Resources, Beijing.
National Development and Reform Commission (NDRC). 2016. *13th Five-Year National Seawater Utilization Plan (全国海水利用十三五规划)*. Beijing: NDRC.

National Development and Reform Commission (NDRC) and Ministry of Commerce (MOC). 2017. Catalogue of Industries for Foreign Investment. June 28, 2017. Available Online: http://fdi.gov.cn. Accessed 15 September 2020.

Oh, Jihye. 2021. Public Private Partnership Projects in the Chinese Water Sector: A Focus on the Jiangsu Province. PhD Dissertation, Graduate School of International Studies, Korea University, February.

Owen, David. 2008. *Pinsent Masons Water Yearbook 2008–2009*. London: Pinsent Masons.

Qian, Neng, Schuyler House, Alfred Wu, and Xun Wu. 2020. Public-Private Partnerships in the Water Sector in China: A Comparative Analysis. *International Journal of Water Resources Development* 36 (4): 631–650.

Shen, Dajun, and Juan Wu. 2017. State of the Art Review: Water Pricing Reform in China. *International Journal of Water Resources Development* 33 (2): 198–232.

Spooner, Simon. 2018. The Water Sector in China: Market Opportunities and Challenges for European Companies. The EU Horizon 2020 Report.

Wang, Anya. 2016. New Year Present for China Seawater Desalination: The National 13th Five-Year Plan. UMORE Consulting. Available Online: http://umoregroup.com/project.aspx?id=2097. Accessed 21 September 2020.

Wang, H., and X. Zhao. 2004. PPP Case Study on China's Urban Water Supply Industry—Factor B of Veolia Chengdu No. 6 Waterworks. Unirule Institute of Economics, China Center for Public-Partnerships.

WaterWorld. 2020. Suez Recognized Among China's Top 10 Most Influential Water Companies. *WaterWorld*, May 11. Available Online: http://www.waterworld.com/print/content/14175707. Accessed 10 October 2020.

World Bank. 2006. *China Water Quality Management—Policy and Institutional Considerations*. Washington, DC: World Bank.

World Bank. 2020. *Private Participation in Infrastructure (PPI) Annual Report 2019*. Washington, DC: World Bank.

World Bank and PPI. 2020. Private Participation in Infrastructure (PPI) Database. Available Online: http://ppi.worldbank.org. Accessed 24 September 2020.

Xinhua. 2020. China's Urbanization Rate Hits 60.6 pct. *Xinhua Net*, January 19. Available Online: http://xinhua.net. Accessed 13 March 2021.

Zhong, Lijin, and Arthur Mol. 2010. Water Price Reforms in China: Policy-Making and Implementation. *Water Resources Management* 24: 377–396.

9

Transboundary Rivers

Introduction

The chapter investigates China's transboundary rivers and their political, socioeconomic, and environmental implications. The complexity of water-related challenges has pushed the central and local governments to seek for additional water sources beyond its domestic freshwater systems, endeavoring to tap in more water resources in transboundary rivers. China seems to have sustained its hegemonic stance related to transboundary rivers so far thanks to its exclusive position as the upstream country in 12 out of its 15 major transboundary rivers, except for the Yalu-Amnok, the Tumen-Dooman, and the Amur- Heilong Rivers. Perhaps it would be natural for China's diplomats or anyone, who are involved in transboundary river matters, to adhere to the principle of absolute sovereignty rather than the principle of 'prior use, no harm and equitable utilization' which has been regarded as an internationally accepted norm associated with transboundary water issues.

The country has opened its door to the outside world since the late 1970s and has actively been engaged in global political, economic, and environmental affairs, taking full advantage of joining the bandwagon of globalization. Nevertheless, China has shown a reticent attitude toward its neighbors, particularly related to natural resource exploitation and protection and has pursued its socioeconomic goals through the unilateral development of water resources. As Ho (2014) stresses, China has almost no comprehensive policy or strategy for managing its transboundary rivers and regards the rivers as a subset of its regional politics. This connotes that transboundary river issues have not drawn major attention of top leaders in Beijing. China's silence about international criticism on the upstream dams in the Lancang-Mekong River is, therefore, not necessarily strategic or intentional, but reflecting the ignorance of the central government about Mekong issues regardless of socioeconomic, environmental, and security-related significance of the river.

However, the study will unfold that China has not been inactive in establishing the relationships with its neighbors and have caused no major trouble or conflict except for mounting tensions with India pertinent to the Yarlung Tsangpo-Brahmaputra River. This is partly because there is no pressure for China to be involved with tough negotiations or bargaining with its neighbors, and in most cases, China overwhelms its riparian counterparts in political, economic, security, and military senses. China's stance can be mirrored as the behavior of hegemon in a transboundary river, and non-hegemonic riparian states are pressurized to follow the terms and conditions for compliance, imposed by its mighty riparian partner, China.

Such a particular pattern of Chinese behavior has gradually been evolving in favor of mutual understanding and cooperation with its riparian countries in recent few decades. Water has increasingly become a security interest in China's foreign policy agendas since the new millennium. The notion of water as a non-traditional security issue was introduced first in 2002 by President Hu Jintao and furthered in 2013 by President Xi Jinping as part of environmental security (Xie and Jia 2018). In addition, China's hegemonic behaviors are not necessarily found in some cases if strategic socioeconomic, political, military, or security interests and benefits are created based on bilateral agreements with China's counterparts. After close encounters with its downstream neighbors,

China began to be more cooperative since the onset of the new millennium in the Lancang-Mekong River Basin. China's Lancang-Mekong Cooperation (LMC) mechanism since 2015 has emerged as a regional cooperative platform competing with the Mekong River Commission that has served as a disappointing cooperative mechanism in the river basin.

The first part of the study highlights the overall situation of transboundary rivers in China, showing China's advantageous position as the upstream superpower in most of its international rivers. In the second part, attention is paid to water diplomacy strategies of China, discussing the concept of water diplomacy coupled with its applicability to China's engagement with transboundary rivers. The third part focuses on China's participation in the architecture of international freshwater laws and a collection of analyses on transboundary river cases, i.e., the Yalu-Amnok, the Tumen-Dooman, the Heilong-Amur, and the Illi and the Irtysh Rivers. In the fourth part, the study evaluates the relationship between China and its riparian countries in the Lancang-Mekong River as a case study, emphasizing the evolving attitude of China toward the downstream countries for maximizing its socioeconomic benefits as well as security advantages, particularly through the creation of the Lancang-Mekong Cooperation mechanism in 2015.

Transboundary Rivers in China

Patterns of transboundary river interactions between riparian countries are affected by various factors, including political, socioeconomic, legal, and environmental circumstances, and more specifically, water-related factors are associated with water demand and use patterns, water use efficiencies, and economic and management capacities. Zhang and Li (2018) classify transboundary rivers into three major types: (1) transboundary rivers; (2) border rivers; and (3) mixed rivers, i.e., a major river with many tributaries and multiple crossings at the border. This classification is significant because riparian countries in rivers running across the border may be confronted with resource-scarcity conflict and countries in which the river becomes the border may be subject to conflict that can be prompted by fuzzy borders. Conflicts related to water quality are likely to occur between countries that share a transborder river rather than border river. Nevertheless, a major river includes numerous tributaries, and some of them are transborder rivers, and others are border rivers in many transboundary rivers.

According to this classification, the Chinese transboundary rivers are largely divided into two groups. First, there are transboundary rivers in which China is located upstream, i.e., the Nu-Salween River, the Yarlung Tsangpo-Brahmaputra River, the Yuan-Red River, the Ganges River, the Indus River, the Lancang-Mekong River, and the Irtysh River. Second, there are border or transborder rivers in which China is situated downstream and mixed rivers. The majority of transboundary rivers in Northeast China is categorized as either border or mixed rivers, such as the Heilong-Amur River, the Yalu-Amnok River, and the Tumen-Dooman River. The cases of China as downstream country are associated with the Kherlen River, stemming from Mongolia. With regard to the Ili River, the largest source of the river originates from Kazakhstan, and another significant tributary of the river is Khorgos River which demarcates the border between China and Kazakhstan with a length of 150 km (Zhang and Li 2018).

Transboundary water resources include not only transboundary rivers but also lakes, inland waters, and aquifers. The study sheds light on transboundary rivers since there have been active interactions between China and its transboundary countries in terms of resource exploitation, inland navigation, environmental protection, and other issues. In addition, together with Chile, China is ranked third regarding the number of transboundary rivers on the globe (Zhang and Li 2018). It is worth noting that the total amounts of water resources flowing into China are estimated at 20.57 billion m^3 in 2018 whereas those flowing out of China into neighboring countries, 610.91 billion m^3, including the volumes of transboundary rivers, 125.55 billion m^3 (MWR, 2019).

There are 15 major transboundary rivers in China and 12 of them originate from the country (see Table 9.1). 15 countries are bordered with China related to the transboundary rivers, which are North Korea, Russia, Mongolia, Kazakhstan, Kyrgyzstan, Tajikistan, Bhutan, Myanmar, Thailand, Laos, Nepal, Pakistan, Afghanistan, India, and

Vietnam. The neighboring countries stretch over 22,000 km along the Chinese border and are adjacent to nine provinces and 132 prefectures impacting upon the socioeconomic relationships with more than 22 million people in China (Chen 2012, 2017; Tang and He 2000).

The 15 major transboundary rivers are concentrated in Northeast, Southwest, and Northwest China. The northeastern river basins are composed of the four rivers, the Heilong-Amur, the Yalu-Amnok, the Tumen-Dooman, and the Suifen-Razdolnaya Rivers, and the three rivers except for the Suifen-Razdolnaya River serve as the borders between China and North Korea, Russia, and Mongolia. The total border length along the three rivers stretches at more than 5,000 km.

China has the largest number of major transboundary rivers in the southwestern parts of its territory, and the river basins are situated within the Qinghai-Tibetan Plateau which is as high as 4,000 m above the sea level. There are eight rivers, i.e., the Irrawaddy, the Nu-Salween, the Lancang-Mekong, the Pearl, the Yalurng Tsangpo-Brahmaputra, the Heng-Ganges, the Sengezhangpu-Indus, and the Yuan-Red Rivers. The plateau is often called as 'Water Tower of China' or 'Water Tower of Southeast Asia or South Asia'. The northwestern river basins are largely situated within the territory of Xinjiang Province, of which natural conditions are characterized with the dry climate, high mountains, and little people. The major rivers are the Irtysh-Ob, the Ili, and the Aksu Rivers (Tang and He 2000).

China's stance toward the 15 transboundary rivers and its neighboring countries was consistent, paying little attention to relevant issues. The country had no reason to be serious about the rivers thanks to its unparalleled position of upstream hegemon in 12 transboundary rivers out of 15 except for the Yalu-Amnok, the Tumen-Dooman, and the Heilong-Amur Rivers. This extremely advantageous position of 'Upstream Superpower' has led China to be unaware of the urgent need to cooperate with the other riparian countries (Chen 2012; Nickum 2008). Kettelus et al. (2015) point out that China's inactive engagement with its riparian neighbors are, to some extent, determined by geography, because transboundary river areas are situated in remote border areas in which less people are around so that no socioeconomic and political attention has been paid to the transboundary rivers.

Table 9.1 15 Major transboundary rivers in China

Location	River name	Size of river basin (10,000 km²)		Length of river (km)		Origin of river	River basin countries
		Total size	Inside China	Total length	Inside China		
Northeast	Heilong-Amur	184.3	88.3	3,420	2,854[a]	Mongolia	China, Russia, Mongolia
	Yalu-Amnok	6.45	3.25	816	816[a]	Jilin, China	China, North Korea
	Tumen-Dooman	3.32	2.29	505.4	490.4[a]	Jilin, China	China, North Korea, Russia
	Suifen-Razdolnaya	1.73	1	443	258	Jilin, China	China, Russia
Southwest	Irrawaddy	43.1	4.33	2,150	178.6	Xizang, China	China, Myanmar
	Nu-Salween	32.5	14.27	3,200	1,540	Xizang, China	China, Myanmar, Thailand
	Lancang-Mekong	80	16.7	4,880	2,129	Qinghai, China	China, Myanmar, Laos, Thailand, Cambodia, Vietnam
	Pearl	45.37	45.37	2,214	2,214	Yunnan, China	China, Vietnam
	Yarlung Tsangpo-Brahmaputra	93.8	23.92	2,900	2,229	Xizang, China	China, Bhutan, India, Bangladesh

Location	River name	Size of river basin (10,000 km²)		Length of river (km)		Origin of river	River basin countries
		Total size	Inside China	Total length	Inside China		
	Heng-Ganges	107.3	0.23	2,700	49	Xizang, China	China, Nepal, India, Bangladesh
	Sengezhangpu-Indus	116.6	2.49	2,880	419	Xizang, China	China, India, Pakistan, Afghanistan
	Yuan-Red	11.3	7.4	1,280	677	Yunnan, China	China, Vietnam, Laos
Northwest	Irtysh-Ob	292.9	5.7	4,248	633	Xinjiang, China	China, Kazakhstan, Russia
	Ili	15.12	5.67	1,237	442	Kazakhstan	China, Kazakhstan
	Aksu	5	3.1	589	449	Kyrgyzstan	China, Kyrgyzstan

Remark [a]This indicates the length of the border rivers between China and the other riparian countries
Source Modified based on Tang and He (2000) and Zhang and Li (2018)

The hegemonic position of China in transboundary rivers seems to have had a far-reaching influence on the country's diplomatic attitudes toward its riparian countries, which is epitomized in its preference to bilateral treaties or agreements rather than participating in multilateral agreements or frameworks as seen from the case of the Lancang-Mekong River (Chen 2012). In recent decades, China has demonstrated its willingness to take an active part in multilateral dialogues and negotiations. The Greater Mekong Subregion (GMS) Program, which was embarked on by the Asian Development Bank (ADB) in 1992, was a good example of China's active involvement with a multilateral cooperative framework, particularly in the fields of transport, energy, and hydropower (Lee 2015). The Lancang-Mekong Cooperation (LMC) is a brand-new vehicle which can accommodate all the Mekong riparian countries on board with the Chinese rules of the game.

A primary motivation of China's changing attitudes toward transboundary rivers is that the country would like to maximize multifaceted benefits from multilateral cooperation rather than unilateral development, i.e., socioeconomic, security, and energy benefits. For instance, although it may take more time to stabilize relations between South Korea, North Korea, and US, the thawing mood of Northeast Asia can entail the countries involved in the Yalu-Amnok and the Tumen-Dooman River Basins to consider a variety of projects in sustainable development (Lee 2020). Socioeconomic and security benefits have been gained for China through the GMS program as well as China's gradual engagement with the Lower Mekong countries through the Mekong River Commission (Lee 2015). The era of the LMC will be likely to accelerate the enlargement of multilateral benefits for China as well as the other riparian countries if the new cooperative framework works for collective interests rather than the hydro-hegemon's unilateral interests.

It is worthwhile to pay attention to the implications of the case of the Lancang-Mekong Rivers from the perspective of China's global development strategy, the Belt and Road Initiative (Eder 2018; Chen 2017). The Lancang-Mekong River is regarded as a starting point or a connecting route between the Mainland Southeast Asia and Southwest China. China will possibly focus more on the area for pushing forward aggressive policies and strategies for bolstering economic growth and having more political influences.

Water Diplomacy Strategies of China

At the global level, tensions have been mounting in a series of transboundary river basins, such as the Tigris-Euphrates, the Nile, and the Jordan River Basins for a few decades, and these cases have helped transboundary water management issues securitized so that water challenges have increasingly become political and national security agendas. The countries involved in transboundary river basins have regarded transboundary water management as part of their foreign policy. Consequently, a new concept of 'water diplomacy' (or hydrodiplomacy) has emerged as a useful tool to address transboundary water cooperation or conflict. The United Nations agencies have actively promoted the approach of water diplomacy, including UNESCO (Cuppari 2017). In accordance with the trend, water diplomacy has increasingly been adopted in official dossiers and academic publications of China (Zhang and Li 2020).

Water diplomacy is defined as the accommodation of water issues into a foreign policy domain with long-term objectives. All measures are considered and implemented by state and non-state actors for preventing or peacefully resolving conflicts on water availability, allocation, or use between and within states and public and private stakeholders. Broader scopes of water diplomacy embrace the enhancement of regional security and stability, the promotion of regional integration, the scale-up of trade relations, and the increase of political influence. This concept emphasizes the magnitude of a multilayered process, which advocates complex interactions between non-state stakeholders, such as sub-national governments, regional and international organizations, and NGOs (Cuppari 2017; Hague Institute for Global Justice 2016; Molnar et al. 2017; Zhang and Li 2020).

Chinese foreign policy has primarily been led and dominated by the central government. The trajectory of China's engagement with transboundary rivers can be divided into the two major periods: (1) the Mao period (1949–1976); and (2) the reform period (1978–now). In the Mao period, the most salient issues for the country were national security and modernization, and therefore, the cooperation of China with its riparian countries was decided primarily based on the extent to which the country kept favorable relations with them in political and ideological senses, such as the Sino-North Korea, the Sino-Soviet Union, and the Sino-Vietnam relations.

Since 1978, China has witnessed the transformation of the global political economy landscape toward the regional economic integration and has been committed to accelerating reform policies in coastal as well as border areas. The mega trend has given profound impacts on China's transboundary river strategy, which has gradually been transformed from bilateral to multilateral cooperation and from a single natural resources development or engineering project to benefit sharing on transboundary natural resources and market development through a regional cooperative framework. This benefit sharing approach is a new norm for transboundary river cooperation, which has been accepted in China as a new pathway which China's water diplomacy should pursue (Tang and He 2000).

Relevant cases have been observed. China was committed to scaling up its trade volumes with Myanmar since the 1980s, which is related to the Nu-Salween River, and in 1994, the total volumes of export from China to Myanmar amounted to more than RMB 4 billion (US$ 615 million). With regard to the Tumen-Dooman River Basin, China was one of the founding countries for the Tumen River Area Development Plan (TRADP) in 1991 together with North Korea, South Korea, Russia, and Mongolia, which was aimed at facilitating trade and achieving regional economic development with the investment of US$ 30 billion. Another case is China's engagement with the Lancang-Mekong River Basin. The Greater Mekong Subregion (GMS) Program was initiated by the Asian Development Bank (ADB) in 1992 for all the Mekong riparian countries, and on the Chinese side, the two provinces were committed to various projects of the GMS program, which were Yunnan as well as Guangxi Provinces (Lee 2015, 2020; Tang and He 2000).

These cases imply China's changing approach to transboundary rivers toward more cooperation without unilateral decisions over natural resources development, such as hydropower dam development. However, it seems that China lacks its comprehensive transboundary water strategy and would like to tackle shared water issues with its riparian neighbors via the so-called one river, one-country approach. In addition, the main player for bilateral or multilateral negotiations in China, the Ministry of Foreign Affairs, does not appear to play an important role in infrastructure developments in transboundary rivers (Ho 2014; Moore 2018). China has demonstrated its reactive attitude toward the low utilization of shared water resources in transboundary rivers due to the insignificance of those rivers. However, pertaining to the major border and transborder rivers where China is situated downstream, the country has adopted a proactive diplomacy because the degree of vulnerability would be higher. Related cases are associated with the Sino-Russia and the Sino-North Korea water agreements, and the Sino-Kazakhstan's Friendship Joint Water Diversion Project on the Khorgos River, a tributary of the Ili River (Zhang and Li 2018).

This state-centric evaluation, however, misses complicated relations between the center and local governments on foreign policy and transboundary river issues, particularly since the late 1970s. One of the major features in the reform period, decentralization, has paved the broad way for local governments to contribute significantly to national-level foreign policies in the course of globalization. Zhang and Li (2020) maintain that local governments of China have emerged as foreign policy agents as well as partners of the national government. As foreign policy agents, local governments are often at the forefront of international negotiations in the fields of cross-border trade, foreign investment and labor exchange, cross-border security, joint police action against crime, and patrolling shared rivers or lakes. Interestingly, local governments with delegated power from the central government do not necessarily behave solely based on the orders from the center, and the pursuit of local interests are prioritized than national interests or international commitments.

Substantial financial resources and economic interests overseas have pushed provincial governments to undertake their own projects without little support from the central government, which demonstrates the emergence of a new type of partnership between the center and local governments. The central government can rely on various activities of local governments for its foreign policy while the endorsement of the national government is particularly useful for local governments' international activities, as seen from the cases in Africa and Southeast Asia. In particular, Yunnan Province has played a pivotal role in reshaping China's foreign policy toward Mainland Southeast Asia and the Lancang-Mekong River Basin (Chen and Jian 2009; Zhang and Li 2020).

China and the 1997 UN Water Convention and Its Transboundary River Cases

It is necessary to discuss the architecture of the international freshwater law and explore the extent to which hydraulic, socioeconomic, and environmental strategies of China have been evolving in contact with other riparian countries. The international law regime for freshwater issues was first codified under the auspices of the League of Nations in 1923 Convention Relative to the Development of Hydraulic Power Affecting More than One State, which provided simple but essential articles on the respect of other parties' interests when building major hydraulic infrastructures. In 1966, the International Law Association (ILA) drafted the Helsinki Rules on the Use of the Waters of International Rivers of 1966 (1966 Helsinki Rules), which describes basic requirements for the state regarding navigational and non-navigational use of international watercourses (ILA 1966).

The International Law Commission under the UN succeeded in making the draft of the 1997 United Nations Convention on the Law of Non-Navigational Uses of International Watercourses (1997 UN Water Convention), which eventually entered into force on 17 August 2014 (UN 2014). In the meantime, the Convention on the Protection and Use of Transboundary Watercourses and International Lakes (1992 Helsinki

Convention) was introduced in Helsinki, 1992, which entered into force in 1996, putting more emphasis on the protection of water ecosystems (UNECE 1992). The ILA endeavored to update what had been done for the last forty years and produced the 2004 Berlin Rules on Water Resources (2004 Berlin Rules), which intended to make sure a management regime for the equitable and reasonable use of international drainage basins that would not cause significant harm to other basin states (ILA 2004). In 2008, the ILC adopted the 2008 Draft Articles on the Law of Transboundary Aquifers (2008 Draft Articles) that is aimed at addressing the issue of water pollution in an aquifer that trigger challenges in groundwater management (UN 2008). All these international instruments were the outcomes of global efforts for sustainable use and protection of international watercourses with no-harm rule, the principle of equitable and reasonable utilization, and a general obligation to cooperate. Also, transboundary water cooperation works on the basis of the principles of limited sovereignty and community of interests (Fry and Chong 2016).

China has demonstrated its overall support of environmental sustainability through its ratification of numerous environmental treaties, i.e., the UN Framework Convention on Climate Change, the Convention on Biological Diversity, and the UN Convention to Combat Desertification. However, the country has never signed or ratified the 1992 Helsinki Convention or the 1997 UN Water Convention. In particular, China has refused to sign the 1997 UN Water Convention, arguing that the convention would significantly harm its territorial sovereignty over water resources within the country. Referring to the UN General Assembly Resolution on Permanent Sovereignty over Natural Resources and the Charter of Economic Rights and Duties of States, China stresses its sovereignty over natural resources (Fry and Chong 2016; Xie and Jia 2018).

Such an argument, however, is not necessarily persuasive, because the state's permanent sovereignty over natural resources should be guaranteed based on mutual respect of another state's permanent sovereignty based on the principle of sovereign equality. In addition, a violation of another state's permanent sovereignty over its natural resources gives negative impacts on the rights of individuals, which is against the spirit of the UN Charter. Therefore, the dilemma for China would be to insist its permanent sovereignty over its natural resources and to respect the same rights of its riparian countries (Fry and Chong 2016).

China was one of the three countries (Turkey and Burundi) that refused to vote for the 1997 UN Water Convention. The country withdrew its commissioner in 1998 from the World Commission on Dams, opposing to the outcomes from the final report of the commission. Considering more than half of the world's 48,000 large dams (above 15 m) existent in China, such an attitude explicitly demonstrates China's conventional and traditional approach of reluctance or resistance to multilateral frameworks regarding international freshwater issues. China's distrust of multilateralism is attributed to the possible intervention against its sovereign rights and freedom of action for managing natural resources, including water resources (Ho 2014).

China's relationship with its riparian states encapsulates the way how China manages transboundary river issues. For instance, the camaraderie tie between China and North Korea for the security of Northeast Asia has spawned smooth cooperation in the Yalu-Amnok and the Tumen-Dooman River Basins. The joint hydropower development and management of the four dams in the Yalu-Amnok River Basin has been a backbone for bilateral cooperation between the two communist countries through the North Korea-China Hydropower Corporation since 1955. Cooperation over the Tumen-Dooman River Basin shows an instance of China's engagement with the other countries involved in the Greater Tumen Initiative (GTI, the changed name of the precursor, the Tumen River Area Development Plan, TRADP) (1991–now), including the riparian countries, North Korea, China, and Russia, and the participating countries, South Korea and Mongolia (Fry and Chong 2016; Lee et al. 2019; Lee 2020).

Another good practice is observed in the case of the Heilong-Amur River between China and Russia. In 2004, China and Russia officially agreed the demarcation of the border in the river basin and reached the agreement about the territorial sovereignty of disputed islands. In addition, the joint-use agreement over the watercourse has paved the way for both countries' vessels to navigate without interruption, and the two countries are now prepared to construct hydropower plants along the river coupled with the creation of a free-trade zone along the Suifen River (Fry and Chong 2016; Lukin 2013; Nickum 2008).

Cooperative interactions between China and other riparian countries have not well occurred in transboundary river basins of Central Asia, particularly Kazakhstan, Kyrgyzstan, and Tajikistan regarding the Ili and the Irtysh Rivers. Large-scale water diversion projects have been undertaken by China in the Irtysh River for delivering water to Xinjiang's oil fields and western China's commercial and industrial development. As for the Ili River, the Chinese used 38% and the Kazakhs 62% of the volumes of the river in the 2000s, however, these figures were changed into 43 and 57% in 2014, respectively. This phenomenon occurred thanks to the rapid expansion of irrigated fields on the Chinese side from 702,100 ha in 2004 to 1,322,700 ha in 2014.

Kazakhstan has been seriously wary of long-term impacts of Chinese heavy withdrawal and water pollution upstream in the Ili and the Irtysh Rivers, such as damage on fisheries, agricultural production, and deforestation. Although Kazakhstan has not been able to challenge China over these matters, transboundary river conflicts would take place in the future without bilateral agreements for mutual benefits between the two countries (Fry and Chong 2016; Kukeyeva et al. 2018).

Ho (2017), however, argues that increasing interdependence between China and Kazakhstan has led to cooperation related to transboundary water issues, and in particular, Kazakhstan has become an important neighbor to help pacify domestic instability associated with the Uighur minorities in Xinjiang Province and to help expand socioeconomic and political influence in Central Asia competing with Russia and the US. This changing pattern of relationship with Kazakhstan is related to China's good-neighborliness policy. Nevertheless, such a bilateral cooperation has not necessarily developed smoothly. China has demonstrated its commitment to bilateral cooperation projects with Kazakhstan without its intention to expand cooperation into a regional level inviting Russia although Kazakhstan would like to take the lead for the regional cooperation framework. The two countries signed the Agreement Concerning Cooperation in the Use and Protection of Transboundary Rivers in September 2001 and established the Sino-Kazakh Joint Commission as a result. However, progress has been very slow, and there will be no mechanisms of compliance and dispute settlement together with the lack of specified volume intakes by both sides. Regarding water allocation, the two countries have not made a consensus on a water-sharing agreement for the Ili and the Irtysh Rivers.

Territorial disputes between China and its riparian countries are observed in many border areas, particularly the border with India. The two countries have remained in dispute about a territory of 83,000 km^2 in the Ganges River, that of 25,900 km^2 in the Indus River, and the Yarlung Tsangpo-Brahmaputra River. These cases demonstrate the lack of effective legal regime applied to the disputed areas. Although each country possesses its own water-related laws, i.e., China's Water Law and India's Environmental Protection Act, water resources in the disputed international rivers cannot adequately be monitored and managed without an institutionalized cooperative framework (Fry and Chong 2016).

Case Study: The Lancang-Mekong River Basin

Special focus is placed on the Lancang-Mekong River Basin as the case study. There have been the evolving patterns of China's engagement in various transboundary river basins. The case of the Lancang-Mekong River Basin is particularly worth paying attention to because the country has recently been committed to creating new rules of the game through the Lancang-Mekong Cooperation mechanism (LMC). It is argued that the architecture of transboundary water cooperation in the river basin has gradually been reshaped, and the pendulum of transboundary water cooperation seems to swing by China's political and economic teeth. Transboundary water cooperation in the Lancang-Mekong River Basin can be divided into the two different phases—before and after 2015 when the LMC was established as the major cooperative vehicle of China over the river basin, and the waning of the MRC's influence began to accelerate thanks to the emergence of the new competitor.

Complex River System with the Primary Interest, Hydropower

The Lancang-Mekong River rises in the Himalayas at an elevation of around 5,000 m. As the world's 12th longest river, the river flows 4,763 km from the Qinghai-Tibetan Plateau in China to the Mekong Delta in Vietnam. The whole river basin stretches over the six riparian countries, namely China, Myanmar, Thailand, Laos, Cambodia, and Vietnam. The Chinese have called the river as the Lancang River, which indicates the upper reaches of the river, and the people in the downstream areas have called it as the Mekong River. China and Myanmar belong to the upper river basin whereas the rest of the four riparian countries are located downstream. The total amounts of discharge flow are as large as 446 km^3, and the size of the river basin is estimated at 810,000 km^2. The river basin is home for more than 70 million people who are heavily relying on the river for their livelihoods, such as freshwater fisheries, irrigation, aquaculture, tourism, and trade of foods (MRC 2020b) (see Fig. 9.1).

Snowmelt from the Tibetan Plateau contributes to the dry season discharge down to the northern part of Vientiane, Laos, and the wet season from June to October entails a discharge pulse from the Mekong tributaries. The discharge of wet months is five to ten times greater than that of dry months. Such wet season peak flows trigger the large flow reversal up to Tonle Sap Lake in Cambodia, facilitating fish movement and delivering a pulse of sediment and nutrients to the floodplain. The pulse system plays a key role in maintaining water and ecosystems in the river basin and supports rice production around the lake. Countless species of flora and fauna flourish throughout the river basin, and the mainstream river as well as tributaries serve as major routes for migrant fish species which ply up and down the river over a few hundred km. Biodiversity is a significant asset for the Lancang-Mekong River (FAO 2011; MRC 2020b). This hydrologically unique regime of the river, however, is threatened by pervasive anthropogenic behaviors, including dam building on the mainstream as well as tributary rivers, and water diversion projects for irrigation, and exacerbated by climate change.

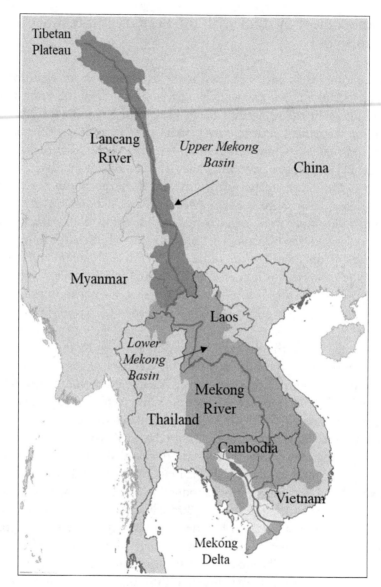

Fig. 9.1 Lancang-Mekong River Basin (*Source* Author)

It is necessary to highlight the issue of hydropower in the river basin, since hydropower development has been, is, and will be the most significant as well as controversial subject. The reason why such a topic is always on top of the agendas between the riparian countries is that hydropower is the most viable option to achieve energy security, poverty eradication, and socioeconomic well-being. This is also regarded as a green option for energy generation and an icon that can beat negative impacts triggered by climate change (Lee 2015, 2020; MRC 2017, 2020b).

Whereas all the countries have endeavored to optimize large potential of hydropower development in the river basin, China and Laos have been the leading countries for building a number of hydropower dams and exporting hydroelectricity to their neighboring countries. A total of 11 hydropower dams of China have been completed in the Lancang River as of December 2020, and 11 additional dams with the capacity of more than 100 MW will be constructed in the further upstream of the Lancang River. The installed hydropower capacity on the Lancang River is as large as 19,285 MW with the planned total increasing to 31,300 MW. The development of hydropower dams in Myanmar, the other upstream country, has been sluggish, and the first dam on Myanmar's Mekong tributaries was commissioned in 2017. The country plans to construct at least six more small storage dams (Eyler et al. 2020; MRC 2020b).

Laos is ambitious to become 'the Asia's Battery' which aims to export hydroelectricity to Thailand, China, Vietnam, and Cambodia. The Xayaburi and the Don Sahong Dams in Laos have been constructed on the mainstream of the Lower Mekong River which is no longer regarded as 'Free Flow' although numerous hydropower dams have already been in operation in Mekong tributaries (Lee 2018; MRC 2020b).Thailand keeps a low profile in terms of hydropower development, however, many Thai consortiums, consisting of Thai developers, construction companies, and banks, have made contracts with the Lao government for exporting large portions of hydroelectricity to Thailand, often, 90% (Lee 2018). Whereas there is almost no potential of hydropower development in Vietnam, Cambodia has gradually developed its hydropower plants in the river with the substantial aid from China (Matthews and Motta 2015; Shin et al. 2020).

In addition to hydropower development, there have been several key issues for socioeconomic development and ecosystem protection in the river basin, such as fishery industries, biodiversity and ecosystems, and food security that is jeopardized by the sea level rise of the Mekong Delta. Millions of people are depending upon freshwater fish caught in the river basin, which account for approximately 50% of the total amounts of protein consumption for the Cambodians, and fishery industry provides innumerable jobs for people in the downstream countries of the river.

Biodiversity and ecosystems in the river basin have drawn major attention of many conservationists who aim to protect rare species of flora and fauna within the river basin which have badly been affected by a variety of anthropogenic activities, including illegal logging, expansion of irrigation fields, rife poaching of animals, and serious wars. In recent decades, water pollution threatens ecosystems of the downstream areas, compounded by population growth, urbanization, and industrialization.

Farmers in the Mekong Delta are increasingly worried about the salinization of arable land thanks to the gradual encroachment of seawater into delta areas and less water volumes flowed in the country due to a number of upstream dams. More seawater intrudes delta areas if less water flows from upstream, which results in decreasing rice production in the delta. Vietnam is one of the largest rice production and export countries in the world, and therefore, it is vital to ensure sufficient amounts of freshwater flowed downstream for both rice production and ecosystem protection (MRC 2020b).

Considering such a complexity of issues within the river basin, the riparian countries have sought after a cooperative pathway to achieve sustainable development. Cooperative efforts embarked on in the 1950s, however, the countries had to wait until 1995 when the agreement on sustainable development in the Lower Mekong River Basin was reached between the four downstream countries, Thailand, Laos, Cambodia, and Laos. China and Myanmar decided to join the cooperative mechanism as dialogue partners in 1996 because of their own concerns and vested interests. As a brainchild of the agreement in 1995, the Mekong River Commission was established as an international river organization and has played an imperative role in tackling water and environmental issues in the river basin between the riparian countries. However, there has been criticism about political legitimacy, roles, and capabilities of the institution from 1995 to 2015, for about 20 years.

The watershed for the political economy landscape of the Lancang-Mekong River Basin was the launch of the Lancang-Mekong Cooperation mechanism (LMC), created by China. Although it is a bit early to judge the way how the LMC gives an influence on complicated interactions between the riparian countries, it is necessary to investigate what has been achieved in the MRC era (1995–2015) and what can be expected in the MRC-LMC era from 2015 onwards.

The Era of the Mekong River Commission from 1995 to 2015

The Mekong River Commission (MRC) was officially launched in 1995 on the basis of the Agreement on the Cooperation for the Sustainable Development of the Mekong River Basin on 5 April 1995 (MRC 1995). The MRC has functioned as a regional platform for water diplomacy and a center of knowledge of water resources management for sustainable development of the region. In addition to the four member countries of the MRC, there are two dialogue partners, China and Myanmar, and these countries have kept this status since the onset of the MRC system. Myanmar is more interested in other rivers, such as the Irrawaddy and the Nu-Salween Rivers and contributes little to the annual water flow of the mainstream of the Lancang-Mekong River. China has been reluctant to join the MRC system because of its own development agendas upstream, such as hydropower dam building.

The mission of the MRC is to promote and coordinate sustainable management and development of water and related resources for the mutual benefits of the Lower Mekong countries. The functions of the commission are as follows: (1) basin monitoring; (2) data and information systems and services; (3) modeling and assessments; (4) basin planning; (5) flood and drought forecasting; (6) implementing MRC Procedures; (7) dialogue and partnership; and (8) stakeholder engagement and communication. The MRC Procedures encompass the Procedures for Data and Information Exchange Sharing; Notification,

Prior Consultation and Agreement; Water Use Monitoring; Maintenance of Flows on the Mainstream; and Water Quality (MRC 2017).

Whereas the MRC has played a pivotal role in facilitating cooperation between the Lower Mekong countries for sustainable development, the institution embraces its critical limitations since the beginning. First, the MRC is not independent in many aspects, particularly financial one. The majority of funding for the institute has stemmed from various donors, such as European Union countries, Australia and Japan. In the period between 2016 and 2020, the necessary budget for the MRC was a total of US$ 65 million, and the four Lower Mekong countries contributed to only US$ 15 million. Adding the leftover amount of US$ 9 million, the total amount of available budget was US$ 24 million, which means that additional US$ 41 million should be channeled from the donor community (Hunt 2016).

Second, there has been a lack of consensus on thorny issues between the riparian countries. The MRC is supposed to facilitate dialogues and discussions between the member countries for tackling critical issues that affect the river basin. The reality is, however, that each member country has put forward its own national interests and often refuses to follow the principles or agreed matters for the sake of its own vested interests. For instance, Laos has often been criticized for pursuing its ambitious plan to build numerous hydropower dams, including large dams on the mainstream of the Lower Mekong River, such as the Xayaburi and the Don Sahong Dams. Thailand diverted water from the mainstream of the Mekong River and filled up 30 water storage facilities in the middle of the historic drought between November 2015 and March 2016. Disappointed with the lethargy of the MRC about state-centric events, donor countries decided to cut their funding for the MRC from US$ 115 million to US$ 53 million from 2011 to 2015 (Kossov and Samean 2016; Zhang and Li 2018).

Third, the structure of the MRC is problematic. The member countries of the MRC are Thailand, Laos, Cambodia, and Vietnam and do not include China and Myanmar. The absence of the two upstream countries has resulted in the lack of political legitimacy and has prevented the riparian countries from reaching any legally binding agreement regarding the Mekong River. The dearth of political legitimacy also limits the roles of the MRC that just conduct scientific and technical works, not mediating conflicts or facilitating talks with law enforcement power (Zhang and Li 2018).

It is critical to examine the reasons why China has not been inclined to join the MRC as a formal member but keeping the dialogue partner status. Zhang and Li (2020) point out that China's Mekong River policy prior to the launch of the LMC was based on the two principles, 'rights protection' and 'stability maintenance.' Rights protection means exclusive rights of China for developing water resources in the Lancang River whereas stability maintenance means China's intention to sustain stable relations with the other riparian countries, particularly preventing any possible conflicts with them, which might hamper the construction of hydropower development upstream.

The notion of rights protection is closely linked to China's firm and consistent attitude toward the 1997 UN Water Convention. The country has refused to sign the convention, being concerned about Part III, Planned Measures that would refrain from its building dams on transboundary rivers. Similar restrictions on dam building are imposed for the member countries of the MRC, which is one of the major reasons why China has never intended to be a full member of the MRC (Ho 2014; Lee 2015; Zhang and Li 2020).

The two top priorities of China about the river are hydropower development as well as the development of inland navigation route. China has kept the three strategies in water diplomacy toward the downstream countries in the Lancang-Mekong River: (1) limited multilateral cooperation; (2) preference for bilateral agreement; and (3) unilateral approach in dam building. The first strategy has primarily been used for promoting inland navigation, which is vital for the take-off of Yunnan Province's economy, particularly regarding international trade with Thailand, Myanmar, and Laos. China signed the Lancang-Upper Mekong River Commercial Navigation Agreement in 2001 with the three downstream countries for promoting trade and tourism. Substantial investments have been made by China for the blasting work to remove rapids, shoals, and reefs along a 330-km stretch between the China-Myanmar border and Ban Houei Sai in Laos. Another agreement about inland navigation was made in 2006 linked to the Greater Mekong

Subregion (GMS) program between China, Thailand, Myanmar, and Laos for a trial program shipping oil along the river during the wet season (Ho 2014).

The second strategy, bilateral agreement, has been a preferential method to be engaged with China's riparian neighbors, which paves the way for international trade between Yunnan Province, China and Vietnam, Laos, Cambodia, and Thailand. As for Vietnam, China has been the largest trade partner since 2004 and is an important market for the export of consumer products and raw materials and the import of essential components for industries. The trade volume between China and Vietnam had increased by almost 30 times from US$ 3.68 billion in 2002 to US$ 106 billion in 2018 with an average growth rate of 23.95% per annum (Huong and Phuong 2019).

Foreign aid is an important issue for China's relationship with Laos and Cambodia. China's foreign aid has been offered for the two countries without major terms and conditions attached even though China's aid packages allow Chinese investors to have easy access to the countries' markets. Large investments through China's foreign aid are made in a wide range of sectors, from transport, communications, health, education, and human resources development to construction.

China's deep and close engagement with the two countries is easily found in many hydropower building sites in which various Chinese companies have been involved as investors or developers. The central government of China has supported the advancement of Chinese companies in hydropower dam building in Laos and Cambodia through the China Exam Bank and Sinosure. Although China's presence is not conspicuous in Vietnam compared with in Laos and Cambodia, hydroelectricity export of China toward Vietnam was planning to increase from 200 MW in 2014 to 2,000 MW in 2020 although the figure in October 2020 reached just 450 MW. Considering power surplus in South China, 150 GW in 2020, China appears to make efforts to sell hydroelectricity to Vietnam as well as the other downstream countries of the river (Ho 2014; Phu 2020).

Thailand exported various goods, such as plastic pallets, rubber products, and fresh produce, to China, which took 12% of the total exports in 2019, the largest export destination of Thai products, followed by the

US and Japan (Boonbandit 2020). In 2018, China made vast amounts of investment to Thailand estimated at Baht 55.48 billion (US$ 1.86 billion), which accounted for 9.5% of the total FDI in the country, competing with the US and Japan (Ministry of Commerce 2019).

The most conspicuous example of the third strategy is China's unilateral development of hydropower in the Lancang River. Hydropower development has been spearheaded by Yunnan Province whose vested interests are closely linked to local economic growth in addition to strategic interests from the central government, such as energy security, cleaner sources of energy, reforms in the energy sector, and the 'Go West' policy. The province firmly established itself as the mecca of hydropower in China based on its substantial contribution to hydropower development, 18% of the total hydropower generation in China, 57.8 GW out of 319.4 GW, in 2015. Yunnan's hydropower potential is as large as 101,939 MW with an annual generation of 491.9 TWh, ranked third in China (No.1 Sichuan, and No 2. Tibet) with the three major rivers, the Jinsha, Lancang-Mekong, and the Nu-Salween Rivers (Liu et al. 2018).

Yunnan Province initially planned to build eight cascade dams in the Lancang River in the 1980s and has expanded its plan for building a total of 22 hydropower dams. The province completed 11 dams as of December 2020, which are Wunonglong, Lidi, Huangdeng, Dahuaqiao, Miaowei, Gongguqiao, Xiaowan, Manwan, Dachaoshan, Nuozhadu, and Jinghong Dams from upstream to downstream (Eyler et al. 2020). In the course of constructing the dams, little information has been shared with the downstream countries regarding hydrological data, water-level monitoring and flood forecasting and mitigation. The downstream countries have expressed their grave concerns to China about diverse impacts possibly caused by the upstream dams, including river ecosystems and local livelihoods related to Tonle Sap and the Mekong Delta. China has been reluctant to provide significant information, i.e., data from hydrologic stations, because there is the ministerial regulation on the information regarding transboundary rivers, which bans public access to the information (MWR and NASSP 2000). Policymakers and elites of China have expressed their concerns about its cooperative manners toward its riparian countries due to this reason (Xie and Jia 2018).

Against complaints and criticism, the Chinese government has counter-argued about positive impacts of the upstream dams on the downstream communities. The Chinese government stresses that considering the complexity of river systems with the length of more than 4,800 km and hundreds of tributaries, a handful of Chinese dams upstream do not necessarily give major impacts on the Lower Mekong countries. In response to the concerns about trapped sediment due to the Chinese dams, certain amounts of sediment can be trapped because of the dams upstream, however, some scholars maintain that such an influence could be reached to the north of Vientiane, Laos, not further down to Cambodia and Vietnam (Lee 2015; Li et al. 2011; Lu and Siew 2006).

Even though the Lancang-Mekong River was hardly one of the most significant matters for top leaders in China, the central government intended to keep stable relations with the downstream countries in a variety of fields, such as hydropower development, water conservation, irrigation, and inland navigation for trade promotion. In addition, China signed a memorandum of understanding with the MRC on the provision of daily river flow and rainfall data from two monitoring stations in Yunnan Province during the wet season in 2002. Other activities, such as technical training, human resources exchange, and field visits, were undertaken with the MRC, however, issues on the Lancang-Mekong River were not regarded as significant ones for Chinese water policy-makers and foreign policy until recently (Nickum 2008; Zhang and Li 2020).

Emergence of the Lancang-Mekong Cooperation (LMC) Since 2015

The Lancang-Mekong Cooperation (LMC) mechanism of China was officially launched in November 2015 for leading transboundary water cooperation in the Lancang-Mekong River Basin. The three major foci of the LMC are political and security issues, economic prosperity, and sustainable development, and socio-cultural aspects and people-to-people exchanges. There are five main areas: (1) agriculture and poverty reduction; (2) water resources management; (3) production capacity; (4)

cross-border economic cooperation; and (5) connectivity. Putting the three foci and the five major areas, China dubs them as the 3 + 5 Cooperation Framework (LMC 2017; Zhang and Li 2020).

China's increasing pragmatic measures toward its riparian countries in the Lancang-Mekong River Basin are featured in three ways: (1) trade; (2) Foreign Direct Investment (FDI); and (3) Official Development Assistance (ODA). These approaches are geared toward removing negative diplomatic and security disputes in Southeast Asia, but whether these have been effective are questionable.

In March 2016, China hosted the first LMC summit and suggested a list of cooperative water resources projects that can be achieved in a short term and the creation of joint working group which would be in charge of planning and monitoring the implementation of the projects. All the member countries agreed to establish the Lancang-Mekong water resources and environment cooperation centers and decided to promote technical cooperation, human and information exchanges, and green, cooperative, and sustainable development (Zhang and Lu 2016).

The topic of water resources development between China and the other riparian countries, particularly hydropower development, has long been regarded as a political taboo that should not be openly discussed in inter-governmental meetings, because the issue can provoke unproductive accusations between the riparian countries, and China may become a main target for being criticized because of its dam building upstream. The LMC is prepared to water down complaints and criticism about the upstream dams and create the new image of China pertinent to transboundary water cooperation in the Lancang-Mekong River through hardware as well as software. As for hardware, China has been strongly committed to providing technical and financial support for water infrastructure projects of the downstream countries, particularly focusing on Laos and Cambodia. Laos has been closely collaborating with China in terms of the creation of regional power hub for the markets of mainland Southeast Asia as well as Yunnan Province. Chinese hydropower and construction companies have scrambled to Laos for implementing a considerable number of hydropower and infrastructure projects in the 2010s. Another instance is Cambodia's active engagement with China in the field of hydropower dam construction. The amount of investment was as large as US$ 2.4 billion from the Chinese side, and the capacity of power generation of the dams will be able to reach 2 trillion watts by 2050. In addition, China is envisaging to develop a regional power grid that connect China with all the other riparian countries (Zhang 2019).

The software areas of the LMC are water-related institution building, technology transfer, the exchange of rules and regulations, and the promotion of water-related data and narrative. Several water-related institutions have been established, including the Joint Working Group on water resources, the Lancang-Mekong Water Resources Cooperation Center, the Environmental Cooperation Center, and the Global Center for Mekong Studies. The first Five Year Action Plan on Lancang-Mekong Water Resources Cooperation is under preparation by the Joint Working Group, and almost 100 water officials and university students from the Mekong countries have participated in training programs offered by the LMC (Zhang and Li 2020).

As the first mission, the LMC played a key role in discharging water downstream for the Lower Mekong countries which suffered from the unprecedented level of drought in late 2015. Thanks to the El Nino effect, the average amount of rainfall in the river basin was recorded historically low so that China and the downstream countries had to be confronted with the serious drought. For instance, the Mekong Delta of Vietnam was badly hit by the drought, which was regarded as the worst in 90 years and affected the arable land of 139,000 km^2 and entailed the lack of access to drinking water for more than 575,000 people. Despite the dire situation, Thailand unilaterally diverted water resources from tributaries for easing the water shortage in irrigation and fisheries without consent with the other Lower Mekong countries. Another factor to make the situation worse would be attributed to a number of large dams built by Laos (Zhang and Lu 2016).

Under these circumstances, the Jinghong Dam of China discharged the emergency water for the downstream countries with an amount of 2,000 m^3/sec every day between 15 March and 10 April 2016. This immediate action was welcomed by the downstream countries, especially from Myanmar, Vietnam, and Cambodia which required substantial amounts of water for rice production as well as fisheries. Laos also decided to discharge the emergency water release for the downstream countries with an amount of 1,136 m³/sec every day, which resulted in alleviating the adverse impacts of the drought in the Mekong Delta, including the seawater intrusion into irrigation fields (Viet Nam News 2016).

It is interesting to notice that the favorable action of China toward the downstream countries has not necessarily received positive responses. The motivation for China's cooperative stance toward the Lower Mekong countries is closely related to the regional strategy under the Belt and Road Initiative (BRI). Nguyen (2016) argues that the impacts of China's water discharge from the Jinghong Dam would not be able to resolve drought conditions of the Mekong Delta since sediments along the river absorb most of the water flowed from the dam. Large amounts of water released from the dam would flow into Tonle Sap Lake in Cambodia before flowing down to the delta. Referring to a scholar from Vietnam, there was no conspicuous change of water level in the Mekong Delta after eight days of discharge from the dam. It would not be feasible to envisage that water from the Chinese dam would reach the Mekong Delta for 4,000 km southward without interruption, because China also suffered from the severe drought.

Zhou (2016) interprets China's water release against the drought in 2015–2016 as the country's strategy to accelerate its trade with the other riparian countries through inland navigation and to improve its diplomatic relationship with Vietnam. On 14 March 2015, anti-Chinese street demonstrations occurred in the city center of Hanoi commemorating the Sino-Vietnamese Sea Battle in South China Sea in 1988. Although the entrenched anti-Chinese sentiment in Vietnam will not fade away soon, China may regard this water release as an opportunity to improve the bilateral relationship (Wu 2015).

From a strategic point of view, China appears to utilize the LMC as an effective game-changer that can water down the troubled image due to its upstream dams and enhance its relationships with the downstream countries for security, economic, and environmental aspects. There are several reasons why China needs its own cooperative framework instead of joining the MRC. First, the country intends to play a leading role in many fields in the Lancang-Mekong River Basin, such as building large dams, agricultural development, aquaculture, and inland navigation.

Pursuing socioeconomic cooperation with the other riparian countries through the LMC, China would like to eliminate the image of troublemaker, especially related to the upstream dams that may cause negative impacts on local livelihoods and ecosystems downstream.

Second, cooperation through the LMC will provide a sound platform for the active implementation of the BRI in the region. The Mekong River Basin is strategically located in one of the major southern routes from China that connects pathways to South China Sea and Maritime Southeast Asia. Vietnam is sensitive to China's policies in South China Sea so that the stable and positive relationship with Vietnam through the LMC would pave the way for China to consolidate its influence over South China Sea.

Third, with regard to China's global strategy, the successful cooperation through the LMC will confirm China's capability to reinvigorate socioeconomic development in the Lancang-Mekong River Basin through large investments in infrastructure, including hydropower dams, agricultural development, aquaculture, and international trade through inland navigation. Such a commitment of China to the region through the LMC will be compared with the limited roles and incapability of investments of the MRC, which demonstrates the viability of the LMC as a replaceable cooperative mechanism in the river basin.

Fourth, the LMC will play an imperative role in securing China's right to promote potential of hydropower development. The Lower Mekong countries have so far been accusing China of its unilateral development and building of hydropower dams upstream. The new cooperative framework under the LMC will help the other riparian countries develop their own hydropower projects with the financial and technical aid of China. In the course of joint projects, the other riparian countries will have a better understanding of China's hydropower development plans and strategies so that unnecessary misunderstandings or accusation can fade away (Zhang and Lu 2016).

In the similar context, Zhang and Li (2020) stress a paradigm shift of China's Lancang-Mekong River policy by adopting water diplomacy through the utilization of the LMC instead of passive multilateral cooperation. In order to achieve the two strategic goals, rights protection as well as stability maintenance, China introduces transboundary river cooperation to attract the other riparian countries on board together with the BRI, which promotes building infrastructures, transfer of industrial capacity, and international trade and serves as a vehicle to consolidate its dominance in Southeast Asia.

Having discussed the evolving patterns of cooperation in the Lancang-Mekong River Basin with special reference to the roles and activities of the MRC and the LMC, it is worth paying particular attention to the relationship between the two institutions and the outlook of cooperation in the MRC-LMC era since 2015. The LMC appears to emerge as an alternative for facilitating cooperation between the riparian countries focusing on various fields, not only water and environmental issues but also socioeconomic development. Although the GMS program since 1992 has enabled all the riparian countries to discuss a variety of development issues, especially hydropower and transport, the program does not necessarily serve as a cooperative framework that can encourage the member countries to make treaties or agreements for sustainable development beyond socioeconomic achievements.

China-led LMC can be transformed as an international river basin organization because the mechanism has already established its several special centers focusing on water resources, the environment, and training. However, Biba (2018) and Williams (2020) argue that the LMC is project-based and focuses on economic cooperation considering policy statements and speeches from press conferences as well as various projects undertaken. It may be too early to make a judgment on this matter but even if the LMC were becoming an international river organization, the institution would not be able to be out of China's far-reaching influence.

The river basin is replete with a myriad of cooperative mechanisms in addition to the MRC and the LMC. The other mechanisms such as ASEAN, the Greater Mekong Sub-region (GMS) initiative, the Lower Mekong Initiative (LMI), the Ayeyarwady-Chao Phraya-Mekong Economic Cooperation Strategy (ACMECS) and the Mekong Initiatives of Japan and the Republic of Korea have turned out to be ineffective in or contribute little to tackling political, socioeconomic, or environmental issues so far. Such a crowded scene on transboundary water cooperation in the river basin can be disrupted or shaken by the emergence of the

LMC thanks to its political, economic, and security teeth which could engender the fundamental change of political economy landscape over the river basin (Williams 2020).

The Chinese government would like to put the LMC as a game-changer for regional cooperation and has expedited its funding of the LMC for developing planning and management policies in the river basin. There are three different measures of China's aid to the other riparian countries: (1) direct financial flows; (2) infrastructure investment; and (3) technology capacity building. Direct financial support has been made for poverty relief. A wide range of aid programs were provided, including the infrastructure interconnectivity program with the investment of US$ 10 billion, poverty alleviation with US$ 490 billion, and loans for the improvement of China's export production capacity with US$ 1.6 billion. In addition, specific financial assistance was given to each riparian country, and Laos received a grant from China worth US$ 4 million for poverty reduction, particularly focusing on suburban areas of the capital, Vientiane, and rural communities of Luang Prabang Province (Chandra 2016; Xie and Jia 2018).

The Mekong River Commission (MRC) has often expressed its willingness to collaborate with the LMC, particularly with regard to water resources management. Responding to the formation of LMC, the MRC explicitly delivered a welcoming message, calling the LMC as the 'important new initiative for regional cooperation' (MRC 2016). After several occasions of visits and encounters in various events, in December 2019, the MRC signed the Memorandum of Agreement with the LMC Water Center. Prior to the event, the MRC made the data and information exchange agreements with China in July 2019 with a renewal of past agreements in 2002, 2008, and 2013 (MRC 2020a).

It is imperative for the LMC to keep a positive relationship with the MRC for now. It may take time for the LMC to function as a cooperative framework with substantial capacity to tackle various issues and problems in the river basin because the LMC requires more scientific and technical capacities to evaluate the complexity of water and ecosystems in the Lower Mekong River Basin. Such an expertise can be pulled out from the MRC which embraces more than 20-year experiences and expertise. Simultaneously, the MRC's wisdom of mediating differences of opinions and conflicting ideas between the Lower Mekong countries is an invaluable asset which China would like to learn.

The rationale behind the creation of the LMC is associated with China's own strategic reasons rather than benefit sharing or anything reflected through regional perspectives. Aggressive investments and joint projects from China to the other riparian countries through the LMC can divert attention of the other countries from possible negative impacts of Chinese upstream dams to development projects in the Lower Mekong River Basin. The title of the mechanism accommodates China's subtle intention to include the names, 'Lancang' indicating the upstream part, and 'Mekong,' the downstream part. In fact, the name, Lancang, was rarely used by non-Chinese experts out of China before 2015 but the new naming repositions and strengthens China's presence and influence over the river basin. The other implication embedded in the naming is that China seems to differentiate between the LMC and the other regional initiatives that have heavily been influenced by donors and foreign governments (Williams 2020).

Future works and projects initiated by the LMC are closely related to the BRI. The Lancang-Mekong River Basin is the salient route which allows the country to have access to the Maritime Silk Road. It is vital for China to consolidate the relationships with its Mekong neighbors in socioeconomic, political, and security senses (MFA 2017; Williams 2020).

Conclusion

The chapter has analyzed an array of critical issues pertinent to transboundary rivers in China. Previously, top leaders in the country were rarely interested in transboundary river issues because of its less significance for the country's socioeconomic development as well as security concerns. However, such a trend has been changed in a recent couple of decades as water becomes a security concern, not only as a resource for optimal use and protection but also a variety of socioeconomic and security interests entangled with its transboundary rivers.

China's little engagement or ignorance with its riparian neighbors has incrementally been changed so that the country's usual diplomatic preference, bilateral agreement, is no more primary option to choose but multilateral cooperation is increasingly becoming a favorable option pertinent to transboundary river management as seen from the Lancang-Mekong River. Although the approach of water diplomacy appears to be adopted by the Chinese authorities for being involved with its riparian countries, whether a set of important norms in international freshwater laws will be recognized and respected by China is unclear. China's refusal to sign the 1997 UN Water Convention is a good reference pertinent to this.

Nevertheless, the cases of the establishment of cooperative relationships with its riparian countries have been found, e.g., the Yalu-Amnok River, the Tuman-Dooman River, the Illi River, and the Irtysh River. The in-depth case study of the Lancang-Mekong River explicitly assures China's proactive engagement with the downstream countries for socioeconomic and security benefits. The Mekong River Commission has played a key role in facilitating dialogue and making substantial scientific and engineering contributions to the region. However, it is disappointing to admit that the MRC has not been effective in mediating or preventing conflicts on thorny issues such as hydraulic infrastructures on the mainstream of the Lower Mekong River, mainly due to its paucity of political legitimacy.

On the contrary, China has gradually changed its engagement strategy with the downstream countries of the river, dealing with each riparian country through bilateral agreements, such as its relationship with Laos and Cambodia and has pushed forward its more aggressive regional development strategies through the creation of the Lancang-Mekong Cooperation (LMC) mechanism. The global development strategy of China, the Belt and Road Initiative (BRI), is neatly connected to the Lancang-Mekong River Basin through the extensive influence of China through the LMC in due course. It will be interesting to observe the extent to which the LMC can accommodate a complexity of vested interests from the downstream countries and shows a path to the achievement of socioeconomic prosperity and environmental sustainability through the leadership of China.

References

Biba, Sebastian. 2018. China's 'old' and 'new' Mekong River Politics: The Lancang-Mekong Cooperation from a Comparative Benefit-Sharing Perspective. *Water International* 43 (5): 622–641.

Boonbandit, Tappanai. 2020. Sino-Thai Trade Continues to Grow Despite Pandemic. *Kahosod English*, August 24, 2020.

Chandra, P. 2016. China Funds Poverty Reduction in Laos. *Vientiane Times*, 8 December 2016. Available Online: http://lotiantimes.com/2016/12/08/china-funds-poverty-reduction-laos (accessed 18 February 2021).

Chen, Huiping. 2012. The 1997 UN Water Convention and China's Treaty Practice on Transboundary Waters. UN Watercourses Convention Global Initiative.

Chen, Huiping. 2017. The Status and Promotion of Transboundary Water Cooperation between China and its Neighbors. Presented at the Regional Workshop on Transboundary Water Cooperation in the context of the SDGs in South Asia and beyond, Pokhara, Nepal, 23–24 May.

Chen, Z. and J. Jian. 2009. Chinese Provinces as Foreign Policy Actors in Africa. South African Institute of International Affairs. SAIIA Occasional Paper No. 22. Available Online: https://media.africaportal.org/documents/SAIIA_Occasional_Paper_no_22.pdf (accessed 26 October 2020).

Cuppari, Rosa. 2017. Water Diplomacy. Policy Brief No. 1. Paris, UNESCO.

Eder, Thomas. 2018. Mapping the Belt and Road Initiative: This Is Where We Stand. The Mercator Institute for Chinese Studies. 7 June. Available Online:https://www.merics.org/en/bri-tracker/mapping-the-belt-and-road-initiative. Accessed 16 January 2019.

Eyler, Brian, Alan Basist, Claude Williams, Allison Carr, and Courtney Weatherby. 2020. Mekong Dam Monitor Lifts the Veil on Basin-Wide Dam Operation. Stimson Center, December 14. Available Online https://www.stimson.org/2020/mekong-dam-monitor-lifts-the-veil-on-basin-wide-dam-operations/. Accessed 16 February 2021.

Food and Agriculture Organisation (FAO). 2011. *Mekong Basin*. Rome: Aquastat.

Fry, James, and Agnes Chong. 2016. International Water Law and China's Management of Its International Rivers. *Boston College International and Comparative Law Review* 39 (2): 227–266.

Hague Institute for Global Justice. 2016. The Multi-Track Water Diplomacy Framework. The Hague Institute for Global Justice, The Hague.

Ho, Selina. 2014. River Politics: China's Policies in the Mekong and the Brahmaputra in Comparative Perspective. *Journal of Contemporary China* 23 (85): 1–20.

Ho, Selina. 2017. China's Transboundary River Policies Towards Kazakhstan: Issue-Linkages and Incentives for Cooperation. *Water International* 42 (2): 142–162.

Hunt, Luke. 2016. Mekong River Commission faces radial change. *The Diplomat*, January 22. Available Online http://thediplomat.com. Accessed 20 October 2020.

Huong Thanh, Vu, and Nguyen Thi Lan Phuong. 2019. Changes in Vietnam-China Trade in the Context of China's Economic Slowdown: Some Analysis and Implications. *VNU Journal of Science: Economic and Business* 35 (2): 11–22.

International Law Association (ILA). 1966. The Helsinki Rules on the Use of the Waters of International Rivers of 1966. The Fifty-second Conference of the ILA, Helsinki, August 1966. Available Online https://www.internationalwaterlaw.org/. Accessed 29 October 2020.

International Law Association (ILA). 2004. The Berlin Rules on Water Resources. Berlin Conference 2004 on Water Resources Law. Available Online: https://www.unece.org/. Accessed 29 October 2020.

Kettelus, Mirja, Matti Kummu, Marko Keskinen, Aura Slmivaara, and Olli Varis. 2015. China's Southbound Transboundary River Basins: A Case of Asymmetry. *Water International* 40 (1): 113–138.

Kossov, Igor and Lay Samean. 2016. Donors Slash Funding for MRC. *The Phnom Penh Post*, January 14. Available Online http://www.phnompenhpost.com. Accessed 20 October 2020.

Kukeyeva, Fatima, Tolganay Ormysheva, Kuralay Baizakova, and Malik Augan. 2018. Is Ili/Irtysh Rivers: A 'Casualty' of Kazakhstan-China Relations? *Academy of Strategic Management Journal* 17 (3): 1–13.

Lancang Mekong Cooperation (LMC). 2017. The 3+5 Cooperation Framework. LMC Website, December 14. Available Online http://www.lmcchina.org/eng/. Accessed 19 October, 2020.

Lee, Seungho. 2015. Benefit Sharing in the Mekong River Basin. *Water International* 40 (1): 139–152.

Lee, Seungho. 2018. Hydropower Dams in Laos: A Solution to Economic Development? The Case of the Nam Theun 2 Dam. *Southeast Asia Research* 28 (2): 145–178.

Lee, Seungho, Ilpyo Hong, and Jongpil Chung. 2019. Transboundary Water Cooperation in the Tumen River Basin: A Focus on the Greater Tumen Initiative. *Journal of Asia Pacific Studies* 26 (4): 21–56.

Lee, Seungho. 2020. Chapter 4: Transboundary River Cooperation Between North Korea and China: The Yalu and Tumen Rivers. In *Chinese People's Diplomacy and Developmental Relations with East Asia: Trends in the Xi Jinping Era*, ed. Lai To Lee. London and New York: Routledge.

Li, Zhiguo, Daming He, and Yan Feng. 2011. Regional Hydropolitics of the Transboundary Impacts of the Lancang Cascade Dams. *Water International* 36 (3): 328–339.

Lu, X.X., and R.Y. Siew. 2006. Water Discharge and Sediment Flux Changes over the Past Decades in the Lower Mekong River: Possible Impacts of the Chinese Dams. *Hydrology and Earth System Science* 10: 181–195.

Liu, Benxi, Shengli Liao, Chuntian Cheng, Fu. Chen, and Weidong Li. 2018. Hydropower Curtailment in Yunnan Province, Southwestern China: Constraint Analysis and Suggestions. *Renewable Energy* 121: 700–711.

Lukin, Alexander. 2013. Territorial Issues in Asia: Drivers, Instruments, Way Forward. Presented at 7th Berlin Conference on Asian Security (BCAS), Berlin, July 1–2. Available Online https://www.swp-berlin.org/fileadmin/contents/products/projekt_papiere/BCAS2013_Alexander_Lukin.pdf. Accessed 28 October 2020.

Matthew, Nathanial, and Stew Motta. 2015. Chinese State-Owned Enterprise Investment in Mekong Hydropower: Political and Economic Drivers and Their Implications across the Water, Energy, Food Nexus. *Water* 7: 6269–6284.

Mekong River Commission (MRC). 1995. *Agreement on the Cooperation for the Sustainable Development of the Mekong River Basin*. Vientiane: MRC.

Mekong River Commission (MRC). 2016. *Lancang-Mekong Cooperation: MRC welcomes the new initiative for regional cooperation by six countries in the Mekong River Basin*. Vientiane: MRC.

Mekong River Commission (MRC). 2017. *1995 Mekong Agreement and Procedures*. Vientiane: MRC.

Mekong River Commission (MRC). 2020a. *Annual Report 2019. Part 1: Progress and Achievements*. Vientiane: MRC.

Mekong River Commission (MRC). 2020b. *Basin Development Strategy for the Mekong River Basin (2021–2030)*. Complete Second Draft of Part I and First Draft of Part II. 4 March.

Ministry of Commerce. 2019. Statistics on China-Thailand Trade in January-December 2018. Ministry of Commerce, People's Republic of China Statistics. March 19. Available Online http://english.mofcom.gov.cn. Accessed 26 November 2020.

Ministry of Foreign Affairs (MFA). 2017. Foreign ministry spokesperson Geng Shuang's regular press conference on March 10, 2017. Beijing, MFA.

Ministry of Water Resources. 2019. *China Water Resources Bulletin (中国水资源公报) 2018*. Beijing: China Water Power Press.

Ministry of Water Resources (NWR) and National Administration of State Secrets Protection (NASSP). 2000. Regulations on State Confidentiality in Water Resources Management and its Specific Scope and Security Classification (水利工作中国家秘密及其密级具体范围的规定), 29 December. Available Online http://www.nwskglj.cn/. Accessed 16 February 2021.

Molnar, K., R. Cuppari, S. Schmeier, and S. Demuth. 2017. Preventing Conflicts, Fostering Cooperation: The Many Roles of Water Diplomacy. UNESCO International Center for Water Cooperation at SIWI.

Moore, Scott. 2018. China's Domestic Hydropolitics: An Assessment and Implications for International Transboundary Dynamics. *International Journal of Water Resources Development* 34 (5): 732–746.

Nguyen, Nam. 2016. China's Water Release Unlikely to Slake Vietnam's Thirst. *Radio Free Asia*, March 23.

Nickum, James. 2008. The Upstream Superpower: China's International Rivers. In *Management of Transboundary Rivers and Lakes*, ed. O. Varis, A. Biswas, and C. Tortajada, 227–244. Berlin: Springer.

Phu, Nhu. 2020. Vietnam to Increase Electricity Imports from Laos, Cambodia, China. *Saigong Times*, November 26. Available Online http://sgtimes.net. Accessed 26 November 2020.

Shin, Nayeon, Seungho Lee, and Ilpyo Hong. 2020. Chapter 9: Regional cooperation through the Greater Mekong Subregion programme: focus on hydropower development and the Mekong Power Grid. In *Opportunities and Challenges for the Greater Mekong Subregion*, ed. Charles Johnston, and Xin Chen. Abingdon: Routledge.

Tang, Qicheng, and Daming He. 2000. Chapter 1: International Rivers of China and Their Characters of Distribution (中国国际河流及其分布特征). In *International Rivers of China (中国国际河流)*, ed. Daming He and Qicheng Tang. Beijing: Science Press.

UN. 2008. Draft Articles on the Law of Transboundary Aquifers. Available Online https://legal.un.org/. Accessed 29 October 2020.

UN. 2014. Convention on the Law of Non-Navigational Uses of International Water Courses 1997. https://legal.un.org/ilc/texts/instruments/english/conventions/8_3_1997.pdf.

UNECE. 1992. The Convention on the Protection and Use of Transboundary Watercourses and International Lakes. Available Online https://unece.org. Accessed 29 October 2020.

Viet Nam News. 2016. Laos Helps Viet Nam Deal with Drought, Salt Intrusion. *Viet Nam News*, March 26. Available Online https://vietnamnews.vn/. Accessed 19 October 2020.

Williams, Jessica. 2020. Is Three a Crowd? River Basin Institutions and the Governance of the Mekong River. *International Journal of Water Resources Development*. https://doi.org/10.1080/07900627.2019.1700779.

Wu, Shou. 2015. Anti-Chinese Demonstration Escalated While China Decided to Release Water from Its Dam for Easing the Historic Drought Downstream in 100 Years (越南反华升级中国开闸放水缓解越百年旱灾). *Duo Wei News*, 15 March. Available Online http://dwnews.com. Accessed 20 October 2020.

Xie, Lei, and Shaofeng Jia. 2018. *China's International Transboundary Rivers*. Abingdon: Routledge.

Zhang, Hongzhou. 2019. Water Conflicts in Mekong. Presented at the 2019 International Specialty Conference, Jointly Organized by the American Water Resources Association and Center for Water Resources Research, Chinese Academy of Sciences. Beijing, September 16.

Zhang, Hongzhou, and Mingjiang Li. 2018. Chapter 1: Thirsty China and its transboundary waters. In *China and Transboundary Water Politics in Asia*, ed. Hongzhou Zhang and Mingjiang Li. Abingdon and New York: Routledge.

Zhang, Hongzhou, and Mingjiang Li. 2020. China's Water Diplomacy in the Mekong: A Paradigm Shift and the Role of Yunnan Provincial Government. *Water International* 45 (4): 347–364.

Zhang, Li., and Guangsheng Lu. 2016. Transboundary Water Resources Cooperation Under the Lacang-Mekong Cooperation Mechanism Through the View of Emergency Water Allocation from Chinese Dams (从应急补水看澜湄合作机制下的跨境水资源合作). *International Prospect (国际展望)* 5: 95–112.

Zhou, Margaret. 2016. China and the Mekong Delta: Water Savior or Water Tyrant? *The Diplomat*, March 23. Available Online https://thediplomat.com/2016/03/china-and-the-mekong-delta-water-savior-or-water-tyrant/. Accessed 27 April 2021.

10
Conclusion

The book has evaluated China's water resources management and policy, especially over the last two decades. An insight on the interconnectedness between water and sustainable development has been the first discussion point for unpacking the far-reaching contribution of water to China's remarkable socioeconomic development and environmental sustainability. China's own targets to achieve SDGs, particularly SDG6 Water and Sanitation, have confirmed its willingness to achieve the 2030 development agenda not only following global norms as well as creating its own path that can be emulated to other countries. The promotion of ecological civilization for achieving 'Beautiful China' is one of the newest editions to ignite the boom of ecosystem-centered policy direction that has been advocated by President Xi Jinping in recent few years.

China has made good progress in terms of various dimensions of water security, i.e., water supply, water quality, and resilience to water-related disasters. Nevertheless, such water challenging issues appear to look like a riddle that is not puzzled out with the recognition of the challenges in water resources management. Climate change fundamentally shakes predictions and preparedness of multi-year efforts against flood and drought as well as plans and projects on water resources development, water supply, water quality, and transboundary rivers. Under the era of climate uncertainties, the country has endeavored to introduce a series of brand-new and innovative plans, programs, and projects for effectively tackling formidable challenges in water resources management.

For resolving the water challenges, Chinese hydraulic engineers have conventionally resorted to physical infrastructures, i.e., dams, inter-basin transfer projects, wastewater treatment plants, long-range sewers, embankments, and agricultural reservoirs. These measures, however, have turned out to be too rigid to cope with climate change-triggered challenges in water resources management. Demand management policies and programs have increasingly been advocated and emphasized not only at the center but also at the local level as seen from the emergence of the River (Lake) Chief System that was initiated in Wuxi City, Jiangsu Province in the Tai River Basin. In addition, the rise of water and wastewater service tariffs in major cities of China has demonstrated a gradual acceptance of the principle of full cost recovery although more sociopolitical and economic concerns are prioritized in the decision of water and wastewater service tariffs. Similar with the carbon emission trading system, the water rights trading and the water pollution trading have been promoted as effective tools of demand management. The virtual water and water footprint approach can serve as an innovative policy tool to resolve food security as well as the imbalance of water resources allocation between North and South China coupled with the recognition of China's contribution to virtual water export and import at the global level.

It is true that top leaders and water engineers' preference to hydraulic engineering projects in China are still significant as the exhaustive assessment of the Three Gorges Dam and the South North Water Transfer Project. The dam has been the symbol of China's modernization that traces back to the early 1910s when the project was first envisaged by Sun Yat-sen and plays an impressive role in preventing floods in the middle of the Yangtze River as seen from the recent 2020 summer flood. More emphasis has been placed on the role of the dam regarding hydropower generation, which serves as the primary power base to electrify economic growth in major industrial bases alongside the eastern and southern coastal areas.

10 Conclusion

The South North Water Transfer Project is supposed to quench the thirst in North China by delivering vast amounts of water from the Yangtze River to the northern parts of China, especially around Beijing and Tianjin. Both projects explicitly demonstrate the interconnectedness between large-scale hydraulic projects and politics because the CCP has reconfirmed its political legitimacy for providing multifaceted dimensions of water security through the projects, water supply, hydropower generation, and flood prevention and control although different voices appear to have been unheard due to a censorship of dissent views and opinions.

Water and wastewater services in urban areas of China have been rapidly evolving thanks to the willingness of the public sector to enhance the quality of relevant services by inviting private players since the late 1990s. Over the two decades, interactions between the public and the private sectors have led to the creation of the Chinese water and wastewater service market with its own features. Chinese players have become more dominant with their advantages in political connections, familiarity with laws and regulations, and financing capacity than foreign counterparts that were previously favored by the central government. Chinese companies are actively collaborating with local governments which would like to improve their water and wastewater services in urban areas.

China's ignorance of transboundary water issues has gradually faded away, and as the country extends its influence beyond its own territory through the implementation of the BRI, it is natural to encounter and bargain with its riparian countries on natural resource exploitation, especially water resources. Transboundary rivers are physically away from the center of power, Beijing, and relevant matters were out of top leaders' concerns for many years. However, water issues are regarded as security matters now, and transboundary river issues are increasingly regarded as critical ones to be discussed in Beijing. The introduction of the Lancang-Mekong Cooperation (LMC) mechanism in 2015 encapsulates the country's recent commitment to participation in multilateral dialogues in transboundary river issues. It will be interesting to observe the extent to which China shapes the political economy landscape in the

Lancang-Mekong River Basin with its own rules, which will be somewhat different from the stalemate in cooperation as seen from the era of the Mekong River Commission between 1995 and 2015.

Although China seems to become more serious about ecological issues, including water resources, it is unlikely that the country would dramatically change its development route in a short term, putting more weight on ecological conservation rather than economic development. Numerous plans and policies on water resources management of the country have demonstrated more infrastructure-focused and top-down projects in pipeline with the titles of 'green or ecological.' The sustainability of those innumerable projects is still dubious, primarily related to the projects in less developed areas, and therefore, as long as the country would prioritize economic development, an array of water challenges would not be adequately tackled. More worrying is related to the spread of such water crises beyond China's borders, not only associated with transboundary water conflicts but also global water shortage due to China's greedy demands of virtual water (Wu and Edmonds 2017).

China's water resources management and policy seems to follow the trajectory of what has been done in environmental management as expressed by Mr Zhenhua Xie, who was the ex-minister of the State Environment Protection Administration in 1998, the precursor of the Ministry of Ecology and Environment. Strong emphasis has been placed on regulation and administrative management for environmental issues. But more work and commitment will be needed for the adoption of economic instruments that can facilitate the application of market mechanisms as well as the encouragement of public and stakeholder participation for the establishment of collaborative governance. Such a collaborative governance mechanism will augur neatly based on mutual support and supervision between the state, local governments, companies, expert groups, and the general public (Xie 2020).

By analogy, the future water resources management and policy direction in China should put an emphasis on the application of economic instruments or soft approaches as well as governance-based management systems at the national and local levels. An array of market-based regulatory programs, i.e., water rights trading and water pollution trading, will serve as the major platforms on which environmental agencies and regulators should rely, and public and stakeholder participation will draw attention of environmental agencies for the establishment of collaborative governance in water resources management and policy in the coming decades. The top-down and the centrally planning economy style plans and programs continue to be imperative as before for managing water resources in the country, not to mention.

The key to achieving sustainable water resources management and policy for Beautiful China by 2035 and the middle of this century will depend upon the extent which the state will be willing to leave room for other stakeholders' involvement in decision making and implementation. Water and environment agencies should pay less attention to command-and-control management of water issues but emphasize the role of market instruments focusing on demand management rather than supply management that are entrenched with hydraulic infrastructures and engineering solutions. China's long march to sustainability in water resources management requires its top-down approaches and open-mindedness to the general public and its neighbors.

References

Wu, Fengshi, and Richard Edmonds. 2017. Chapter 7: Environmental Degradation. In *Critical Issues in Contemporary China*, 2nd ed., ed. Czeslaw Tubilewicz, 105–119. London: Routledge.

Xie, Zhenhua. 2020. China's Historical Evolution of Environmental Protection Along with the Forty Years' Reform and Opening-Up. *Environmental Science and Ecotechnology* 1 (100001): 1–8.

Index

A

Aksu River 341
Ammonia Nitrogen 37, 76, 110, 212, 214, 249
Amur-Heilong River 9, 337, 339–341, 350
Anglian Water 314
Anhui Guozhen 329
Aqueducts 4, 41, 177, 231
Aquifer salinization 64, 99
'The Asia's Battery' 355
Asia Environment, Singapore 318
Asian Infrastructure Development Bank (AIIB) 98
Asia Water Technology, Singapore 318

B

Baiji dolphin (white dolphin) 251

'Beautiful China' 4, 16, 18, 37, 44, 99, 320, 377, 381
Beijing 2, 6–8, 20, 25, 29, 30, 34, 43, 63, 97, 111, 148, 154, 156–158, 162–166, 172, 178, 181, 182, 184, 196, 203, 230, 247, 256, 258, 267, 269, 270, 273, 278, 280–282, 298, 309, 338, 379
Beijing Capital Group 297, 324, 329, 330
Beijing Enterprises Water Group 297, 324
Beijing No.10 Water Project 314
Beijing Origin Water Technology 324
Beijing-Tianjin-Hebei Region 31–33, 44, 114, 157
Beijing Urban Construction 330

Belt and Road Initiative (BRI) 9, 22, 31–33, 44, 98, 175, 241, 344, 365–367, 369, 370, 379
Biodiversity 3, 19, 196, 219, 248, 250, 353, 356
Blue green algae 8, 10, 143, 193, 201, 212, 217, 218, 221, 223, 225
Brownfield projects 315, 316
Build-Own-Operate (BOO) 316
Build-Renovate-Operate-Transfer (BROT) 316, 317
Build-Transfer 317

C

Cambodia 9, 353, 355, 356, 358, 360, 362–365, 370
'Cancer Villages' 83, 99
Catastrophe insurance 41, 42, 53, 86
Cathay International Holdings, Hong Kong 318
Chao Lake 52, 80
Chemical Oxygen Demand (COD) 37, 76, 78, 79, 82, 110, 142, 212–214, 221, 249
Chengdu No. 6 Build-Operate-Transfer (BOT) plant 296
China, 15 major international river basins 9
China Datang Corporation 236
China Everbright 330
China Huadian Corporation 236
China Huaneng Corporation 236
China Hydropower Engineering Consulting Group (Hydrochina) 237
China Lesso Group 324

China State Grid Corporation 236
China Three Gorges Corporation 236, 237
China Water Affairs Group, Hong Kong 318
China Water Exchange 172, 175, 184, 215
China Wuzhou Engineering 330
Chinese Community Party (CCP) 7, 16, 20, 28, 32, 39, 116, 129, 130, 133, 149, 168, 175, 202, 203, 205, 206, 208, 224, 230, 239, 257, 283, 293, 302, 325, 326, 379
Chinese paddlefish 251
Chinese sturgeon 251
Chongqing 20, 35, 41, 86, 243, 245–247, 250, 298
Circular on Accelerating the Reform of Water Price, Promoting Water Saving and Protecting Water Resource, 2004 304
Climate change 3, 6, 16–18, 20, 22, 41, 53, 64, 67, 85, 88, 91, 93–98, 100, 122, 130, 131, 229, 230, 251, 278, 281, 282, 353, 355, 377, 378
Climate change impacts 51, 53, 93
Climate finance 98
Compensated Use of Tradable Emission Permit Program (CUTEPP) 215–217
Cooperative Joint Venture (CJV) 314, 317
COVID-19 28, 333
Cultural Revolution (1966–1976) 256

Index

D

Da Chang BOT water project, Shanghai 314
Dadu River 273, 275
Danjiangkou Reservoir 247, 267, 270, 272, 283
Decoupling 29, 30
Deeper aquifers 63
Deforestation 95, 351
Demand management 6, 7, 10, 114, 131, 158, 159, 168, 281, 378, 381
Desalination 5, 107, 112, 114, 266, 295, 312, 313
Desertification 43, 61, 195, 349
Dianchi Lake 52, 80
Dongting Lake 52, 60, 66, 67, 245, 250
Don Sahong Dam 355, 358
Drought-prone areas 53, 90, 92
Drought risk management 53, 92
Droughts 3, 5, 16, 20, 40, 41, 51, 53, 84, 86, 90–93, 95, 99, 100, 106, 130, 131, 141, 156, 251, 282, 357, 377
Dujiangyan 2

E

Eastern route 253, 256, 257, 260, 262–267, 269, 271, 272, 276, 280, 281, 284
Ecological civilization 4, 21, 22, 32, 44, 99, 121, 130, 148, 163, 225, 294, 302, 303, 311, 377
Ecological degradation 2, 254, 278
Ecological Red Lines 4, 27, 191–193, 195, 196

Ecology and Environment Bureaus (EEBs) 80, 139, 142, 192, 198, 199, 211, 215, 308
Economic instruments 4, 32, 153, 183, 192, 199, 380
Ecosystem rehabilitation 19, 27, 229
Ecosystems 3, 5, 6, 17, 18, 22, 26, 37, 42–45, 51, 61, 62, 66, 76, 98, 99, 114, 127, 131, 148, 154, 156, 172, 179, 206, 209, 219, 251, 254, 255, 263, 265, 272, 275–278, 284, 349, 353, 356, 361, 366, 368
Electric Power Law 236
Environmental Pollution Tax 160, 192
Environmental protection 19
Environmental Protection Bureaus (EPBs) 80, 82, 142, 192. *See also* Ecology and Environment Bureaus (EEBs)
Environmental Protection Law 106, 115, 137, 138, 193, 198, 207, 302, 305
Environmental Protection Tax Law 139, 199
Environmental sustainability 9, 15, 35, 106, 108, 148, 159, 160, 168, 219, 349, 371, 377
Environment Impact Assessment (EIA) 192, 194
Environment Impact Assessment (EIA) Law 137, 138
Equity Joint Venture (EJV) 314, 317
Estuary salinization 265
Eutrophication 52, 79, 99, 218, 219, 221, 222, 251

Index

F

Fertilizers and pesticides 52, 82, 83, 222
Fisheries Law 137, 144
Five Energy Giants 237. *See also* China Datang Corporation; China Huadian Corporation; China Huaneng Corporation; China State Grid Corporation; China Three Gorges Corporation
Five Year Environmental Plans (FYEPs) 195
Flood control 26, 33, 85, 89, 97, 129, 134, 144, 229, 231, 242, 244, 245, 248, 282, 309
Flood Control Law 86, 106, 132, 137, 144
Flood control levees 53, 85
Flood prevention 10, 43, 85, 86, 95, 144, 201, 206, 229, 283, 284, 333, 379
Flood Prevention Regulation 86
Flood prone areas 84
Floods 2, 3, 5, 16, 40, 41, 44, 45, 51, 53, 84–88, 93, 94, 97–100, 106, 130, 131, 140, 141, 229, 242, 248, 251, 357, 361, 377, 378
14th Five Year Plan (2021–2025) 4, 273, 284

G

Ganges River 340, 352
General Water of China 324, 325
Gezhouba Dam 251
Glacier melting 93
Golden State Environment Group 315
Golden waterway 35, 243
'Go West' campaign 273, 361
Grand Canal 2, 264, 265
Grassland Law 137, 144
Greater Mekong Subregion (GMS) program 344, 346, 360, 367
Great Leap Forward Movement (1958–962) 256
Greenfield projects 315, 316, 327
Groundwater resource management 42
Groundwater resources 5, 17, 19, 22, 26, 34, 42, 43, 45, 51, 52, 54, 61–63, 68–70, 78, 84, 98, 99, 108, 120, 153, 155, 156, 163, 198, 224, 263, 269, 270, 278
Guangdong 20, 70, 113, 135, 171–173, 181, 182, 200, 242, 245, 247, 313
Guangzhou 29, 30, 298

H

Hai River 2, 57, 61, 64, 69, 93, 132, 156, 178, 255, 256
Han River 111, 112, 250, 254, 267, 272
Heilongjiang Inter China Water 329
The Helsinki Rules on the Use of the Waters of International Rivers of 1966 (1966 Helsinki Rules) 348
Heng-Ganges River 36, 341
Hindu Kush Himalayan region 36
Household wastewater discharge 52, 80

Huai River 2, 52, 59, 64, 69, 84, 93, 132, 156, 252, 256
Hua Yan Water, Hong Kong 320
Hydraulic bureaucracy 2
Hydraulic mission 169, 229, 257, 283
Hydrodiplomacy. *See* Water diplomacy
Hydropower dams 33, 131, 229, 231–233, 236, 237, 355, 358, 361, 366
Hydropower installed capacity 232
Hydropower potential 232, 237, 361
Hyflux, Singapore 318, 320

Ili River 339–341, 347, 351
Increasing block tariffs 163, 164
Indus River 36, 340, 352
Industrial wastewater discharge 52, 77, 80–82
Inland navigation 10, 27, 35, 92, 229, 243, 246, 248, 265, 283, 284, 340, 359, 362, 365, 366
Institutional framework 4, 136, 201
Integrated Water Resources Management (IWRM) 97, 106, 137, 139, 144, 282, 309
Inter-basin water transfer projects 7, 10, 69, 131, 229, 230, 252, 254–256, 283
International Law Association (ILA) 348, 349
Inter-provincial coordination 134
Irrawaddy River 341, 357
Irrigation project 2, 27, 231
Irtysh-Ob River 339–341, 351, 352, 370

Jiabao, Wen 155, 241
Jiangsu Province 8, 83, 143, 157, 178, 194, 201, 204, 214, 217, 219, 222, 223, 263, 264, 282, 325, 378
Jiangsu Zhongnan Construction 330
Jinghong Dam 361, 364, 365
Jinping, Xi 32, 34, 130, 175, 241, 324, 338, 377
Jintao, Hu 175, 241, 338
Joint Stock Company (JSC) 315

Keqiang, Li 16, 18, 192, 193, 257, 258
Kherlen River 340

Lancang-Mekong Cooperation (LMC) mechanism 9, 26, 339, 344, 352, 357, 362, 379
Lancang-Mekong River 9, 10, 36, 60, 236, 338–341, 344, 346, 348, 352, 353, 357, 359, 361–363, 365, 366, 369, 370
Lancang-Mekong River Environment Cooperation Center 26
Lancang-Mekong Water Resources Cooperation Center 364
on Five Year Action Plan 364
Land subsidence 43, 52, 63, 99
Laos 9, 340, 353, 355, 356, 358–360, 362–364, 368, 370
Large dams 10, 231, 275, 283, 350, 358, 364, 365
Liao River 52, 56, 64, 69, 76, 156

Loess Plateau 61
Lower Mekong countries 9, 344, 357, 358, 362, 364–366, 369
Lower Mekong River 355, 356, 358, 368–370

M

Marubeni 314
Measures on the Guarantee of Fixed Profit Margins for Foreign Investment Projects, 2002 304
Mekong Delta 353, 356, 361, 364, 365
Mekong River Agreement, 1995 357
Mekong River Commission (MRC) 9, 339, 344, 352, 353, 355–359, 362, 365–368, 370, 380
Middle East routes 230
Middle route 34, 97, 157, 158, 163, 178, 247, 252–254, 256, 257, 259, 260, 265, 267, 269–271, 273, 276–278, 280–284
Ministry of Agriculture and Rural Affairs 120, 136, 237
Ministry of Ecology and Environment (MEE) 5, 7, 54, 74–77, 79–81, 93, 95–98, 106, 116, 120–123, 127, 130–132, 134–136, 147, 148, 220, 237, 238, 258, 302, 307, 308, 380
Ministry of Emergency Management 87, 106, 120, 127, 129, 130
Ministry of Environmental Protection 19, 105, 116, 120–123, 213, 222, 311
Ministry of Housing and Urban-Rural Development 120
Ministry of Natural Resources (MNR) 106, 120, 123, 127, 128, 131, 303, 313
Ministry of Water Resources (MWR) 7, 54–58, 61, 62, 64, 66, 68–70, 73, 85, 90, 106, 109, 116, 120, 122, 123, 128, 130–132, 136, 147, 148, 156, 161, 162, 168–172, 206, 207, 233, 237, 258, 273, 303, 305, 309
Mitsubishi 314
Multi-purpose and agricultural dams 4

N

National Assessment Report on Climate Change, 2015 53, 94
National Climate Change Adaptation Strategy 96
National Development and Reform Commission (NDRC) 120, 122, 129, 147, 210, 237, 280, 307
Nationally Determined Contributions (NDCs) 97
National People's Congress 17, 18, 135, 144, 147, 239
New World Development Company 314
NH_3-N (ammonia nitrogen). *See* Ammonia Nitrogen
1995 BOT Circular 304
1997 United Nations Convention on the Law of Non-Navigational Uses of International Watercourses (1997 UN Water

Convention) 348–350, 359, 370
No.1 Central Document 2011 107
Non-Point Source (NPS) 5, 37, 52, 77, 80, 99, 113, 123, 191
North China Plain 28, 42, 61, 62, 64, 78, 82, 83, 99, 132, 163, 231, 269, 273
Nu-Salween River 36, 60, 236, 340, 341, 346, 357, 361
NWS Holdings, Hong Kong 318

O

Oligotrophication 52, 79, 99
Open-door policy 2, 15, 70, 137, 229, 231, 256
Operation and Maintenance (O&M) 81, 131, 166, 295
Overseas Chinese companies 296, 318, 320

P

Paris Agreement 97
Pearl River 52, 56, 57, 60, 64, 69, 76, 87, 93, 113, 132, 173, 341
Peng, Li 238, 239
Per capita water consumption 155, 156
Per capita water resources 5, 29, 54, 56, 158, 270
Permanganate 76, 79, 218, 249
Physical infrastructure 4, 22, 45, 185, 378
Pollution Discharge Fee System 192, 193, 199, 216
Poyang Lake 52, 66, 67, 251

Precipitation 5, 41, 42, 51, 54, 56–62, 79, 88, 90, 91, 94, 95, 132, 157, 179, 276
Public participation 114, 140, 142, 144, 207, 209, 210, 224
Public Private Partnership (PPP) 294, 295, 297–299, 303, 305, 307–310, 312, 314, 315, 317, 324–330, 332, 333

Q

Qinghai-Tibetan Plateau 59, 274, 341, 353

R

Rainwater harvesting 5, 7, 68–70, 312
Renewable water resources 16, 54, 55, 68, 98, 155, 156
  average volume of 54
Renovate-Operate-Transfer (ROT) 316, 317
'Reservoir-induced seismicity' 248
Reservoirs 2, 41, 53, 85, 95, 110, 157, 158, 163, 174, 260, 270, 271, 283, 378
Resettlement 238, 247, 265, 269, 271, 272, 279
Resettlement with Development 247
River and freshwater lake basins 51
River basins 2, 22, 26, 60, 64, 76, 78, 79, 99, 108, 113, 114, 135, 141, 142, 233, 253, 255, 341, 345, 351, 352
River (Lake) Chief System 8, 10, 26, 115, 143, 192, 193, 200–207, 223, 224, 305, 378

S

Salcon, Malaysia 318
Sanchuan Wisdom Technology 330
Sanitation and hygiene 21
Schistosomiasis 251, 265
SDG 6 Water and Sanitation 17, 19, 21
Sea level rise 53, 93, 94, 356
Sembcorp Utilities, Singapore 320
Sengezhangpu-Indus River 341
Seven River Basin Commissions 131, 132, 172, 237
Shallow aquifer 70
Shanghai 20, 29, 30, 35, 63, 94, 111, 157, 163, 192, 199, 202, 210, 214, 217, 219, 223, 224, 242, 245, 247, 265, 282, 297, 298, 309, 314, 315
Shanghai Pudong Water Supply Corporation 315
Shenzhen Water Group 325
Shuangliang Eco-Energy Systems 329
Silk Road Economic Belt 31, 32
Siltation 231, 249, 250
Sime Darb, Malaysia 318
Sino-French Company 314
Socialist market economy 28, 160, 294, 301, 324, 325
Soil erosion 16, 26, 61, 195
Songhua River 52, 56, 64, 69, 76
Sound Group 297
South North Water Transfer Project 7, 10, 25, 34, 36, 53, 70, 97, 123, 157, 163, 168, 175–179, 184, 207, 230, 247, 252, 255, 257, 258, 260–262, 264, 268, 274, 277, 284, 313, 378, 379
South-South cooperation 26
Sponge City Initiative 43, 45, 85, 312, 329, 333
State Environment Protection Administration (SEPA) 220, 222, 380
State Oceanic Administration 123
State-Owned Assets Supervision and Administration Commission 238
State-Owned Enterprises (SOEs) 7, 210, 236, 238, 258
Stringent Water Resources Management (SWRM) 107, 108
Suez Group 314
Suifen-Razdolnaya River 341
Surface water 34, 52, 62, 69, 75, 77, 82, 116, 139, 141, 156, 162, 163, 197, 217, 218, 263, 271, 281
Surface water resources 5, 54, 56–59, 62, 68, 98, 184
Sustainable development 3, 15
Sustainable Development Goals (SDGs) 4, 9, 17–19, 21, 22, 33, 44, 195, 377

T

Tai Basin Authority (TBA) 223, 225
Tai Lake 8, 10, 52, 60, 67, 132, 143, 193–195, 201, 210, 212–214, 217, 219, 221–223, 225
Target Responsibility System 203
Tarim River 36, 254
Thailand 9, 340, 353, 355, 356, 358–361, 364
Thames Water 39, 296, 303, 314, 318

Index 391

13th Five Year Plan (2016–2020) 4, 17, 19, 20, 25, 31, 33, 37, 42, 44, 53, 96, 97, 100, 156, 192, 195–197, 257, 302, 311, 312, 320, 329
Three Gorges Dam 7, 10, 85, 92, 95, 99, 229–232, 237, 239, 240, 243, 244, 259, 277, 279, 283, 378
3H Rivers. *See* Hai River; Huai River; Yellow River (Huang River)
Three Red Lines 4, 9, 22, 25, 27, 31, 105–107, 111, 112, 143, 148, 154, 184, 207, 266, 301, 305, 311
Three Synchronization System 192, 193, 198, 199, 215, 224
Tianjin Water Group 330
Tiao-Kuai (条块) lines 120, 133, 201
Tongtian River 273, 275
Tonle Sap Lake 353, 361, 365
Total Phosphorus (TP) 76, 79, 110, 214, 218, 222
Total Pollution Load Control (TPLC) 142, 194, 210, 212
Township Village Enterprises (TVEs) 52, 77, 82, 83, 99
Transboundary river cooperation 26, 346, 367
Transboundary river issues 4, 10, 338, 347, 350, 369, 379
Transboundary rivers 8, 9, 22, 60, 237, 276, 337–342, 344, 346, 347, 359, 361, 369, 378, 379
Transfer-Operate-Transfer (TOT) 298, 315, 317
Tsinghua Holdings Ltd 330

Tsinghua Tongfang 325
Tumen-Dooman River 9, 337, 339–341, 344, 346, 350
Tumen River Area Development Plan (TRADP) 346, 350
12th Five Year Plan (2011–2015) 37, 53, 96, 109, 203, 207, 232, 257, 266, 299, 312, 320
21st-Century Maritime Silk Road 31–33
2004 Berlin Rules 349
2018 Water Pollution Prevention and Control Law 115
2018 administrative reform 7, 106, 120, 130, 136, 258, 305, 309
2030 Sustainable Development Agenda 18, 39

U

Upstream hegemon 9, 341
Upstream Superpower 339, 341
Urbanization 1, 3, 8, 16, 43, 67, 70, 219, 233, 256, 277, 295, 299, 356
Urban water and wastewater service charges 154, 155

V

Veolia Environment 296, 297, 315
Vietnam 9, 341, 346, 353, 355, 356, 358, 360, 362, 364–366
Virtual water 30, 153, 155, 176–185, 380
Virtual water and water footprint approach 6, 154, 175–177, 180, 378

W

War on pollution 192, 193, 199
Wastewater 30, 35, 39, 43, 52, 68–70, 76, 80–82, 123, 154, 160, 161, 163, 166, 167, 184, 221, 249, 295, 298, 299, 303, 305, 312, 315, 329–331, 378
Wastewater treatment plants 25, 81, 142, 157, 163, 266, 281, 300, 311, 312, 315, 320, 331, 378
'Water Tower of China' 341
'Water Tower of Southeast Asia or South Asia'. *See* 'Water Tower of China'
Water abstraction 72, 108, 109, 112, 139, 161, 169–172
Water allocation 58, 74, 97, 98, 108, 112, 132–134, 140, 167, 170, 171, 178, 180, 181, 183, 277, 282, 352
Water and wastewater PPP market 294, 330, 333
Water and wastewater service market 10, 293, 295, 332, 379
Water and wastewater services 4, 154, 167, 184, 293–296, 299, 303, 307, 309, 310, 324, 331–333, 379
  in private sector participation 8, 10, 162, 293
  in public sector participation 8
Water and wastewater treatment facility 4
Water availability 54, 61, 108, 153, 156, 180, 182, 185, 345
Water availability per capita 54, 60
Water demand and availability 5, 51, 60
Water diplomacy 33, 339, 345, 346, 357, 359, 366, 370
Water-Energy-Climate nexus 36
Water footprint 10, 30, 153, 155, 175, 176, 178, 180, 182, 183
Water Function Zones 110, 112
Water governance 7, 121, 144, 168, 295
Water Law 6, 86, 106, 132, 134, 137, 139, 140, 160, 161, 169, 184, 194, 305, 352
Water Pollutant Discharge Permit System (WPDP) 215, 216
Water pollution 3, 6–8, 10, 16, 18, 31, 33, 39, 53, 74, 76, 77, 79, 82–84, 97, 99, 107, 114, 122, 134, 135, 139, 142, 143, 191, 192, 194, 198, 202, 205, 207, 210, 212, 214, 215, 219, 220, 222–225, 265, 266, 349, 351, 356, 378, 380
Water Pollution Prevention and Control Action Plan (Water Ten Plan) 10, 22, 27, 105–107, 113, 115, 117, 148, 205, 302, 320, 329, 333
Water Pollution Trading Scheme 8, 10, 192, 200, 210, 211, 214, 223
Water Prevention and Control Law 106
Water pricing 6, 10, 39, 114, 154, 155, 158–162, 167, 171, 183, 184, 256, 280, 281, 295, 301, 303, 305, 332
Water productivity 71, 72, 179, 183
Water quality 5, 8–10, 22, 25, 26, 31, 35, 37, 43, 51–53, 63, 74–80, 82, 84, 99, 107, 110,

112–114, 120, 122, 123, 130, 131, 136–138, 142, 143, 163, 191–193, 196–203, 205, 210, 214, 218, 219, 221–225, 248, 249, 266, 271, 272, 294, 301, 305, 308, 310, 311, 320, 333, 340, 377, 378
Water-related disasters 3, 6, 7, 9, 16, 18, 40, 41, 53, 84, 95, 96, 106, 130, 131, 136, 377
Water resources 1, 5, 6, 9, 15, 17, 19, 21, 27, 34, 51, 53, 61, 68, 69, 97, 105, 108, 116, 120, 130, 138, 148, 154, 162, 166–168, 170, 173, 174, 178, 206, 364, 367, 380
   total amount of 51, 176, 255
Water resources development 7, 10, 107, 109, 131, 132, 148, 177, 229, 230, 283, 284, 363, 377
Water resources districts 54, 56, 58, 69
Water resources fee 112, 139, 160, 161, 163, 304
Water resources management 2–4, 7–9, 15–18, 20–22, 26, 27, 31, 32, 35–37, 39, 44, 45, 51, 53, 97, 98, 105–107, 109, 112, 113, 120–124, 130–137, 139, 141, 144, 148, 161, 167, 168, 172, 175, 176, 185, 195, 201, 202, 205, 206, 210, 225, 252, 280, 295, 301, 302, 307, 309, 332, 357, 362, 368, 377, 378, 380, 381
Water Resources Tax Reform, 2016 160
Water reuse and recycle 5, 34, 312

Water rights trading 6, 8, 10, 108, 109, 153–155, 167–175, 183, 184, 193, 214, 224, 378, 380
Water Saving Society 96, 168, 169
Water security 33, 62, 97, 253, 270, 377, 379
Water shortage 3, 5, 6, 16, 31, 33, 39, 51, 53, 57, 59, 84, 90, 93, 95, 100, 107, 135, 148, 153–158, 163, 174, 176, 177, 180, 182–184, 230, 253, 254, 256, 263, 269, 277, 294, 301, 380
Water Soil and Conservation Law 106
Water stress 25, 34, 113, 155, 157, 176, 177
Water supply and use 9, 34, 51
Water supply engineering fee 160, 161, 304
WaterSure Plus 39
Water Ten Plan. *See* Water Pollution Prevention and Control Action Plan (Water Ten Plan)
Water Towers of Asia 36
Water transfer projects 157, 177, 230, 252, 255, 277, 282
Water use efficiency 20, 22, 25, 28, 29, 31, 32, 34, 36, 37, 52, 71, 72, 107–109, 114, 148, 161, 176–180, 183, 281, 301
Water withdrawals 20, 61, 181, 183
Wenchuan 512 earthquake 41
Western route 230, 256, 257, 259, 273–276, 284
Wholly Foreign Owned Enterprise (WFOE) 315, 317

X

Xayaburi Dam 358
Xiaolangdi Dam 232
Xiaoping, Deng 2
Xie, Zhenhua 380

Y

Yalong River 273, 275
Yalu-Amnok River 9, 337, 339–341, 344, 350, 370
Yangtze River 2
Yangtze River Economic Belt (YREB) 31–37, 44, 196
Yangtze sturgeon 251

Yarlung Tsangpo-Brahmaputra River 36, 60, 276, 277, 284, 338, 340, 352
Yat-Sen, Sun 239, 378
Yellow River Conservancy Commission 132, 133, 135, 256
Yellow River (Huang River) 2, 5, 61, 64, 69, 84, 87, 99, 113, 132, 140, 156, 231, 253–256, 262, 267, 273, 275, 283
Yuan-Red River 340, 341
Yunnan Province 348, 359–363

Z

Zhangmu Dam 276

CPSIA information can be obtained
at www.ICGtesting.com
Printed in the USA
LVHW081926080123
736723LV00008B/770